SUSTAINABLE AGRICULTURE AND FOOD SUPPLY

Scientific, Economic, and
Policy Enhancements

SUSTAINABLE AGRICULTURE AND FOOD SUPPLY

Scientific, Economic, and Policy Enhancements

Edited by
Kim Etingoff

Apple Academic Press Inc. | Apple Academic Press Inc.
3333 Mistwell Crescent | 9 Spinnaker Way
Oakville, ON L6L 0A2 | Waretown, NJ 08758
Canada | USA

First issued in paperback 2021

Exclusive worldwide distribution by CRC Press, a member of Taylor & Francis Group
No claim to original U.S. Government works

ISBN 13: 978-1-77463-710-4 (pbk)
ISBN 13: 978-1-77188-384-9 (hbk)

Library and Archives Canada Cataloguing in Publication

Sustainable agriculture and food supply: scientific, economic, and policy enhancements / edited by Kim Etingoff.

Includes bibliographical references and index.
Issued in print and electronic formats.
ISBN 978-1-77188-384-9 (hardcover).--ISBN 978-1-77188-385-6 (pdf)
1. Sustainable agriculture. 2. Food supply. I. Etingoff, Kim, editor

S494.5.S86S88 2016   630   C2015-908304-4   C2015-908305-2

Library of Congress Cataloging-in-Publication Data

Names: Etingoff, Kim, editor.
Title: Sustainable agriculture and food supply : scientific, economic, and policy enhancements / editor: Kim Etingoff.
Description: Waretown, NJ : Apple Academic Press, 2015. | Includes bibliographical references and index.
Identifiers: LCCN 2015047104 (print) | LCCN 2016000093 (ebook) | ISBN 9781771883849 (hardcover : alk. paper) | ISBN 9781771883856 ()
Subjects: LCSH: Sustainable agriculture. | Sustainable agriculture--Economic aspects. | Sustainable agriculture--Social aspects. | Sustainable agriculture--Government policy. | Food supply.
Classification: LCC S494.5.S86 S8556 2015 (print) | LCC S494.5.S86 (ebook) | DDC 338.1--dc23
LC record available at http://lccn.loc.gov/2015047104

Apple Academic Press also publishes its books in a variety of electronic formats. Some content that appears in print may not be available in electronic format. For information about Apple Academic Press products, visit our website at **www.appleacademicpress.com** and the CRC Press website at **www.crcpress.com**

About the Editor

Kim Etingoff

Kim Etingoff has a Tufts University terminal master's degree in Urban and Environmental Policy and Planning. Her recent experience includes researching with Initiative for a Competitive Inner City, a report on food resiliency within the city of Boston. She worked in partnership with the Dudley Street Neighborhood Initiative and Alternatives for Community and Environment to support a community food-planning process based in a Boston neighborhood, which was oriented toward creating a vehicle for community action around urban food issues, providing extensive background research to ground the resident-led planning process. She has worked in the Boston Mayor's Office of New Urban Mechanics, and has also coordinated and developed programs in urban agriculture and nutrition education. In addition, she has many years of experience researching, writing, and editing educational and academic books on environmental and food issues.

Contents

List of Contributors

Kalli Anderson
Department of Liberal Studies, Humber College, 205 Humber College Boulevard, Toronto, ON M9W 5L7, Canada

Richard D. Bardgett
Faculty of Life Science, The University of Manchester, Michael Smith Building, Oxford Road, Manchester, M13 9PL, UK

John R Beddington
Government Office of Science, London, UK

Jenny Bell
Pepsico UK and Ireland, Theale Head Office, 1600 Arlington Business Park, Theale, Reading, RG7 4SA, Berkshire, UK

Tim G. Benton
Global Food Security Programme, University of Leeds, Leeds, LS2 9JT, UK

Karin Edvardsson Björnberg
Division of Philosophy, KTH Royal Institute of Technology, Brinellvägen 32, SE-100 44 Stockholm, Sweden

Jan J. Boersema
Institute for Environmental Studies, VU University, De Boelelaan 1087, 1081 HV Amsterdam, The Netherlands

Angela Booth
AB Agri, 64, Innovation Way, Peterborough, PE2 6FL, Canada

Jan Bouwman
Syngenta Crop Protection B.V., Jacob Obrechtlaan 3a, Bergen op Zoom 4600AM, The Netherlands

Brook O. Brouwer
Northwestern Washington Research and Extension Center, Washington State University, Mount Vernon, WA 98273, USA

Chris Brown
Asda, Asda House, Southbank, Great Wilson Street, Leeds LS11 5AD, UK

Ann Bruce
Science, Technology and Innovation Studies & Innogen, University of Edinburgh, Old Surgeons' Hall, High School Yards, Edinburgh, EH1 1LZ, UK

Paul J. Burgess
Department of Environmental Science and Technology, Cranfield University, Cranfield, Bedfordshire, MK43 0AL, UK

Simon J. Butler
School of Biological Sciences, University of East Anglia, Norwich Research Park, Norwich, NR4 7TJ, UK

Bruce M. Campbell
CGIAR Research Program on Climate Change, Agriculture and Food Security, Cali, Colombia; International Center for Tropical Agriculture (CIAT), Cali, Colombia

Vincenzina Caputo
Department of Food and Resource Economics, Korea University, Anamdong-5-1, Seongbukgu, Seoul 136-701, Korea

Ian Crute
Agriculture and Horticulture Development Board, Stoneleigh Park, Kenilworth, Warwickshire, CV8 2TL, UK

Joop de Boer
Institute for Environmental Studies, VU University, De Boelelaan 1087, 1081 HV Amsterdam, The Netherlands

Fabrice DeClerck
Bioversity International, Montpellier, France

Alessandro De Pinto
International Food Policy Research Institute (IFPRI), Washington, DC, USA

Lynn V. Dicks
Conservation Science Group, Department of Zoology, University of Cambridge, Cambridge, CB2 3EJ, UK

Frances Dixon
Welsh Government, Sustainable Land Management, Rhodfa Padarn, Llanbadarn Fawr, Aberystwyth, Ceredigion, Wales, SY23 3UR, UK

Caroline Drummond
LEAF Chief Executive, Stoneleigh Park, Stoneleigh, Warwickshire, CV8 2LG, UK

Robert P. Freckleton
Department of Animal & Plant Sciences, University of Sheffield, Sheffield, S10 2TN, UK

Maggie Gill
Department for International Development, 22 Whitehall, London, SW1A 2EG, UK

Mary V. Gold
Alternative Farming Systems Information Center, National Agricultural Library, U.S. Department of Agriculture, 10301 Baltimore Avenue, Beltsville, MD 20705-2351, US

Andrea Graham
National Farmers Union, Agriculture House, Stoneleigh Park, Stoneleigh, Warwickshire, CV8 2TZ, UK

Jay Gulledge
Environmental Sciences Division, Oak Ridge National Laboratory, Oak Ridge, TN, USA

Rosie S. Hails
Centre for Ecology and Hydrology, Maclean Building, Benson Lane, Crowmarsh Gifford, Wallingford, Oxfordshire, OX10 8BB, UK

James Hallett
British Growers Association Ltd, BGA House, Nottingham Road, Louth, Lincolnshire, LN11 0WB, UK

Beth Hart
Sainsbury's Supermarkets Ltd, 33 Holborn, EC1N 2HT, UK

Jonathan Hellin
International Maize and Wheat Improvement Center (CIMMYT), Texcoco, Mexico

Mario Herrero
Commonwealth Scientific and Industrial Research Organisation (CSIRO), Brisbane, Australia

Jon G. Hillier
School of Biological Sciences, University of Aberdeen, 23 St Machar Drive, Aberdeen, Scotland, AB24 3UU, UK

John M. Holland
Game and Wildlife Conservation Trust, Fordingbridge, Hampshire, SP6 1EF, UK

Jonathan N. Huxley
University of Nottingham, School of Veterinary Medicine and Science, Sutton Bonington Campus, Loughborough, Leicestershire, LE12 5RD, UK

John S. I. Ingram
Environmental Change Institute, Oxford University Centre for the Environment, South Parks Road, Oxford, OX1 3QY, UK

Molly Jahn
Department of Agronomy and Laboratory of Genetics, University of Wisconsin, Madison, WI, USA

Andy Jarvis
CGIAR Research Program on Climate Change, Agriculture and Food Security, Cali, Colombia; International Center for Tropical Agriculture (CIAT), Cali, Colombia

Elisabeth Jonas
Department of Animal Breeding and Genetics, Swedish University of Agricultural Sciences, P.O. Box 7023, SE-750 07 Uppsala, Sweden

Stephen S. Jones
Northwestern Washington Research and Extension Center, Washington State University, Mount Vernon, WA 98273, USA

Vanessa King
Unilever UK Central Resources Ltd, Unilever House, Blackfriars, London, EC4Y 0DY, UK

Maggie Kisaka-Lwayo
Department of Agricultural Economics & Extension, University of Fort Hare, Alice, South Africa

Michal Kulak
Life Cycle Assessment group, Institute for Sustainability Sciences, Agroscope Reckenholzstrasse 191, CH-8046 Zurich, Switzerland

Les Levidow
Centre for Technology Strategy, Open University, Milton Keynes MK6 6AA, UK

David LeZaks
Department of Agronomy and Laboratory of Genetics, University of Wisconsin, Madison, WI, USA

Luanne Lohr
Food and Specialty Crops, Economic Research Service, USDA, 1800 M. St. NW, Washington, DC 20036-5831, USA

Fiona Louden
Faculty of Environmental Studies, York University, 4700 Keele St., Toronto, ON M3J 1P3, Canada

Tom MacMillan
Soil Association, South Plaza, Marlborough Street, Bristol, UK

Rod MacRae
Faculty of Environmental Studies, York University, 4700 Keele St., Toronto, ON M3J 1P3, Canada

Mads V. Markussen
Center for BioProcess Engineering, Department of Chemical and Biochemical Engineering, Technical University of Denmark DTU, DK-2800 Kgs. Lyngby, Denmark

Håkan Marstorp
Department of Soil and Environment, Swedish University of Agricultural Sciences, P.O. Box 7014, SE-750 07 Uppsala, Sweden

Daniel F. McGonigle
Department for the Environment, Food and Rural Affairs, Nobel House, 17 Smith Square, London, SW1P 3JR, UK

Carmel McQuaid
Marks & Spencer, 5 Merchant Square, Paddington Basin, London, W2 1AS, UK

Holger Meinke
Tasmanian Institute of Agriculture, University of Tasmania, Hobart, Australia; Centre for Crop Systems Analysis, Wageningen University, Wageningen, The Netherlands

Kevin M. Murphy
Department of Crop and Soil Sciences, Washington State University, Pullman, WA 99164-6420, USA

Rodolfo M. Nayga Jr.
Department of Food and Resource Economics, Korea University, Anamdong-5-1, Seongbukgu, Seoul 136-701, Korea; Department of Agricultural Economics and Agribusiness, University of Arkansas, Fayetteville, AR 72701, USA; Norwegian Agricultural Economics Research Institute, Storgata 2/4/6, Oslo NO-0155, Norway

Thomas Nemecek
Life Cycle Assessment group, Institute for Sustainability Sciences, Agroscope Reckenholzstrasse 191, CH-8046 Zurich, Switzerland

Henry Neufeldt
World Agroforestry Centre (ICRAF), Nairobi, Kenya

Tim Nevard
Conservation Grade, Gransden Park, Abbotsleigh, Cambridgeshire, PE19 6TY, UK

Steve Norman
Dow AgroSciences, Milton Park, Abingdon, Oxfordshire, OX14 4RN, UK

Ken Norris
Centre for Agri-Environmental Research, University of Reading, Earley Gate, Reading, RG6 6AR, UK

Ajuruchukwu Obi
Department of Agricultural Economics & Extension, University of Fort Hare, Alice, South Africa

Hanne Østergård
Center for BioProcess Engineering, Department of Chemical and Biochemical Engineering, Technical University of Denmark DTU, DK-2800 Kgs. Lyngby, Denmark

Timothy Park
Food Marketing Branch, Economic Research Service, USDA, 1800 M. St. NW, Washington, DC 20036-5831, USA

Catherine Pazderka
British Retail Consortium, Westminster, SW1H 9BP, UK

Inder Poonaji
Nestle UK Ltd, 1 City Place, Gatwick, RH6 0PA, UK

Katerina Psarikidou
Department of Sociology, Lancaster University, Lancaster, LA1 4YT, UK

Athina-Evera Qendro
Institute for Management, Governance and Society (IMaGeS), Robert Gordon University, Garthdee Road, AB10 7QB Aberdeen, UK

Claire H. Quinn
Sustainability Research Institute, School of Earth and Environment, University of Leeds, Leeds LS2 9JT, UK

Stephen J. Ramsden
School of Biosciences, University of Nottingham, Sutton Bonington Campus, Loughborough, Leicestershire, LE12 5RD, UK

Todd Rosenstock
World Agroforestry Centre (ICRAF), Nairobi, Kenya

Giovanna Sacchi
Department of Management, Ca' Foscari University of Venice, San Giobbe—Cannaregio 873, Venice 30121, Italy

Mary Scholes
School of Animal, Plant and Environmental Science, University of Witwatersrand, Johannesburg, South Africa

Robert Scholes
Council for Scientific and Industrial Research (CSIR), Cape Town, South Africa

Hanna Schösler
Institute for Environmental Studies, VU University, De Boelelaan 1087, 1081 HV Amsterdam, The Netherlands

Duncan Sinclair
Waitrose, John Lewis plc, 171 Victoria Street, London, SW1E 5NN, UK

Gavin M. Siriwardena
British Trust for Ornithology, The Nunnery, Thetford, Norfolk, IP24 2PU, UK

Laurence G. Smith
The Organic Research Centre, Elm Farm, Hamstead Marshall, Newbury, Berkshire RG20 0HR, UK

Crystal Snyder
Department of Agricultural, Food & Nutritional Science, 4-10 Agriculture/Forestry Centre, University of Alberta, Edmonton, Alberta, T6G 2P5, Canada

Dean Spaner
Department of Agricultural, Food & Nutritional Science, 4-10 Agriculture/Forestry Centre, University of Alberta, Edmonton, Alberta, T6G 2P5, Canada

William J. Sutherland
Conservation Science Group, Department of Zoology, University of Cambridge, Cambridge, CB2 3EJ, UK

Michelle Szabo
Faculty of Environmental Studies, York University, 4700 Keele St., Toronto, ON M3J 1P3, Canada

Pernilla Tidåker
JTI—Swedish Institute of Agricultural and Environmental Engineering, P.O. Box 7033, SE-750 07 Uppsala, Sweden

Sandi Trillo
Graduate Program in Interdisciplinary Studies, York University, 4700 Keele St., Toronto, ON M3J 1P3, Canada

Sonja Vermeulen
CGIAR Research Program on Climate Change, Agriculture and Food Security, Cali, Colombia; Gund Institute for Ecological Economics, University of Vermont, Burlington, VT, USA

Juliet A. Vickery
RSPB, the Lodge, Sandy, Bedfordshire, SG19 2DL, UK

Andrew P. Whitmore
Rothamsted Research, Harpenden, Hertfordshire, AL5 2JQ, UK

Eva Wollenberg
CGIAR Research Program on Climate Change, Agriculture and Food Security, Cali, Colombia; Department of Plant and Environmental Sciences, Faculty of Science, University of Copenhagen, Copenhagen, Denmark

William Wolmer
Blackmoor Estate Ltd, Blackmoor, Liss, Hampshire, GU33 6BS, UK

Robert Zougmoré
CGIAR Research Program on Climate Change, Agriculture and Food Security, Cali, Colombia; International Crops Research Institute for the Semi-Arid Tropics (ICRISAT), Bamako, Mali

Acknowledgment and How to Cite

The editor and publisher thank each of the authors who contributed to this book. The chapters in this book were previously published in various places in various formats. To cite the work contained in this book and to view the individual permissions, please refer to the citation at the beginning of each chapter. Each chapter was read individually and carefully selected by the editor; the result is a book that provides a nuanced look at sustainable agriculture. The chapters included are broken into five sections, which describe the following topics:

Part 1: What Is Sustainable Agriculture?
- Chapter 1 provides an overview of sustainable agriculture, including similarities and differences in various actors' definitions.

Part 2: Sustainably Connecting Producers and Consumers
- Chapter 2 uses qualitative methods to investigate consumers' attitudes to organic food outlets in two different countries.
- Chapter 3 explores the determinants of both the production and consumption aspects of organic agriculture in South Africa in order to promote more effective relevant public policy.

Part 3: Localizing the Food System
- Chapter 4 outlines potential consumer information systems that enhance the ability of "citizen-consumers" to make healthy and sustainable choices.
- Chapter 5 uses a case study in the UK to highlight important food relocalization initiatives, namely greater governmental support of short food supply chains and intermediaries that can expand markets for producers.
- Chapter 6 argues for the need for localizing plant breeding to account for regional differences in food culture, and proposes farmer-breeder-chef collaborative models as possible methods for achieving this goal.
- Chapter 7 examines the effects of deciding to engage in local selling on the earned income of organic farms.

Part 4: Organic Food and the Human Element: Consumers & Farmers

- Chapter 8 looks at cost-effective alternatives to organic certification—specifically the Brazilian model of Participatory Guarantee Systems—and their effect on consumers interested in purchasing organic food.
- Chapter 9 compares multiple dimensions of sustainability of a case organic farm system in the UK with two model organic food supply systems, using energy accounting and Life Cycle assessment.
- Chapter 10 discusses the challenges that face organic grain production in Canada, including agronomic, environmental, and economic factors.

Part 5: The Future of Sustainable Agriculture

- Chapter 11 identifies twenty-six priority "knowledge needs" in the UK concerning the the creation and maintenance of environmentally sustainable agricultural systems based on input from practitioners and scientists.
- Chapter 12 provides a look at Swedish perceptions of sustainability in the food supply chain and the potential of biotechnology to increase food production and promote agricultural sustainability within that system.
- Chapter 13 argues for a clarified understanding and definition of climate-smart agriculture, which would include indicators and metrics that more easily measure sustainable food systems over time.

Introduction

Sustainable agriculture is an increasingly crucial concept and practice as we realize the ways in which our food systems contribute and respond to climate change. The sustainability of agriculture is a complicated, multi-dimensional issue, which should be considered from a variety of angles. This book engages with sustainable agriculture from the necessary perspectives of science, economics, sociology, and policy in order to move farming forward into a more promising and healthy future.

Part 1 tackles the deceptively difficult task of defining sustainable agriculture. Although there are many global initiatives that seek to increase the sustainability of farming and food chains, there is not always agreement on goals or methods. In chapter 1, Mary Gold discusses the meaning of the term "sustainable agriculture." She attempts to answer important questions such as, "In such a quickly changing world, can anything be sustainable?" "What do we want to sustain?" "How can we implement such a nebulous goal?" "Is it too late?" Her discussion forms a cogent and necessary foundation for the following chapters.

Part 2 considers both sides of sustainable agriculture: the producers and the consumers.

The purpose of chapter 2 is to elicit UK and Albanian consumers' perceptions of food outlets in order to understand their views on supermarkets and farmers' markets as outlets for organic food. Qendro chose a qualitative research methodology was chosen as the best way to get an in-depth understanding of how consumers of these two different countries understand and evaluate buying organic food from two different food outlets. This exploratory research is a first step to find out how and why organic food is being bought in supermarkets and farmers' markets. The results show that respondents associated organic with vegetables and fruit, that taste good, are healthy, and are free of pesticides and hormones. The importance of motives varies between the outlets they prefer for buying organic food. An interesting finding is the fact that Albanian respondents refer to the farmers' markets as the villagers' market.

The objective of chapter 2 is to provide, through an exploratory analysis of data from farm and households surveys, empirical insights into determinants of organic farming adoption, differentiating between fully-certified organic, partially-certified organic and non-organic farmers; eliciting farmers risk preferences and management strategies and; exploring consumer awareness, perceptions and consumption decisions. By exploring a combination of adoption relevant factors in the context of real and important land management choices, Kisaka-Lwayo and Obi provide an empirical contribution to the adoption literature and provides valuable pointers for the design of effective and efficient public policy for on-farm conservation activities. Similarly, achieving awareness and understanding the linkage between awareness and purchasing organics is fundamental to impacting the demand for organically grown products. Consumer awareness of organic foods is the first step in developing demand for organic.

In Part 3, the prominent topic of localizing food systems is addressed. Local agriculture is a visible and key part of the discussion surrounding sustainable agriculture.

In chapter 5, Levidow and Psarikidou outline key short-, medium-, and long-term initiatives to facilitate the citizen-consumer phenomenon and better support consumers in their efforts to promote health and sustainability in the Canadian food system. Both health and sustainability are stated public policy objectives in Canada, but food information rules and practices may not be optimal to support their achievement. In the absence of a stated consensus on the purposes of public information about food, the information provided is frequently determined by the marketers of product. No institution or agency has responsibility for determining the overall coherence of consumer food messages relative to these broader social goals of health and sustainability. Individual firms provide information that shows their products to best advantage, which may contradict what is provided about the product by another firm or government agency. Individual consumers do not have the resources to determine easily the completeness of any firm's messages, particularly in light of the size of food industry advertising budgets. Government rules confound this problem because there is also little coherence between the parts of government that have responsibility for point of purchase, advertising rules, and labelling. The healthy eating messages of health departments are often competing with contradictory messages permitted by the regulatory framework of other arms of government. Investments in programs that successfully promote environmental stewardship in agriculture are undercut in the market because consumers cannot support those efforts with their dollars. This problem exists despite the emergence of "citizen-consumers" who have a broader approach to food purchasing than individual maximization. Only recently have some health professionals and sustainable agriculture proponents turned their attention to these factors and designed interventions that take them into account.

The rapid growth and co-option of the local agriculture movement highlights a need to deepen connections to place-based culture. Selection of plant varieties specifically adapted to regional production and end-use is an important component of building a resilient food system. Doing so will facilitate a defetishization of food systems by increasing the cultural connection to production and consumption. Today's dominant model of plant breeding relies on selection for centralized production and end-use, thereby limiting opportunity for regional differentiation. On the other hand, end-user-driven selection of heirloom varieties with strong cultural and culinary significance may limit productivity while failing to promote continued advances in end-use quality. Farmer-based selection may directly reflect local food culture; however, increasing genetic gains may require increased exchange of germplasm, and collaboration with trained plant breeders. Participatory farmer–breeder–chef collaborations are an emerging model for overcoming these limitations and adding the strength of culturally based plant breeding to the alternative food movement. These models of variety selection are examined in chapter 6 within the context of small grain and dry bean production in Western Washington.

The primary purpose of chapter 7 is to examine the factors that influence earned income of organic farmers explicitly incorporating farmer decisions to engage in local selling. The stochastic frontier model identifies role model producers who are the most technically efficient in achieving the maximum output that is feasible with a given set of inputs along with farm and demographic factors that enhance efficiency. Organic earnings equations that control for producer and farm characteristics reveal that organic farmers who are involved in local sales achieve lower earnings. Producer involvement in local sales has little impact on observed technical efficiency on organic farms.

Part 4 addresses another important aspect of this topic, the role of organic farming along the producers and consumer dimensions.

Regulatory standards and certification models are essential tools guaranteeing the authenticity of organic products. In particular, third-party certification is useful to consumers since it provides guarantees regarding production processes and food quality. In an attempt to cope with the costs and bureaucratic procedures related to the adoption of such certification, groups of small producers have begun to rely upon alternative quality assurance systems such as Participatory Guarantee Systems (PGS). Chapter 8 contextualizes and analyzes the PGS scheme and describes the Brazilian Rede Ecovida de Agroecologia network. Sacchi and her colleagues then investigate the effect of various factors on Brazilian consumers' purchasing behavior for organic products guaranteed by PGS. The results show that employed and older consumers who live in rural.

In chapter 9, resource use and environmental impacts of a small-scale low-input organic vegetable supply system in the United Kingdom were assessed by emergy accounting and Life Cycle Assessment (LCA). The system consisted of a farm with high crop diversity and a related box-scheme distribution system. Markussen and colleagues compared empirical data from this case system with two modeled organic food supply systems representing high- and low-yielding practices for organic vegetable production. Further, these systems were embedded in a supermarket distribution system and they provided the same amount of comparable vegetables at the consumers' door as the case system. The on-farm resource use measured in solar equivalent Joules (seJ) was similar for the case system and the high-yielding model system and higher for the low-yielding model system. The distribution phase of the case system was at least three times as resource efficient as the models and had substantially less environmental impacts when assessed using LCA. The three systems ranked differently for emissions with the high-yielding model system being the worst for terrestrial ecotoxicity and the case system the worst for global warming potential. As a consequence of being embedded in an industrial economy, about 90% of resources (seJ) were used for supporting labor and service.

Demand for organically produced food products is increasing rapidly in North America, driven by a perception that organic agriculture results in fewer negative environmental impacts and yields greater benefits for human health than conventional systems. Despite the increasing interest in organic grain production on the Canadian Prairies, a number of challenges remain to be addressed to ensure its long-term sustainability. In chapter 10, Snyder and Spaner summarize Western Canadian research into organic crop production and evaluate its agronomic, environmental, and economic sustainability.

Finally, Part 5 provides chapters that consider the future of farming and the aspects missing from sustainable agriculture today.

Increasing concerns about global environmental change and food security have focused attention on the need for environmentally sustainable agriculture. This is agriculture that makes efficient use of natural resources and does not degrade the environmental systems that underpin it, or deplete natural capital stocks. In chapter 11, Dicks and her colleagues convened a group of 29 "practitioners" and 17 environmental scientists with direct involvement or expertise in the environmental sustainability of agriculture. The practitioners included representatives from UK industry, non-government organizations and government agencies. The authors collaboratively developed a long list of 264 knowledge needs to help enhance the environmental sustainability of agriculture within the UK or for the UK market. The authors then refined and selected the most important knowledge needs through a three-stage process of voting, discussion and scoring. Scientists and practitioners identified similar priorities. Finally, the authors present the 26 highest priority knowledge needs. Many of them demand integration of knowledge from different disciplines to inform policy and practice. The top five are about sustainability of livestock feed, trade-offs between ecosystem services at farm or landscape scale, phosphorus recycling and metrics to measure sustainability. The outcomes will be used to guide on-going knowledge exchange work, future science policy and funding.

Researchers have put forward agricultural biotechnology as one possible tool for increasing food production and making agriculture more sustainable. In chapter 12, Björnberg and her colleagues investigated how key actors in the Swedish food supply chain perceive the concept of agricultural sustainability and the role of biotechnology in creating more sustainable agricultural production systems. Based on policy documents and semi-structured interviews with representatives of five organizations active in producing, processing and retailing food in Sweden, an attempt is made to answer these three questions: How do key actors in the Swedish food supply chain define and operationalize the concept of agricultural sustainability? Who/what influences these organizations' sustainability policies and their respective positions on agricultural biotechnology? What are the organizations' views and perceptions of biotechnology and its possible role in creating agricultural sustainability? Based on collected data, the authors conclude that, although there is a shared view of the core constituents of agricultural sustainability among the organizations, there is less explicit consensus on how the concept should be put into practice or what role biotechnology can play in furthering agricultural sustainability.

Finally, agriculture is considered to be "climate-smart" when it contributes to increasing food security, adaptation and mitigation in a sustainable way. This new concept now dominates current discussions in agricultural development because of its capacity to unite the agendas of the agriculture, development and climate change communities under one brand. In the opinion piece in chapter 13, authored by scientists from a variety of international agricultural and climate research communities, Neufeldt and his colleagues argue that the concept needs to be evaluated critically because the relationship between the three dimensions is poorly understood, such that practically any improved agricultural practice can be considered climate-smart. This lack of clarity may have contributed to the broad appeal of the concept. From the understanding that we must hold ourselves accountable to demonstrably better meet human needs in the short and long term within foreseeable local and planetary limits, we develop a conceptualization of climate-smart agriculture as agriculture that can be shown to bring us closer to safe operating spaces for agricultural and food systems across spatial and temporal scales. Improvements in the management of agricultural systems that bring us significantly closer to safe operating spaces will require transformations in governance and use of our natural resources, underpinned by enabling political, social and economic conditions beyond incremental changes. Establishing scientifically credible indicators and metrics of long-term safe operating spaces in the context of a changing climate and growing social-ecological challenges is critical to creating the societal demand and political will required to motivate deep transformations. Answering questions on how the needed transformational change can be achieved will require actively setting and testing hypotheses to refine and characterize our concepts of safer spaces for social-ecological systems across scales. This effort will demand prioritizing key areas of innovation, such as (1) improved adaptive management and governance of social-ecological systems; (2) development of meaningful and relevant integrated indicators of social-ecological systems; (3) gathering of quality integrated data, information, knowledge and analytical tools for improved models and scenarios in time frames and at scales relevant for decision-making; and (4) establishment of legitimate and empowered science policy dialogues on local to international scales to facilitate decision making informed by metrics and indicators of safe operating spaces.

PART 1
What Is Sustainable Agriculture?

CHAPTER 1

Sustainable Agriculture: The Basics

Mary V. Gold

Some terms defy definition. "Sustainable agriculture" has become one of them. In such a quickly changing world, can anything be sustainable? What do we want to sustain? How can we implement such a nebulous goal? Is it too late? With the contradictions and questions have come a hard look at our present food production system and thoughtful evaluations of its future. If nothing else, the term "sustainable agriculture" has provided "talking points," a sense of direction, and an urgency, that has sparked much excitement and innovative thinking in the agricultural world.

The word "sustain," from the Latin *sustinere* (*sus-*, from below and *tenere*, to hold), to keep in existence or maintain, implies long-term support or permanence. As it pertains to agriculture, sustainable describes farming systems that are "capable of maintaining their productivity and usefulness to society indefinitely. Such systems... must be resource-conserving, socially supportive, commercially competitive, and environmentally sound." [John Ikerd, as quoted by Richard Duesterhaus in "Sustainability's Promise," Journal of Soil and Water Conservation (Jan.-Feb. 1990) 45(1): p.4. NAL Call # 56.8 J822]

"Sustainable agriculture" was addressed by Congress in the 1990 "Farm Bill" [Food, Agriculture, Conservation, and Trade Act of 1990 (FACTA), Public Law 101-624, Title XVI, Subtitle A, Section 1603 (Government Printing Office, Washington, DC, 1990) NAL Call # KF1692.A31 1990]. Under that law, "the

Reprinted from "Sustainable Agriculture: The Basics," Alternative Farming Systems Information Center, National Agricultural Library, U.S. Department of Agriculture, Special Reference Briefs Series no. SRB 99-02, 2007.

term sustainable agriculture means an integrated system of plant and animal production practices having a site-specific application that will, over the long term:

- satisfy human food and fiber needs;
- enhance environmental quality and the natural resource base upon which
  the agricultural economy depends;
- make the most efficient use of nonrenewable resources and on-farm
  resources and integrate, where appropriate, natural biological cycles and
  controls;
- sustain the economic viability of farm operations; and
- enhance the quality of life for farmers and society as a whole."

1.1   SOME BACKGROUND

How have we come to reconsider our food and fiber production in terms of sustainability? What are the ecological, economic, social and philosophical issues
that sustainable agriculture addresses?

The long-term viability of our current food production system is being ques-
tioned for many reasons. The news media regularly present us with the paradox
of starvation amidst plenty—including pictures of hungry children juxtaposed
with supermarket ads. Possible adverse environmental impacts of agriculture
and increased incidence of food-borne illness also demand our attention. "Farm
crises" seem to recur with regularity.

The prevailing agricultural system, variously called "conventional farming,"
"modern agriculture," or "industrial farming" has delivered tremendous gains in
productivity and efficiency. Food production worldwide has risen in the past
50 years; the World Bank estimates that between 70 percent and 90 percent of
the recent increases in food production is the result of conventional agriculture
rather than greater acreage under cultivation. U.S. consumers have come to
expect abundant and inexpensive food.

Conventional farming systems vary from farm to farm and from country to
country. However, they share many characteristics: rapid technological innovation; large capital investments in order to apply production and management
technology; large-scale farms; single crops/row crops grown continuously over
many seasons; uniform high-yield hybrid crops; extensive use of pesticides, fertilizers, and external energy inputs; high labor efficiency; and dependency on
agribusiness. In the case of livestock, most production comes from confined,
concentrated systems.

Philosophical underpinnings of industrial agriculture include assumptions that "a) nature is a competitor to be overcome; b) progress requires unending evolution of larger farms and depopulation of farm communities; c) progress is measured primarily by increased material consumption; d) efficiency is measured by looking at the bottom line; and e) science is an unbiased enterprise driven by natural forces to produce social good." [Karl N. Stauber et al., "The Promise of Sustainable Agriculture," in Planting the Future: Developing an Agriculture that Sustains Land and Community, Elizabeth Ann R. Bird, Gordon L. Bultena, and John C. Gardner, editors (Ames: Iowa State University Press, 1995), p.13. NAL Call # S441 P58 1995]

Both positive and negative consequences have come with the bounty associated with industrial farming. Some concerns about contemporary agriculture are presented below. They are drawn from the resources compiled at the end of this chapter. While considering these concerns, keep the following in mind: a) interactions between farming systems and soil, water, biota, and atmosphere are complex—we have much to learn about their dynamics and long term impacts; b) most environmental problems are intertwined with economic, social, and political forces external to agriculture; c) some problems are global in scope while others are experienced only locally; d) many of these problems are being addressed through conventional, as well as alternative, agricultural channels; e) the list is not complete; and f) no order of importance is intended.

1.2   ECOLOGICAL CONCERNS

Agriculture profoundly affects many ecological systems. Negative effects of current practices include the following:

Decline in soil productivity can be due to wind and water erosion of exposed topsoil; soil compaction; loss of soil organic matter, water holding capacity, and biological activity; and salinization of soils and irrigation water in irrigated farming areas. Desertification due to overgrazing is a growing problem, especially in parts of Africa.

Agricultural practices have been found to contribute to non-point source water pollutants that include: sediments, salts, fertilizers (nitrates and phosphorus), pesticides, and manures. Pesticides from every chemical class have been detected in groundwater and are commonly found in groundwater beneath agricultural areas; they are widespread in the nation's surface waters. Eutrophication and "dead zones" due to nutrient runoff affect many rivers, lakes, and oceans. Reduced water quality impacts agricultural production, drinking water supplies, and fishery production.

Water scarcity in many places is due to overuse of surface and ground water for irrigation with little concern for the natural cycle that maintains stable water availability.

Other environmental ills include over 400 insects and mite pests and more than 70 fungal pathogens that have become resistant to one or more pesticides; stresses on pollinator and other beneficial species through pesticide use; loss of wetlands and wildlife habitat; and reduced genetic diversity due to reliance on genetic uniformity in most crops and livestock breeds.

Agriculture's link to global climate change is just beginning to be appreciated. Destruction of tropical forests and other native vegetation for agricultural production has a role in elevated levels of carbon dioxide and other greenhouse gases. Recent studies have found that soils may be sources or sinks for greenhouse gases.

1.3   ECONOMIC AND SOCIAL CONCERNS

Economic and social problems associated with agriculture cannot be separated from external economic and social pressures. As barriers to a sustainable and equitable food supply system, however, the problems may be described in the following way:

Economically, the U.S. agricultural sector includes a history of increasingly large federal expenditures and corresponding government involvement in planting and investment decisions; widening disparity among farmer incomes; and escalating concentration of agribusiness—industries involved with manufacture, processing, and distribution of farm products—into fewer and fewer hands. Market competition is limited. Farmers have little control over farm prices, and they continue to receive a smaller and smaller portion of consumer dollars spent on agricultural products.

Economic pressures have led to a tremendous loss of farms, particularly small farms, and farmers during the past few decades—more than 155,000 farms were lost from 1987 to 1997. This contributes to the disintegration of rural communities and localized marketing systems. Economically, it is very difficult for potential farmers to enter the business today. Productive farmland also has been pressured by urban and suburban sprawl—since 1970, over 30 million acres have been lost to development.

1.4 IMPACTS ON HUMAN HEALTH

As with many industrial practices, potential health hazards are often tied to farming practices. Under research and investigation currently is the sub-therapeutic use of antibiotics in animal production, and pesticide and nitrate contamination of water and food. Farmer worker health is also a consideration in all farming practices.

1.5 PHILOSOPHICAL CONSIDERATIONS

Historically, farming played an important role in our development and identity as a nation. From strongly agrarian roots, we have evolved into a culture with few farmers. Less than two percent of Americans now produce food for all U.S. citizens. Can sustainable and equitable food production be established when most consumers have so little connection to the natural processes that produce their food? What intrinsically American values have changed and will change with the decline of rural life and farmland ownership?

World population continues to grow. According to recent United Nations population projections, the world population will grow from 5.7 billion in 1995 to 9.4 billion in 2050, 10.4 billion in 2100, and 10.8 billion by 2150, and will stabilize at slightly under 11 billion around 2200. The rate of population increase is especially high in many developing countries. In these countries, the population factor, combined with rapid industrialization, poverty, political instability, and large food imports and debt burden, make long-term food security especially urgent.

Finally, the challenge of defining and dealing with the problems associated with today's food production system is inherently laden with controversy and emotion. "It is unfortunate, but true, that many in the agriculture community view sustainable agriculture as a personal criticism, or an attack, on conventional agriculture of which they are justifiably proud. ‹I guess that the main thing people get defensive about when you say sustainable,' explained one agent, 'is that it implies that what they've been doing is not sustainable. And that's the biggest issue.'" [Judy Green, "Sustainable Agriculture: Why Green Ideas Raise a Red Flag," Farming Alternatives Newsletter (Cornell) (Summer 1993).

1.6 A SAMPLING OF PERSPECTIVES

"It's easy to understand why key individuals and organizations in agriculture have flocked to this term. After all, who would advocate a 'non-sustainable agriculture?'" [Charles A. Francis, "Sustainable Agriculture: Myths and Realities," Journal of Sustainable Agriculture (1990) 1(1): p.97. NAL Call # S494.5.S86S8] Despite the appeal of a sustainable agriculture philosophy, however, discussions about how best to define and achieve sustainability present some controversy. Supporters of sustainable agriculture come from diverse backgrounds, academic disciplines, and farming practices. Their convictions as to what elements are acceptable or not acceptable in a sustainable farming system sometimes conflict. They also disagree on whether sustainable agriculture needs defining at all.

"Wes Jackson, geneticist and co-founder of the Land Institute, was probably the first to use the term 'sustainable agriculture' in recent times (Jackson, 1978) ["Toward a Sustainable Agriculture," Not Man Apart, p. 4-6. Friends of the Earth, 1978]. Since natural ecosystems have stood the test of time, Jackson argued, they should serve as models for sustainable agriculture." [Greg McIsaac, Sustainable Ag. Definition (SANET-mg post, March 1994). Available at SANET-mg Archives Website: http://www.sare.org/sanet-mg/archives/html-home/4-html/0101.html (8/23/07)]

"Sustainable agriculture is a philosophy based on human goals and on understanding the long-term impact of our activities on the environment and on other species. Use of this philosophy guides our application of prior experience and the latest scientific advances to create integrated, resource-conserving, equitable farming systems. These systems reduce environmental degradation, maintain agricultural productivity, promote economic viability in both the short and long term, and maintain stable rural communities and quality of life." [Charles Francis and Garth Youngberg, "Sustainable Agriculture — An Overview," in Sustainable Agriculture in Temperate Zones, edited by C.A. Francis, C.B. Flora and L.D. King (New York: Wiley, 1990), p.8. NAL Call # S494.5.S86S87]

"Sustainable agriculture does not mean a return to either the low yields or poor farmers that characterized the 19th century. Rather, sustainability builds on current agricultural achievements, adopting a sophisticated approach that can maintain high yields and farm profits without undermining the resources on which agriculture depends." ["Frequently Asked Questions About Sustainable Agriculture," in Sustainable Agriculture–A New Vision (Union of Concerned Scientists, 1999). Available at UCS Website: http://www.ucsusa.org/food_ and_environment/sustainable_food/questions-about-sustainable-agriculture. html (8/23/07)]

"A systems approach is essential to understanding sustainability. The system is envisioned in its broadest sense, from the individual farm, to the local ecosystem, and to communities affected by this farming system both locally and globally... A systems approach gives us the tools to explore the interconnections between farming and other aspects of our environment." [University of California Sustainable Agriculture Research and Education Program (SAREP), What Is Sustainable Agriculture? (SAREP, 1998). Available at SAREP Website: http://www.sarep.ucdavis.edu/concept.htm (8/23/07)]

"Environmental sustainability implies the following:

- meeting the basic needs of all peoples, and giving this priority over meeting the greeds of a few
- keeping population densities, if possible, below the carrying capacity of the region
- adjusting consumption patterns and the design and management of systems to permit the renewal of renewable resources
- conserving, recycling, and establishing priorities for the use of nonrenewable resources
- keeping environmental impact below the level required to allow the systems affected to recover and continue to evolve.

"An environmentally sustainable agriculture is one that is compatible with and supportive of the above criteria." [Stuart B. Hill, Environmental Sustainability and the Redesign of Agroecosystems (Ecological Agriculture Projects (EAP), McGill University, 1992). Available at EAP Website: http://eap.mcgill.ca/publications/EAP34.htm (2/24/2009)] Dr. Hill goes on to explain: "To help recognize these real issues I distinguish between shallow (short-term, symbolic) and deep (long-term, fundamental) sustainability. Shallow sustainability focuses on efficiency and substitution strategies with respect to the use of resources. It usually accepts the predominant goals within society without question, and aims to solve problems by means of curative solution. Deep sustainability, in contrast, re-evaluates goals in relation to higher values and redesigns the systems involved in achieving these goals to that this can be done within ecological limits." [Ibid.]

Sustainable agriculture is "a way of practicing agriculture which seeks to optimize skills and technology to achieve long-term stability of the agricultural enterprise, environmental protection, and consumer safety. It is achieved through management strategies which help the producer select hybrids and varieties, soil conserving cultural practices, soil fertility programs, and pest management programs. The goal of sustainable agriculture is to minimize adverse impacts to the immediate and off-farm environments while providing a sustained level of production and profit. Sound resource conservation is an integral part of the means to achieve sustainable agriculture." [USDA Natural Resource Conservation Service (NRCS) General Manual (180-GM, Part 407). Available at USDA Website: http://www.info.usda.gov/default. aspx?l=176 Select Title 180; Part 407 - Sustainable Agriculture; Subpart A - General. (10/20/09)]

"Today, sustainable farming practices commonly include:

- crop rotations that mitigate weeds, disease, insect and other pest problems; provide alternative sources of soil nitrogen; reduce soil erosion; and reduce risk of water contamination by agricultural chemicals
- pest control strategies that are not harmful to natural systems, farmers, their neighbors, or consumers. This includes integrated pest management techniques that reduce the need for pesticides by practices such as scouting, use of resistant cultivars, timing of planting, and biological pest controls
- increased mechanical/biological weed control; more soil and water conservation practices; and strategic use of animal and green manures
- use of natural or synthetic inputs in a way that poses no significant hazard to man, animals, or the environment.

"This approach encompasses the whole farm, relying on the expertise of farmers, interdisciplinary teams of scientists, and specialists from the public and private sectors." [Paul F. O'Connell, "Sustainable Agriculture, a Valid Alternative," Outlook on Agriculture (1992) 21(1): p.6. NAL Call # 10 Ou8]

From NGO Sustainable Agriculture Treaty, Global Forum at Rio de Janeiro, June 1-15, 1992:

- "Sustainable agriculture is a model of social and economic organization based on an equitable and participatory vision of development which recognizes the environment and natural resources as the foundation of economic activity. Agriculture is sustainable when it is ecologically sound, economically viable, socially just, culturally appropriate and based on a holistic scientific approach.

- "Sustainable agriculture preserves biodiversity, maintains soil fertility and water purity, conserves and improves the chemical, physical and biological qualities of the soil, recycles natural resources and conserves energy. Sustainable agriculture produces diverse forms of high quality foods, fibers and medicines.
- "Sustainable agriculture uses locally available renewable resources, appropriate and affordable technologies and minimizes the use of external and purchased inputs, thereby increasing local independence and self sufficiency and insuring a source of stable income for peasants, family and small farmers and rural communities. This allows more people to stay on the land, strengthens rural communities and integrates humans with their environment.
- "Sustainable agriculture respects the ecological principles of diversity and interdependence and uses the insights of modern science to improve rather than displace the traditional wisdom accumulated over centuries by innumerable farmers around the world." [These excerpts are from NGO Sustainable Agriculture Treaty (Global Forum at Rio de Janeiro, June 1-15, 1992). Available at Information Habitat Website: http://habitat.igc.org/treaties/at-20.htm (8/23/07]
- "Sustainable agriculture does not refer to a prescribed set of practices. Instead, it challenges producers to think about the long-term implications of practices and the broad interactions and dynamics of agricultural systems. It also invites consumers to get more involved in agriculture by learning more about and becoming active participants in their food systems. A key goal is to understand agriculture from an ecological perspective—in terms of nutrient and energy dynamics, and interactions among plants, animals, insects and other organisms in agroecosystems—then balance it with profit, community and consumer needs."

[Sustainable Agriculture Research and Education (SARE), Exploring Sustainability in Agriculture: Ways to Enhance Profits, Protect the Environment and Improve Quality of Life." (SARE, 1997). Available at SARE Website: http://www.sare.org/publications/exploring.htm (8/23/07)]

"Sustainable agriculture: A whole-systems approach to food, feed, and other fiber production that balances environmental soundness, social equity, and economic viability among all sectors of the public, including international and intergenerational peoples. Inherent in this definition is the idea that sustainability must be extended not only globally, but indefinitely in time, and to all living organisms including humans.

"Sustainable agroecosystems:

- maintain their natural resource base
- rely on minimum artificial inputs from outside the farm system
- manage pests and diseases through internal regulating mechanisms
- recover from the disturbances caused by cultivation and harvest."

[Stephen R. Gliessman, "An Ecological Definition of Sustainable Agriculture," Principles of Agroecology and Sustainability (1998). Available at Agroecology Home Website: http://agroecology.org/Principles_Def.html (6/9/08)]

"Consumers can play a critical role in creating a sustainable food system. Through their purchases, they send strong messages to producers, retailers and others in the system about what they think is important. Food cost and nutritional quality have always influenced consumer choices. The challenge now is to find strategies that broaden consumer perspectives, so that environmental quality, resource use, and social equity issues are also considered in shopping decisions. At the same time, new policies and institution must be created to enable producers using sustainable practices to market their goods to a wider public." [University of California Sustainable Agriculture Research and Education Program (SAREP), What Is Sustainable Agriculture? (SAREP, 1998). Available at SAREP Website: http://www.sarep.ucdavis.edu/concept.htm (8/23/07)]

"... I think the community has reached about as explicit, useful, concrete a definition of sus ag [sic] as now possible, or possible at any given time, given the differences of opinion, world view, etc., that exist. At any point in time, in any society, the definition of any concept like sus ag is going to be a compromise among differing world views, sets of values, etc. no one of which has any way to prove the other wrong, or illegitimate. So the sus ag tent is now relatively stable; its shape and innards perhaps fully pleasing to no one, but I am certain there is no real point in debating the fine points anymore because we will simply document more crisply the differences that are out there, and have been all along.... One of the other realities is that the 'definition' of something like sus ag is going to remain fluid, driven by changes in politics, ideology, science, community values, etc." [Charles Benbrook, More Soil Quality and Def. (SANET-mg post,

Feb. 1995). Available at SANET-mg Archives Website: http://www.sare.org/
sanet-mg/archives/html-home/7-html/0080.html (8/23/07)]

"I concluded some time ago that we didn't need to spend much more time
and effort attempting to define sustainability. We have sufficient commonality
among our different understandings of it to continue moving in the right general
direction, even if we are not yet all moving toward precisely the same destination by the same route. More recently, I have come to the conclusion that we
may never have a generally accepted definition of sustainability, and perhaps, we
don't need one." [John Ikerd, On Not Defining Sustainability (SANET-mg post,
May 1998). Available at SANET-mg Archives Website: http://www.sare.org/
sanet-mg/archives/html-home/25-html/0203.html (8/23/07)]

"'Sustainability' is at once extremely important and practically useless. It consists of a set of concepts which are fundamentally different in nature. That is
why there has been no success in attempt to identify THE definition of sustainability. There can be no satisfactory definition which is not multifaceted. This
poses serious difficulties for the practical application of sustainability as an
objective in real decision-making. We have suggested here that these difficulties
be addressed by focusing on the particular aspects of sustainability which the
decision maker considers to be important, and presenting information about the
trade offs between these aspects within a multiple criteria decision making for-
mula." [David J. Pannell and Steven Schilizzi, "Sustainable Agriculture: A Matter
of Ecology, Equity, Economic Efficiency or Experience?" Journal of Sustainable
Agriculture (1999) 13(4): p.65. NAL Call #: S494.5 S86S8]

1.7   THE FUTURE OF THE SUSTAINABLE
AGRICULTURE CONCEPT

Many in the agricultural community have adopted the sense of urgency and
direction pointed to by the sustainable agriculture concept. Lack of sharp
definition has not lessened its authenticity. Sustainability has become an integral component of many government, commercial, and non-profit agriculture research efforts, and it is beginning to be woven into agricultural policy.
Increasing numbers of farmers and ranchers have embarked on their own
paths to sustainability, incorporating integrated and innovative approaches
into their own enterprises.

This just-do-it attitude is the real force carrying the issue of sustainability into the next century. "The best way to communicate the meaning of sustainable agriculture is through real-life stories of farmers who are developing sustainable farming systems on their own farms," says John Ikerd, describing the 1,000 Ways to Sustainable Farming project funded by USDA's Sustainable Agriculture Research and Education Program. The project sought to explore and refine the definition of sustainable agriculture by profiling successful sustainable farmers and ranchers." SARE continued the project, renaming it The New American Farmer. "In addition to describing successful farming practices, the features in The New American Farmer detail the effects of those practices on farm profitability, quality of life, rural communities and the environment." [see The New American Farmer: Profiles of Agricultural Innovation, 2nd ed. (SARE, 2005). Available at SARE Website: http://www.sare.org/publications/naf.htm (8/23/07)]

Critical discussion of the sustainable agriculture concept will and should continue. Understanding will deepen; answers will continue to come. On-going dialog is important for another reason: with more parties, each with its own agenda, jumping into the sustainable agriculture "tent," only a continued focus on the real issues and goals will keep sustainable agriculture from becoming so all-encompassing as to become meaningless.

Youngberg and Harwood's 1989 statement still holds true: "We are yet a long way from knowing just what methods and systems in diverse locations will really lead to sustainability... In many regions of the country, however, and for many crops, the particular mix of methods that will allow curtailing use of harmful farm chemicals or building crop diversity, while also providing economic success, are not yet clear. The stage is set for challenging not only farm practitioners, but also researchers, educators, and farm industry." [Garth Youngberg and Richard Harwood, "Sustainable Farming Systems: Needs and Opportunities," American Journal of Alternative Agriculture (1989) 4(3 & 4): p.100. NAL Call # S605.5.A3]

REFERENCES

1. Clive A. Edwards, Rattan Lal, Patrick Madden, Robert H. Miller, and Gar House, editors, Sustainable Agricultural Systems (Soil and Water Conservation Society, 1989). Chapters 1–6 and Chapter 38.
2. Committee on the Role of Alternative Farming Methods in Modern Production Agriculture, National Research Council, Alternative Agriculture (National Academy Press, 1989). Chapter 2.

3. Elizabeth Ann R. Bird, Gordon L. Bultena, and John C. Gardner, Planting the Future: Developing an Agriculture that Sustains Land and Community (Iowa State University Press, 1995). Chapter 1, Part 1 and Chapter 2, Introduction.

4. J. Patrick Madden and Scott G. Chaplowe, editors, For All Generations: Making World Agriculture More Sustainable (World Sustainable Agriculture Association, 1996). Chapter 1 and Chapter 10.

5. Raymond Forney, A Common Vision: Evaluating the Farming Industry's Progress Toward Sustainability (Dupont/Chesapeake Farms, 1998).

6. Tracy Irwin Hewitt and Katherine R. Smith, Intensive Agriculture and Environmental Quality: Examining the Newest Agricultural Myth (Henry A. Wallace Institute for Alternative Agriculture, 1995). Available at: http://www.winrock.org/wallace/wallacecenter/documents/iaeq.pdf (8/23/07).

7. Committee on Long-Range Soil and Water Conservation, National Research Council, Soil and Water Quality: An Agenda for Agriculture (National Academy Press, 1993). Executive Summary and Chapter 1.

8. Pesticide Action Network (PAN) Updates Service, World and U.S. Agrochemical Market in 1998 (PAN, July 23, 1999). Available at: http://www.panna.org/resources/panups/panup_19990723.dv.html (8/23/07).

9. Arnold L. Aspelin, Pesticides Industry Sales and Usage: 1994 and 1995 Market Estimates (U.S. EPA, 1997).

10. National Water Quality Assessment, Pesticide National Synthesis Project, U.S. Geological Survey, Pesticides in Ground Water (USGS Fact Sheet FS-244-95), Pesticides in Surface Waters (USGS Fact Sheet FS-939-97), and Pesticides in the Atmosphere (USGS Fact Sheet FS-152-95).

11. David Pimentel et al., "Environmental and Economic Costs of Pesticide Use," in BioScience (Nov. 1992) 42(10): pp.750-760.

12. John Fedkiw, compiler, USDA Working Group on Water Quality, Nitrate Occurrence in U.S. Waters (and Related Questions): A Reference Summary of Published Sources from an Agricultural Perspective (USDA, 1991). Question 1 (To What Extent is Nitrate-N Found in U.S. Waters), Executive Summary.

13. Sandra Postel, Last Oasis: Facing Water Scarcity, rev. ed. (Worldwatch Environmental Alert Series; Norton, 1997). Chapters 1 and 2.

14. Wayne A. Morrissey and John R. Justus, Congressional Research Service Issue Brief for Congress, 89005: Global Climate Change (CRS and the Committee for the National Institute for the Environment, May 1999). 2006 version available at: http://www.ncseonline.org/NLE/CRS/abstract.cfm?NLEid=348 (8/23/07).

15. Farms and Land in Farms (USDA National Agricultural Statistics Service, Feb. 1999). Available at: http://usda.mannlib.cornell.edu/usda/nass/FarmLandIn/1990s/1999/FarmLandIn-02-26-1999.pdf (8/23/07).

16. Marty Strange, Family Farming: A New Economic Vision (University of Nebraska Press, 1988). Introduction and Chapters 1 and 2.

17. Ann Sorensen, Richard P. Greene, and Karen Russ, Farming on the Edge (American Farmland Trust, 1997). Major Findings. Available at: http://www.farmland.org/resources/fote/default.asp (8/23/07).

18. National Commission on Small Farms, USDA, A Time to Act: A Report of the USDA National Commission on Small Farms (USDA Misc. Pub. 1545) (USDA, 1998).

Introduction. Available at: http://www.csrees.usda.gov/nea/ag_systems/pdfs/time_to_act_1998.pdf. Overview: http://www.csrees.usda.gov/nea/ag_systems/in_focus/small-farms_if_time.html(2/5/09).

19. Edward Groth III, Charles M. Benbrook, and Karen Lutz, Do You Know What You're Eating? An Analysis of U.S. Government Data on Pesticide Residues in Foods (Consumers Union of United States, Consumer Reports, 1999).

20. Margaret Reeves and Kristin Schafer, Fields of Poison: California Farmworkers and Pesticides (Pesticide Action Network, 1998?). Executive Summary. Available at: http://www.panna.org/resources/documents/fieldsSum.dv.html (8/23/07).

21. Population Division, Department of Economic and Social Affairs, United Nations Secretariat, World Population Projections to 2150 (United Nations, Feb. 1998).

PART 2
Sustainably Connecting Producers and Consumers

Albanian and UK Consumers' Perceptions of Farmers' Markets and Supermarkets as Outlets for Organic Food: An Exploratory Study

Athina-Evera Qendro

2.1 INTRODUCTION

There has been a recent increase in consumers' interest in local, healthy and environmentally friendly food which has brought about a renewed interest in farmers' markets [1]. Farmers' markets have become a synonym for food quality considering that local produce is often picked when ripe and fresh [2].

Over the past decade, there has been a rise in the interest that UK consumers show in purchasing food from farmers' markets [3]. This is partly due to positive associations with both the products available and the shopping process at farmers' markets [4,5]. However, it may also be due in part to the negative connotations that some green consumers have of more mainstream food outlets, particularly supermarkets [6] (p. 620). In Albania, on the other hand, consumers who have been traditionally reliant on farmers' markets for a wide range of local foodstuffs are embracing supermarkets for the first time. The rise of supermarkets in this post-communist (Post-communist country is a country in transition from a communist regime. Communist is the regime where the private property and a profit-based economy are replaced with public ownership and communal control of at least the major means of production (e.g., mines, mills, and factories) and the natural resources of a society (Encyclopaedia Britannica).) European country has been dramatic [7].

In the UK, the farmers' market is often portrayed as a greener alternative for purchasing food and the rise of the farmers' market is understood to denote the rise of a greener consumer interested in food that is more likely to be local, organic and seasonal. From this perspective, the rise of supermarkets in Albania could be understood to represent a move away from green consumption. However, given the dearth of research into consumer behavior in Albania, it is impossible to tell how this changing food retail landscape is being understood by individual consumers. The study described in this paper tackles these issues for the first time by conducting a qualitative pilot study which asks both UK and Albanian consumers to talk about their food shopping in detail in order to explore consumers' perceptions and better understand these two opposite retail trends and understand how this affects the purchasing of organic food.

There are two reasons why the UK and Albania were chosen for this study. The first reason is that opposing trends can be seen to be at work within the retail sectors in these two contexts. Consumers in the UK have had access to organic food and supermarkets for many decades. However, farmers' markets are emerging as a new food outlet in the country. British consumers' perceptions of organic food and supermarkets are documented in many academic studies [8,9,10,11,12,13,14,15,16,17,18,19,20]. Farmers' markets have recently gained popularity among consumers and researchers [21,22,23,24,25,26]. A report conducted by the Soil Association [27] in the UK showed that respondents purchase organic food mainly from supermarkets and farmers' markets. Sales of organic products through farmers' markets and farm shops increased by 3.5% to £42.7 million in 2013 [27] whereas sales of organic products through supermarkets increased by 1.2% to £1275.6 million. It seems that UK consumers' interest in organic food is increasing and that the growth rate of organic food bought from farmers' markets is higher than for supermarkets. In contrast, in Albania organic food and supermarkets are new concepts among Albanian consumers whereas farmers' markets have been the main food outlets in the country for the last 50 years to provide fruit and vegetables.

The second reason is that UK and Albania are very different in terms of their economies, and the extent to which they have been researched. Albania is an economy emerging from the old communist regime. Thus, cultural inheritances persist alongside institutional and market changes. Moving from a state controlled economy, interest from private investors and private organizations has risen in the country which also involves the emergence of supermarkets and modern terms such as organic food. In terms of consumer behavior research, little is known about the Albanian consumer. The UK, on the other hand, provides a free market context for the retail sector that has been subject to decades of scrutiny from consumer behavior researchers. These differences make these two contexts interesting for asking questions about consumer perceptions of organic food and the outlets that it is purchased from. Therefore, the purpose of this pilot study is to find out whether there are similarities in the perception of organic food provided in supermarkets and farmers' markets by respondents in a developing, post-communist country as opposed to a Western European country.

The paper will begin by outlining the current debates within the literature on supermarkets and then farmers' markets. Following this, the methods used for data collection will be outlined. Data from both countries is then presented and discussed in the light of the extant literature in order to demonstrate how supermarkets and farmers' markets are currently understood by the respondents in each country as outlets for organic food and make recommendations for how researchers can take forward this under-researched area.

2.2 LITERATURE REVIEW

2.2.1 SUPERMARKETS

Supermarkets emerged as a particular retail form in the United States of America (USA) during the 1930s and gradually spread to countries outside the USA, adopting local and global characteristics [28]. The trend characterizing the development of these new retail forms was similar in the United States and the United Kingdom: supermarket stores started as places offering dry food and then full-line fresh foods [29]. Fresh produce was not available in supermarkets until the 1960s due to the belief that it was impossible to move consumers from the wet markets and fruit shops to these new big retail stores [29].

To begin with, the supermarket development began with locations in urban areas in Western countries, and then in the 1990s, an out-of-town initiative was accompanied by the decision to increase the size of the stores as well as building supermarkets at the edge-of-town centers. Latterly, supermarkets began to locate abroad because of the limited opportunities for market growth in the US and Europe [30]. For this reason, supermarket chains began searching for new market opportunities overseas [30].

Supermarkets are the dominant format of food and grocery retailing in Western societies [31]. For example, in the UK there are four major British supermarket chains—Tesco, Asda, Sainsbury's and Morrisons—which together account for a total of 75.4% of all UK food shopping [32].

According to Reardon et al. [33] following the success in Western countries, the "supermarket revolution" took place in three waves in developing countries (A developing country is one in which the majority lives on far less money—with far fewer basic public services—than the population in highly industrialized countries (World Bank).): the "first-wave" started in the mid-1990s in South American and East Asian countries, "the second-wave" encompassed Mexico, much of South-East Asia, Central America, and Southern-Central Europe, and in the "third-wave," which occurred in the late 1990s early 2000s, Eastern/Southern Africa, India and China experienced the "supermarket revolution" [34].

The above findings show that there is a contrast in the supermarket development growth between developed and developing countries. Developed countries have experienced a gradual growth in the number of supermarkets while in developing countries, the supermarket entry and expansion has happened in only a matter of years. In Albania, for example, Euromax was the first supermarket chain that entered the country in 2005 [35]. Other supermarket chains include the Greek chain Marinopoulos with Carrefour introducing the first hypermarket in 2011, which is the largest shopping center in the country [35]. This fact supports Hawke's [30] statement about supermarkets' fast growth in developing countries where traditional outlets, hypermarkets, and discounters coexist at the same time.

Supermarkets have been emerging not only in developing countries but also in post-communist countries in Europe such as in Russia, Croatia, Latvia, Slovakia Hungary, Poland and the Czech Republic. Most of these countries were state controlled economies and the state played an important role in the retail sector [36,37]. Research in post-communist countries is focused on the expansion strategies of large Western supermarket chains that are entering these countries as opposed to analyzing consumers' perceptions [36,37]. There is some evidence of studies investigating consumers' perceptions towards supermarkets exist in Lithuania [38] and another one revealing that young Czechs, Hungarians, and Poles like shopping at supermarkets and are more open to Western shopping ways [39]. As stated earlier, Albania is another post-communist country that is experiencing the supermarket phenomenon. Four international supermarket brands are now operating in the country sharing 20% of the market share with the remaining 80% being spent on local markets and smaller shops [7]. Supermarket stores in Albania are perceived to be trusted sources of food products [40].

Supermarket development came as a response to the changes in the habits, demands and preferences of consumers in developing countries [30]. Urbanization, the entry of women into the workforce outside of the home, and the higher per capita incomes are all contributory factors to this expansion [41]. According to Traill [42], supermarkets in developing countries are no longer places frequented only by rich people. This is partly because of the rising incomes and the desire to imitate Western lifestyles [42]. Moreover, supermarkets in developing countries are perceived as places for entertainment, where families spend their day shopping together [43]. According to Harvey [19], another reason behind the fast supermarket development is the internal niche market they create by bringing together high with low-income consumers with distinctive marques at high prices, along with their own-label discount ranges. However, despite the fact that supermarkets are widely accessible to lower-income consumers, their use is not similar to higher-income consumers [44].

Academic literature presents a plethora of studies around supermarkets discussing a variety of matters such as their entry into various countries, the threat they pose to the local small shops [45,46], prices [47,48,49], nutrition issues of the products available in supermarkets [50,51,52], customer relationship marketing (CRM), strategies and maintaining customer relationships [53,54], and customer service and loyalty [55,56,57,58,59]. In Western countries, the supermarket literature is becoming more specialized, abandoning slowly the concept of a pure consumer perception of supermarkets. These new studies are covering topics such as "recording brain waves at the supermarket: what can we learn from a shopper's brain" [60], "the influence of background music on consumer behaviour [61], "aroma stimuli influencing shopper's behaviour and satisfaction" [62] and obesity issues related to the food bought in supermarkets [63,64]. However, this exploratory study aims to refocus on the traditional consumer perception research. The findings of this study will have managerial implications for farmers' markets and supermarkets on how to develop strategies to promote organic food to consumers in two different countries where two dis-similar trends of consumption are developing.

2.2.2 FARMERS' MARKETS

In the past, farmers' markets used to be held in outdoor settings in which small farmers gathered to sell directly to the public, usually from the backs of pick-up trucks or on makeshift tables [65] (p. 15). Nowadays, farmers' markets are not necessarily held only outdoors. The farmers' market in Cheltenham, for example, is held in a venue dedicated to local farmers and local producers in the heart of the town's shopping center [66]. It seems that the farmers' market structure has changed over the years. Back in the 80s, farmers would sell their products in the streets while nowadays farmers' markets take place in organized venues/spaces.

According to the National Farmers' Retail & Markets' Association [67], the first UK farmers' market started in Bath in September 1997 and since then the term "farmers" market' has stood for locally produced food sold by the people who made it. FARMA accredits certificates to farmers when they comply with the following criteria:

- The food is produced locally
- The stall is attended by the producer or someone involved in production
- All the goods on sale will have been grown, reared or processed by the stallholders.

Farmers' markets have long existed as the prime outlet for such small-scale producers with the appeal of providing "local" produce to urban residents [68]. However, there is a consensus as to their common characteristics:

- "involving direct selling to the consumer by the person who grew, reared, or produced the foods,
- "in a common facility where the above activity is practiced by numerous farmers,
- "who sell local produce" [69] (p. 399)

Typically, the goods sold at a farmers' market would include: fruit and vegetables, meats and meat products, cheeses and other dairy products, fish, honey, bakery products, jams, pickles, dressings and sauces [68]. Farmers' markets are also believed to offer a more diverse range of products compared to supermarkets [70].

The increasing preference for locally produced food has resulted in 30% sales growth in the UK over the last four years [3]. There is also growth in the number of farmers' markets which stands at over 550, representing some 9500 market days and 230,000 stallholders throughout the UK [71]. The growth of farmers' markets in the UK is often linked to consumers' desire to move away from food that was produced using industrialized techniques [72]. They have been associated with a growing interest in rural heritage, culinary traditions and tourism [4]. Farmers' markets have also been identified as part of a larger food system movement that establishes consumers' cultural and personal identity [5]. Increased consumption of fresh fruit and vegetable products of better nutritional content, good taste, and flavour has also contributed to the success of farmers' markets over the years [73].

It has been argued that farmers' markets serve to [68]:

- enable increased profit margins, by shortening the supply chain
- provide a secure and regular outlet for produce
- provide fresh local produce, direct from the producer
- generate new business and boost trade for adjacent shops
- encourage consumers to support local business and local agriculture
- ensure that money stays in the locality
- reduce transport, food miles and packaging and
- encourage people to interact.

Further benefits are associated with farmers' markets such as attracting people into the areas in which they take place and they also contribute to local economic development through the support of local traders [70]. La Trobe [22] argues that farmers' markets may increase local economic sustainability and they do not have a negative impact on the environment in terms of transportation and food miles [68,70]. Farmers' markets may also act as a regeneration initiative for the local economy [22,70,74]. Gerbasi's study [74] showed that farmers themselves believe that the farmers' markets not only increase their household income (from a consumers' perspective, the most common reason for buying from a farmers' market is to support the local farmers) but also stimulate the local economy, add cultural value and improve social networks between farmers and their customers.

Tong et al. [1] examine farmers' markets from a different angle. They suggest that farmers' markets differ from other food stores because they quite often have very limited hours of service and are distributed very sparsely in space.

This means that farmers' markets are not easily accessible by people whose schedule does not fit with the markets' service hours.

Studies in most European countries such as Sweden [75,76], Norway [77,78,79], the Netherlands, and Germany [80,81] indicate that the development of alternative food networks such as the farmers' markets is often based on "modern" and more "commercial" quality definitions, stressing environmental sustainability or animal welfare, and on innovative (and retailer-led) forms of marketing [82]. According to Evans et al. [83], the general move towards "ecological modernization" may reconcile agriculture and environment in ways which continue to support agricultural production.

However, the development of farmers' markets in countries such as Italy, Spain, and France mostly builds on activities of regional quality production and direct selling with long-lasting traditions [12]. Other countries such as Portugal and Greece originate more than 75% of European registered, regionally designated products. This happens due to cultural and structural factors that reinforce links among region, tradition, origin and quality in countries in the south of Europe [84]. These countries include small, diversified, and labour-intensive family farms employing traditional methods of producing food, the prevalence of the convention of "domestic worth," with the market and industrial conventions embedded in robust localistic and civic orders of evaluation [85] (p. 186).

Literature around farmers' markets in Eastern or post-communist Europe is very scarce: Rizov and Mathijs [86] have analyzed farm survival related to the scale of farms in Hungary; Blaas [87] focuses on farm holders and their households in Slovakia. However, in the Czech Republic, the consumers' perceptions of farmers' markets have been investigated [88]. In contrast to the US/Western findings of farmers' market consumer perception [89,90], the survey in Prague showed that consumers of different ages, family sizes and occupation type shop from farmers' markets. These consumers are mainly motivated by the freshness of food and better taste. Many Western scholars aimed to reveal the characteristics of farmers' markets consumers. They concluded that the typical customers are predominantly educated, urban, middle-class, middle aged and well off [91,92,93,94]. In Prague, a variety of consumer types would visit the markets such as urban middle class shoppers, mothers with children, and pensioners, claiming that farmers' markets provide an interesting alternative and a welcomed uniqueness to the shopping experience across the socioeconomic and demographic divide of shoppers to the anonymous, large-scale hypermarkets [88]. This study showed another interesting outcome: Czech consumers seemed to care less about environmental or ethical issues, support for local farmers or community building functions than their fellow shoppers in Western European countries and the USA [3,94,95]. The main motivation of Czech consumers is rather hedonistic; it is about the freshness, taste and enjoyment of the experience [88].

2.2.3   BUYING ORGANIC FOOD

"It is not an easy task to provide precise and universally agreed definition of organic agriculture and organic food" [96] (p. 358). "Organic food products" is a term that began to be used in the 1940s. It refers to food raised, grown and stored and/or processed without the use of synthetically produced chemicals or fertilizers, herbicides, pesticides, fungicides, growth hormones and regulators or generic modification [96]. According to the European Commission's regulation [97]: "Organic production is an overall system of farm management and food production that combines the best environmental practices, a high level of biodiversity, the preservation of natural resources, the application of high animal welfare standards and a production method in line with the preference of certain consumers for products produced using natural substances and processes." Organic foods are defined as products that contain no fertilizers, no chemicals, no pesticides, no antibiotics, no hormones, no genetically modified organisms (GMOs); and are not processed, not over packaged, no injection/no harm for animals, natural, tasty, nutritious, colorful, fresh/stays fresh longer, and are labor intensive [98,99,100].

Defining organic food is with no doubt significant, however, it is also important to distinguish between certified and non-certified organic food. To begin with, in UK there are 11 official certification bodies: Organic Farmers & Growers Ltd., Organic Food Federation, Soil Association Certification Ltd., Biodynamic Association Irish Organic, Farmers and Growers Association (IOFGA), Organic Trust Limited Quality Welsh, Food Certification Ltd. Ascisco Ltd., Global Trust Certification Ltd. and Scottish Food Quality Certification Ltd. (Department for Environment, Food and Rural Affairs, UK, 2012). In Albania, there are three official certification bodies that certificate and appropriately label organic products: BioAdria (founded in 2006), Institute of Organic Agriculture (founded in 2010) and Albinspekt (founded in 2006) [101]. A key distinction between certified and non-certified organic production is the inspections–labeling process [102] Inspections mobilize contractually independent inspectors, and then forward inspection information to labelers who determine whether the product meets standards to be certified as organic [102].

"Organic" has many different meanings and interpretations and is often associated, and sometimes confused, with terms such as "green," "ecological," "environmental," "natural" and "sustainable" [10,103,104]. It is interesting to note that consumers have different meanings for the term "organic" and what is organic to one may not be organic to another [105]. Moreover, producers and regulators of organic food may also interpret the meaning of organic differently from consumers [105].

On the other hand, in the UK consumers consider that farmers provide them with organic, locally produced, fresh, tasty and quality food and there is often an implicit assumption that farmers' market food is organic, which is not always the case [106]. Moreover, the empirical findings of Vanhonacker et al. [107] suggested that consumers perceive organic food to be the same as locally produced food. Motives of British consumers in buying organic food include fair trade, social aspects, and support of local farmers and environmental protection [6,108] as well as ethical and moral issues [109]. Health and food safety are factors that motivate consumers to buy organic products as well as high standards of animal welfare [14]. The factors that may affect consumers' decisions to buy organic food are knowledge, social, cultural and personal factors [15]. "Consumers perceive organic food as a means of achieving individual and social values of which the most important is centered around the health factor for themselves and their families" [110] (p. 351).

As mentioned above, "organic" may be interpreted differently by consumers' interpretations that may vary from country to country. Their primary motivations for buying organic food has been found to be related to health issues, particularly the fear of ingesting pesticide residues and chemical inputs [111,112,113,114,115,116,117]. In Albania for example, despite the fact that very little research has been undertaken to investigate Albanians' perceptions of organic food, a study showed [118] that Albanians believe that organic food comes directly from farmers' markets. However, their findings also showed that Albanians prefer to buy organic food from supermarkets which offers a contradiction to respondents' beliefs that "organic" food comes directly from farmers.

Consumers of organic products also justify their choice by the quality and taste of these foods and by their alignment with an agricultural model that matches their values: animal welfare, support for local producers, small-scale production, and trust in organic producers, who are not perceived as profit-driven [117]. Large companies such as supermarkets that produce and market organically certified foods are therefore somewhat distrusted by consumers [119].

"Organic" is used as a heuristic for naturalness and "greenness" [120] (p. 98). However, a study indicates that the majority of the respondents were unsure of the sources of the ideas they had about organic food and this process may be described as accidental learning [121]. People seem to build up knowledge of the meaning of the term over time via "osmosis" and terms like organic have different meanings to different individuals or groups especially when the international dimensions are concerned [120] (p. 99).

Seyfang's [122] (p. 198) findings showed that consumers expressed a wide range of economic, social, environmental and personal reasons for purchasing local organic food from a farmers' market, and many were quite deliberately avoiding supermarkets where possible, choosing to support the alternative food network instead. On the other hand, a study conducted in Malaysia showed that organic products were mainly bought by organic food buyers from conventional markets followed by natural and whole food supermarkets [123]. Chinnici et al. [124] found that there are four segments of organic consumers: "pioneers" who purchase at the supermarket out of curiosity; "nostalgic" are the ones that associate organic produce with the past; "health conscious" consumers who regularly purchase organic produce due to health concerns and who prefer specialized retailers such as farmers' markets and expect to pay a premium; and "pragmatist" consumers who are knowledgeable, but price-sensitive. In the UK, a study [10] showed that all respondents wanted organic food to be available in supermarkets because they are convenient places to shop.

Weatherell et al. [16] also found that despite consumers' willingness to engage with local food producers in order to buy organic food, supermarkets remain their first point of reference when buying organic food. Supermarkets providing organic food have an advantage compared to other distribution channels because they are easily accessed and consumers believe that they sell a large variety of organic food which is fresh, tasty with good appearance [18]. However, Young [125] argues that there are criticisms of supermarkets overselling organic food for large profit and supplying organic food from overseas markets—adding to food miles—despite the fact that UK-produced supplies are available. Major supermarket retailers use food labels to indicate the quality and the level of sustainability of the products they provide according to their perceptions of what consumers need or prefer [126].

In summary then, this paper seeks to understand how supermarkets and farmers' markets are understood by consumers in the UK and Albania as potential sources of organic food. In the next section, the methods used for the pilot study that was undertaken are outlined.

2.3   METHODS

This is a preliminary, pilot study that employs an exploratory, qualitative research approach. Explanatory research is suitable for complex and profound examinations of issues and makes recommendations for the future [127]. Exploratory research on the other hand is better suited to identifying the real nature of research problems or formulating relevant hypotheses for later tests [128]. Taking into consideration the above, an exploratory research approach has been adopted because very little is known about Albania and its consumers and Albania as a research topic is underdeveloped in the literature [127]. Moreover, as this is a preliminary, pilot study employing exploratory research, it is more suitable in order to make recommendations for future empirical investigations [127]. This sort of exploratory, qualitative approach has been used to great effect by researchers looking at issues such as belief formation [129], consumer perceptions [130], and decision-making processes [100].

According to Strauss and Corbin [131], qualitative research provides an in-depth, rich understanding of the topic under research as well as an understanding of the participant as an individual which is not possible to gain through quantitative methods. Qualitative research is a useful and appropriate tool for investigating perceptions, beliefs, motivations and attitudes [132].

This study is attempting to understand the perceptions of consumers drawn from different cultural backgrounds. A qualitative research methodology is the best way to get an insight on how consumers of two different countries understand their world and life [133].

In-depth open-ended interviews were used to gather the data. As Chisnall [128] (p. 219) proposed: "Basically, in-depth interviews are non-directive interviews in which the respondent is encouraged to talk about the subject rather than to answer 'yes' or 'no' to specific question." This means that the interviewer has the ability to adjust the questions according to the responses that are given by the respondents and open a two-way communication.

2.3.1   DATA COLLECTION

Data in Albania were collected in the urban area of Tirana which is the country's capital city. This area was chosen because purchasing power is concentrated mainly in Albania's capital, and is subsequently a reasonably good representative area of the country due in large part to internal migration. During the last 20 years, Tirana has grown from 200,000 to around 700,000 inhabitants.

As people from all parts of Albania have migrated to Tirana, it represents the whole variety of subcultures within the country [134].

Data in the UK were collected in Edinburgh and Glasgow, the two largest cities in Scotland, UK so as to have a direct assessment with the capital of Albania and because, as with Tirana, these major cities contain diverse populations.

Six interviews were conducted in total, with three in Albania and three in the UK, at farmers' markets and outside supermarkets so as to get immediate data from the consumers' shopping experience.

A topic guide was prepared for the interviews that consisted of four parts (see Appendix): Part 1 included questions about respondents' general food shopping experience; part 2 about food outlet perception; part 3 about shopping decisions; and part 4 included questions about their demographic variables.

The purpose of this outlay of questions was to get a general feeling of respondents' shopping routines, evaluate their perceptions of shopping outlets and the reasons they frequent each one.

Respondents were not asked about their perceptions of organic food directly either in Albania or in the UK. The reason was the fact that previous studies suggest that one of the reasons consumers buy food in farmers' markets is that they provide organic produce [6,22,104]. A general question was asked instead regarding the reason respondents visit a farmers' market in order to find out whether they would give organic food as a reason in their responses without being prompted.

An additional reason that Albanian respondents were not asked any questions about organic food directly was to discover their level of familiarity with organic food.

When respondents did not mention organic food in their responses, then a more specific question was addressed around organic food perception. On the other hand, when respondents mentioned organic food spontaneously then more specific, follow up questions were addressed in order to gain insights into their perception of organic food.

2.4   RESULTS AND DISCUSSION

The following section contains the results of the Albanian and British respondents that reveal some first insights into the respondents' perceptions of organic food, highlighting the reasons behind their purchases, and the outlet of their preference. Table 2.1 and Table 2.2 below show the demographic characteristics of the UK and Albanian respondents:

TABLE 1: Respondents' demographic information in Albania.

Age Range	Gender	Education	Occupation	Household Income Range	No of Children
40-49	Female	High School	Housewife	More than 30,000 Lek	2
70+	Female	Primary	Pensioner	More than 15,000 Lek	0
21-29	Female	BSc	Musician	More than 30,000 Lek	0

TABLE 2: Respondents' demographic information in the UK.

Age Range	Gender	Education	Occupation	Household Income Range	No of Children
30-39	Male	PhD	Engineer	More than 30,000 Lek	2
40-49	Female	MSc	IP Manager Part-time	More than 15,000 Lek	0
40-49	Female	BSc	Accountant Part-time	More than 30,000 Lek	0

2.4.1 PERCEPTION OF ORGANIC FOOD

All respondents were positive about knowing what organic food is. When asked to elaborate more on the meaning of the term, different answers were given by the Albanian respondents as opposed to the British ones. To begin with, the term "bio" was more recognisable among Albanian interviewees than organic. This might be due to the labelling efforts of official certification bodies (BioAdria, Institute of Organic Agriculture, Albinspekt) [101] that promote organic food in Albania. They all supported the notion of organic food, which was mainly associated with fruit and vegetables being fresh and tasty. However, Albanian respondents relate organic food with local food without "hormones" produced by local villagers who have used "local seeds" to grow their products and "organic fertilizer-cattle waste."

British responses are similar to what previous studies have already suggested that [105,106,107,108,109,110,111] consumers buy organic food because it is healthy, tasty, has no pesticides, and that they wish to support their local community. Surprisingly, environmental concerns were not mentioned. However, one of the British respondents seemed to be more aware of the matter and gave a more insightful answer:

"It is not using pesticides excessively. I know there are some that use pesticides. It's more using natural fertilizers, not chemicals to stop bugs, they are using field rotation, and they are using techniques to minimize wastage that don't involve just being crops sprayed with chemicals."

2.4.2 FOOD OUTLET CHOICE SUPERMARKETS VS. FARMERS' MARKETS

2.4.2.1 FARMERS' MARKETS

2.4.2.1.1 Albania

One first point that it is worth mentioning is that Albanian respondents refer to the farmers' markets as the villagers' market. This finding is in contrast to another Albanian study [112] referring to these types of markets as farmers' markets. It is important to mention this new term within the findings of this study, because it shows that these markets are perceived differently by Albanian consumers. This is a social stereotype, a remnant of the communist regime where the fruit and vegetables were produced by villagers who would then sell their products in large cities. This is still the case in Albania 25 years after the change of regime.

Generally, Albanian respondents were positively predisposed towards farmers' markets. One of the reason they frequent these markets is the taste and the quality of the products. Albanians believe that villagers provide them with the freshest, best produce they have as: "they collect as much in the morning as they expect to be sold during the day." Another reason is that villagers take care of their product, provide healthy, local products that stay fresh without the use of preservatives. Throughout the interviews, it was made clear that Albanians trust the villagers about the quality and origin of the food they provide. This is similar to what Marsden et al. [12] found that consumers wish to engage in a relationship with farmers and food producers that will be based on trust and reciprocity. It is also remarkable that two out of three interviewees (the older ones) mentioned that the food they purchase from the villagers market reminds them of the taste of food they had when they were younger: "I choose the taste of my time." This is also similar to what Chinnici [124] found in Italy where he identified the "nostalgic" type of organic consumer. Availability and convenience were among the other reasons that Albanian respondents shop from farmers' markets. These markets take place every weekday, in many streets throughout the major cities of Albania. The relationship built between the producer and the end customer was also mentioned several times: "You build like a spiritual relationship with the villager, we know them for many years and they cannot lie about the quality of their products because if I stop shopping from him he will face difficulties" or "The villagers treat you well."

Albanian respondents buy organic food from farmers' markets because they offer fresh, quality products, without hormones and pesticides. They trust the farmers and are willing to pay premium prices despite their low income. This finding is consistent with findings of Western studies identifying consumers with high incomes, willing to pay premium prices for fresh, quality products, without hormones and pesticides in farmers' markets [14,117,118,119,120,121].

Albanian and UK respondents were also asked to provide a definition of a typical farmers' market consumer. Research indicates that farmers' markets provide organic food, thus their typical consumer may be similar to the organic food consumer. The reason why this question was addressed was in order to identify whether the respondents' descriptions would be similar to what studies have indicated about the organic food consumer in Western contexts.

When Albanian respondents were asked to describe a typical consumer of the farmers' markets, they were unable to give a definition: "The largest population shop from the villagers' market" as they believe that everyone shops from them. They also believe that farmers' market will not be affected by the entrance of supermarkets in the country: "The villagers' market existed, exists and will exist in the future because it is important they continue existing for the people that sell the products and their customers."

Considering that the literature around the farmers' markets in the Western countries is almost exclusively positive, Albanian respondents were also asked to name any negative aspects of these markets in Albania. Hygiene and cleanliness were two of the most important issues that came up: "They sell on the street, dust is one negative, they also provide milk in plastic bottles. The products are not displayed in carts like in supermarkets but they put them on fig leaves to give you a taste of the village."

2.4.2.1.2   UK

Two out of three respondents from the UK sample mentioned the frequency, convenience and the prices of the products provided in the farmers' markets as negatives and reasons for not shopping there: "It tends to be quite expensive and it doesn't happen quite often." The general feeling was that the best deal wins over the quality and organic standards: "Ideally if I could get cheap fresh products directly from the farms that could attract me to go there." This is consistent to similar findings that Seyfang [122] found when consumers of the Eostre farmers' market in Norfolk were asked to express their opinion about the negative aspects of the farmers' markets.

The UK respondents were of a well educated and affluent background, however the best deal was more important than buying organic food generally or buying organic food from farmers' markets: "It is always going to be the price for me, it is the major factor" or "Often at farmers' markets there is an amount for a bunch of carrots but I don't know physically how much pound of carrots it is. I don't like going to shop from farmers market because I am so focused on ensuring that I get the best deal." One of the respondents had used a farm shop before where she bought a chicken on her first trip. When she went back to shop again, she mentioned that the size of the chicken became smaller but the price remained the same: "But the chickens got smaller and the price stayed the same and I thought hang on, I am tripping out here to do this and I don't think it tasted better. When the size changed I thought I am not travelling here for double the price for something less in the size. So I switched back to Aldi's." This indicates that there is no trust between the farmer and the consumer as seen in the Albanian sample. This finding is also inconsistent with what previous studies have found, where farmers providing organic food are seen as trusted food sources [105]. The farmer in Albania would not try and deceive his/her customers because of the fear of losing them. The third UK respondent was a more concerned consumer and more aware of organic farming and farmers' markets. Her reasons for frequenting the farmers' markets are similar to those of Albanians respondents: "you know where the food has come from. The meat is from a farm that is between Kilmarnock and Ayre. So you know you have seen the cows in the field and you know they have been quite happy wondering about and eat nice grass. And that is what you get in the packet and is the same for the pig farm which is fairly local, there is also a poultry farm and you again get to see the hens coming out, it is not a battery farmed hen, you know the source of your food, you know what food you are getting." A sign of mistrust was also pointed out about the farmers taking place in large cities within UK. It was mentioned that they do not provide locally sourced or organically grown food: "I used to live up in Glasgow and go to a farmers market in Queen's Park and there were times there were some fruits and vegetables sold there that made you think there is no way these have been grown by you. You went to the fruit and vegetable market this morning and bought these things. So I think the farmers' market works better when there is a local knowledge."

Farmers' markets that take place in big cities were considered to be a negative aspect of these markets, they were considered to be a "big rip-off." The reason is the fact that in a big city there is no direct understanding of who the farmers are. Concerns were also expressed that farmers selling in farmers' markets in big cities, buy their products from elsewhere, mark them up significantly and then sell them in their markets. So, once more, trust issues arise again between the consumer and the farmer.

When respondents were asked to describe the typical farmers' market consumer, they all agreed that the consumer is a middle age, middle class person whose children have left the home: "I think it is people who are really keen on knowing their food sources, that really matters to them the food chain or very affluent people who like the concept of buying at farmers' markets. I think there is a bit of snobbery about it but I think there is a fair amount of people who care about their food."

In contrast to the Albanian findings, the UK data indicated that farmers' market will be affected by supermarkets who have started advertising about buying as much local "as Scottish as they can," "in some ways I can delude myself that Aldi's care more about Scotland. If people perceive that supermarkets are buying more locally they could mitigate the use for farmers' markets. On the other hand, when supermarkets are getting bad press the farmers' markets do better you'd imagine. But farmers' markets are small, the one in Linlithgow got small the last time I saw it."

2.4.2.2   SUPERMARKETS

2.4.2.2.1   Albania

The interviews with the Albanian consumers showed that there are two different types of supermarkets operating in the country. The local ones, which are small supermarkets existing in neighborhoods, that consumers visit frequently during weekdays. The large ones are built outside the city center and are used for larger scale shopping and as places of entertainment: "I visit the large ones once in two weeks or once a month. We also go there for entertainment purposes. We make an excursion for the kids, have fun and drink a coffee while passing your time." This finding is similar to a previous study suggesting that supermarkets in developing countries are perceived as places of entertainment [43]. Supermarkets are viewed positively in terms of offering a variety of products in one place. They are used for shopping for dairy or other products such as detergents and shampoos. All Albanian respondents agreed that they don't shop for fruit and vegetables there because they don't offer "bio" products and: "You can tell just by looking the vegetables available in supermarkets that are full of hormones. For example tomatoes or cucumbers they may look beautiful outside but when you eat them they are full of nerves. They also taste differently. It's like an old saying "it feels like you are eating hay." They believe that supermarket products are full of pesticides but they are well presented and organized within the stores. They offer lower prices when they make offers in some supermarket stores. However, it was concluded that all respondents buy in supermarkets all the products that the villagers cannot produce. An interesting controversy that was identified between the responses was that two mentioned they trust supermarkets because they control better the product they offer on their shelves while the third one said that there might be a manipulation around the expiration date on the labels of the products: "The sausages for example that are imported products you cannot trust that because it was not consumed at the place where it was produced was sent for consumption in Albania."

When interviewees were asked to describe the supermarket consumer it was much easier than the farmers' market consumer. The general conclusion according to their comments is that a typical supermarket consumer: "is someone of a good economic background, well dressed, who shops after work. I can see them in the afternoon shopping in supermarkets and the appearance (clothing) indicates good economic background."

2.4.2.2.2   UK

In contrast to the Albanian interviewees, it was easier for the UK respondents to speak about supermarkets. They shop mostly in supermarkets, including for fruit and vegetables. These respondents would not go out of their way to buy or search for organic food in supermarkets. Prices, once again, were found to be a more important factor thus finding the best deal was the goal when shopping in a supermarket outlet.

When asked to define the supermarket consumer, they struggled to do so as they believe everyone shops from supermarkets. They are convenient food outlets where you can buy anything you need and are opened for a long time during the day. However, they are considered to be unethical because of their "general attitude, they don't care what they are damaging, it is producing things as cheaply as they can get it."

2.5   CONCLUSIONS

The findings from the interviews conducted between Albanian and UK respondents showed that these two different types of consumers perceive the meaning of organic food slightly differently. This might be due to the fact that the term organic has been present in the UK for quite a long time now and consumers are fairly educated by digital and offline media of what it means/represents. Whereas in Albania, the term has only been used very recently and consumers describe organic according to their personal experiences and understandings of how a "natural" product should be.

The responses also showed a different understanding of farmers' markets and supermarkets as food outlets. Supermarkets in Albania are perceived to be places of shopping and entertainment for the whole family. The typical consumer is an affluent, educated, working individual. The opinions though are divided about whether they should be trusted food outlets or not. The villagers' market as they call the farmers' markets turned out to be better-trusted outlets that provide better quality fruit and vegetables and bio products. For these reasons, Albanian consumers are willing to pay the premium prices of the farmers' markets.

Paradoxically, price seemed to be an important factor for the UK respondents in deciding whether to shop at the farmers' markets. These markets seem to operate differently in these two countries. In the UK, it looks like it is a habit of luxury to visit a farmers' market as they require time, money, transport and prior research as per where they occur and attract a specific type of consumer. Whereas, in Albania they are spread throughout the large cities, occur every day of the week and attract all type of consumers who do not sacrifice food quality over price.

Supermarkets on the other hand are also perceived differently. In the UK, consumers visit them because they offer good deals, are convenient places to shop and are open for longer hours. Supermarkets are not frequented though as places where they can buy organic food. In Albania, on the contrary, supermarkets are seen as places where they can spend a Sunday with their family, have a coffee and do their shopping. However, Albanians, like the UK consumers interviewed, also do not trust supermarkets as providers of organic/good quality fruit and vegetables.

One of the contributions of this study is to add an Albanian strand to the literature as very little is known about the country and its consumers. An examination of previous studies conducted in other contexts showed that both organic food and farmers' markets consumers are middle class individuals, well educated with good household incomes. However, the results of this study challenge this. The Albanian interviewees were of different educational and income backgrounds with one having finished just primary school to another one having university education. All of the three interviewees were shopping from farmers' markets and willing to buy the products despite the higher prices. On the contrary, the UK respondents were all well educated, one had a PhD, and apart from one, they would not go out of their way to visit a farmers' market or spend more to buy good quality food. The fact that this pilot study has offered different results to some of the previous studies underlines the need for a wider study in order to explore these findings further.

The paradox seen in the literature about supermarkets and farmers' markets in Albania and the UK is that there is a reverse trend of the growth and popularity of these two outlets for organic food in these countries. As stated in the literature, farmers' markets started developing in the UK in the mid-late 1990s whereas in Albania this was the traditional way of buying fresh food. On the other hand, supermarkets started developing in Albania in the mid-2000s whereas in the UK supermarkets have been present for many decades.

Another interesting finding from the Albanian data is the perception that the respondents had about organic food. In the literature, it is found that ideas about organic food for consumers from Western countries are not only related to food without pesticides, health and taste but also to animal welfare and environmental sustainability. In Albania, the notion of organic food is related to food that farmers have produced in a traditional way, the way they used to cultivate food in Albania during communism using natural fertilizers and agricultural techniques. This perception is probably related to the fact that many people that were alive in Albania during that time are still alive and are transferring these opinions/perceptions down to their children. Thus, this paper is not only casting light on Albania and its consumers but also recording some consumer perceptions before the country becomes part of a more Westernized Europe.

2.6   FURTHER RESEARCH

This exploratory study has found that the study of supermarkets and farmers' markets as outlets of organic foods in these two differing contexts is certainly worth pursuing. Due to the limitation of the sample size of this study, further research needs to be undertaken in order to cast some more insights into the existing findings. A larger sample size can help examine whether, for example, there is a cultural factor that drives the results found in this study. The results thus far suggest that a further study would benefit from the insights that could be gained by the use of Consumer Culture Theory (CCT) [135] as a theoretical framework to underpin any further investigation.

It is also worth investigating how consumers' perceptions change in Albania in the forthcoming years with Albania's candidacy to enter the EU.

APPENDIX

ALBANIAN AND UK CONSUMERS' PERCEPTIONS OF FARMERS' MARKETS AND SUPERMARKETS AS OUTLETS FOR ORGANIC FOOD: AN EXPLORATORY STUDY

Pilot Interview Protocol
Interview Number:
Part 1: Food shopping experience

- Tell me about your most recent food shopping trip.
- Was there any time pressure on your shopping duration?
- Was this trip planned or spur of the moment?
- When did you go?
- When do you usually prefer/enjoy going for food shopping?
- Who did you go with?
- How much did you spend in your last food shopping trip?
- What transport did you use for your food shopping?
- How much time did you spend on food shopping?
- Was this a normal trip for you?

Part 2: Food outlet perception
Farmers' market

- (if they don't mention FM in part 1): Do you ever shop at a farmers' market?
- (if no): Is there any particular reason for that?
- (if yes): Can you tell me about the last time you bought food in a farmers' market?
- Do you go alone to a farmers' market?

- What do you usually buy when visiting a farmers' market?
- Why would you shop from a farmers' market?
- How frequently do you shop from farmers' markets?
- What types of food do you buy from farmers' markets?
- What are the negative aspects of shopping from a farmers' market?
- What are the positive aspects of shopping from a farmers' market?
- Is buying seasonal food important to you? Do you think farmers' markets provide seasonal food?
- (either yes or no): What kind of people shop from a farmers' market? Could you possibly describe a typical farmers' market customer?
- How do you think farmers' markets will be affected by the new supermarkets?

Supermarkets

- (if they don't mention SM in part 1): Do you ever shop at a supermarket?
- (if no): Is there any particular reason for that?
- (if yes): Can you tell me about the last time you bought food in a supermarket?
- Who do you go with when shopping in a supermarket?
- What do you usually buy when visiting a supermarket?
- Why would you shop in a supermarket?
- How frequently do you shop in supermarkets?
- What types of food do you buy in supermarkets?
- What are the negative aspects of shopping in a supermarket?
- What are the positive aspects of shopping in a supermarket?
- (either): What kind of people shop from a supermarket? Could you possibly describe a typical supermarket customer?
- (for those who use both):
- Do you spend more in supermarkets or in farmers' markets?

Part 3: Shopping decision

- How do you decide where to go shopping?
- Who in your household influences this decision?
- What are the most important differences for you between farmers' markets and supermarkets?
- Are there any constraints on your choice of the food outlet (money, distance, location, parking)?

- What are your top priorities for making a decision on which food outlet to shop from (prices, brands, packaging, convenience of outlet, environment)?
- Which would be your favourite food outlet if there weren't any constraints on your shopping process, why?

Part 4: General info
Gender:
Female   Male
Age Range:
less than 20   ☐
21–29   ☐
30–39   ☐
40–49   ☐
50–59   ☐
60–69   ☐
70+   ☐
Number of Adults in Household:
Number of Children in Household:
Education:
Occupation:
Household Income range:
up to £ 15,000–20,000   ☐
£ 21,000–30,000   ☐
£ 31,000–40,000   ☐
£ 41,000–50,000   ☐
£ 51,000–60,000   ☐
£ 61,000–70,000   ☐
more than 70,000   ☐

REFERENCES

1. Tong, D.; Ren, F.; Mack, J. Locating farmers' markets with an incorporation of spatio-temporal variation. Socio-Econ. Plan. Sci. 2012, 46, 149–156.
2. Connell, D.; Smithers, J.; Joseph, A.E. Farmers' markets and the good food value chain: A preliminary study. Local Environ. 2008, 13, 169–185.]
3. Carey, L.; Bell, P.; Duff, A.; Sheridan, M.; Shields, M. Farmers' Market consumers: A Scottish perspective. Int. J. Consum. Stud. 2011, 35, 300–306.
4. Bessière, J. Local development and heritage: Traditional food and cuisine as tourist attractions in rural areas. Sociol. Rural. 1998, 38, 21–34.

5. Delind, L.B. Of bodies, place, and culture: Re-situating local food. J. Agric. Environ. Ethics 2006, 19, 121–146.

6. Susanne, P.; Foster, C. Exploring the gap between attitudes and behaviour: Understanding why consumers buy or do not buy organic food. Br. Food J. 2005, 107, 606–625.

7. Vorpsi, D. The chaos of Albanian Market. Available online: Http://www.ekon.al/2012/07/01/kaosi-i-tregut-shqiptar/ (accessed on 5 March 2013). (In Albanian).

8. Tregear, A.; Dent, J.B.; McGregor, M.J. The demand for organically grown produce. Br. Food J. 1994, 96, 21–25.

9. Davies, A.; Titterington, A.J.; Cochrane, C. Who buys organic food?: A profile of the purchasers of organic food in Northern Ireland. Br. Food J. 1995, 97, 17–23.

10. Hutchins, R.K.; Greenhalgh, L.A. Organic confusion: Sustaining competitive advantage. Nutr. Food Sci. 1995, 95, 11–14.

11. Wright, S. Europe goes organic. Food Ingred. Eur. 1997, 3, 39–43.

12. Marsden, T.K. Theorising food quality: Some issues in understanding its competitive production and regulation. In Qualities of Food; Harvey, M., McMeekin, M., Warde, A., Eds.; Manchester University Press: Manchester, UK, 2004.

13. Barret, H.R.; Browne, A.W.; Harris, P.J.C.; Cadoret, K. Organic certification and the UK market: Organic imports from developing countries. Food Policy 2002, 27, 301–318.

14. Harper, G.; Makatouni, A. Consumer perception of organic food production and farm animal welfare. Br. Food J. 2002, 104, 287–299.

15. Hill, H.; Lynchehaun, F. Organic milk: Attitudes and consumption patterns. Br. Food J. 2002, 104, 526–542.

16. Weatherell, C.; Tregear, A.; Allison, J. In search of the concerned consumer: UK public perceptions of food, farming and buying local. J. Rural Stud. 2003, 19, 233–244.

17. Seyfang, G. Growing sustainable consumption communities: The case of local organic food networks. Int. J. Sociol. Soc. Policy 2007, 27, 120–134.

18. Wier, M.; O'Doherty, J.K.; Andersen, M.L.; Millock, K. The character of demand in mature organic food markets: Great Britain and Denmark compared. Food Policy 2008, 33, 406–421.

19. Harvey, M. Innovation and competition in UK supermarkets. Supply Chain Manag. Int. J. 2000, 5, 15–21.

20. Burch, D.; Lyons, K.; Lawrence, G. What do we mean by 'Green'? Consumers, Agriculture and the Food Industry. In Consuming Foods, Sustaining Environments; Australian Academic Press: Samford, Australia, 2001; pp. 33–46.

21. Holloway, L.; Kneafsey, M. Reading the space of the farmers' market: A preliminary investigation from the UK. Sociol. Rural. 2000, 40, 285–299.

22. La Trobe, H. Farmers' markets: Consuming local rural produce. J. Consum. Stud. Home Econ. 2001, 25, 181–192.

23. Youngs, J. Consumer direct initiatives in North West England farmers' markets. Br. Food J. 2003, 105, 498–530.

24. Kirwan, J. Alternative Strategies in the UK Agro-Food System: Interrogating the Alterity of Farmers' Markets. Sociol. Rural. 2004, 44, 395–415.

25. Lyon, P.; Collie, V.; Kvarnbrink, E.B.; Colquhoun, A. Shopping at the farmers' market: Consumers and their perspectives. J. Foodserv. 2009, 20, 21–30.

26. MacLeod, M.J. The origins, operation and future of farmers' markets in Scotland. J. Farm Manag. 2007, 13, 177–203.

27. Soil Association. Organic Market Report, 2014. Available online: http://action.soilassociation.org/page/s/marketreportlicensees?utm_campaign=OMR+Update+%26+Export+Support&utm_source=emailCampaign&utm_medium=email&utm_content= (accessed on 5 March 2015).

28. Humphery, K. Shelf Life: Supermarkets and the Changing Cultures of Consumption; Cambridge University Press: Cambridge, UK, 1998.

29. Reardon, T.; Gulati, A. The Rise of Supermarkets and Their Development Implications: International Experience Relevant for India; International Food Policy Research Institute: Washington, DC, USA.

30. Hawkes, C. Dietary Implications of Supermarket Development: A Global Perspective. Dev. Policy Rev. 2008, 26, 657–692.

31. Wrigley, N.; Branson, J.; Murdock, A.; Clarke, G. Extending the Competition Commission's findings on entry and exit of small stores in British high streets: Implications for competition and planning policy. Environ. Plan. 2009, 41, 2063–2085.

32. Thompson, C.; Cumminsa, S.; Brown, T. Understanding interactions with the food environment: An exploration of supermarket food shopping routines in deprived neighborhoods. Health Place 2013, 19, 116–123. [PubMed]

33. Reardon, T.; Timmer, C.P.; Barrett, C.B.; Berdegué, J. The Rise of Supermarkets in Africa, Asia and Latin America. Am. J. Agric. Econ. 2003, 85, 1140–1146.

34. Reardon, T.; Timmer, C.P. Transformation of markets for agricultural output in developing countries since 1950: How has thinking changed? Handb. Agric. Econ. 2007, 3, 2807–2855.

35. Likmeta, B. Carrefour Expands Retail Network in Albania, 2013. Available online: http://www.balkaninsight.com/en/article/carrefour-expands-its-retail-network-in-albania (Accessed on 20 August 2014).

36. Reardon, T.; Swinnen, J.F. Agrifood Sector Liberalisation and the Rise of Supermarkets in Former State-controlled Economies: A Comparative Overview. Dev. Policy Rev. 2004, 22, 515–523.

37. Dries, L.; Reardon, T.; Swinnen, J.F. The rapid rise of supermarkets in Central and Eastern Europe: Implications for the agrifood sector and rural development. Dev. Policy Rev. 2004, 22, 525–556.

38. Pileliené, L.; Grigaliūnaitė, V. Determination of customer satisfaction with supermarkets in Lithuania. Organizacijų Vadyba Sisteminiai Tyrimai 2013, 66, 99–114.

39. Money, R.B.; Colton, D. The response of the 'new consumer' to promotion in the transition economies of the former Soviet bloc. J. World Bus. 2000, 35, 189–205.

40. Kapaj, A.; Chan-Halbrendt, C.; Deci, E.; Kapaj, I. Milk consumer's preferences in urban Albania. Adv. Manag. Appl. Econ. 2011, 1, 197–215.

41. Gaiha, R.; Thapa, G. Supermarkets, smallholders and livelihood prospects in selected Asian countries. Available online: Http://www.ifad.org/operations/projects/regions/pi/paper/4.pdf (accessed on 15 October 2014).

42. Traill, W.B. The rapid rise of supermarkets? Dev. Policy Rev. 2006, 24, 163–174.

43. Reardon, T.; Henson, S.; Berdegué, J. Proactive fast-tracking' diffusion of supermarkets in developing countries: Implications for market institutions and trade. J. Econ. Geogr. 2007, 7, 399–431.

44. Goldman, A.; Hino, H. Supermarkets vs traditional retail stores: Diagnosing the barriers to supermarkets' market share growth in an ethnic minority community. J. Retail. Consum. Serv. 2005, 12, 273–284.

45. Igami, M. Does Big Drive Out Small? Rev. Ind. Organ. 2011, 38, 1–21.
46. Borraz, F.; Dubra, J.; Ferrés, D.; Zipitría, L. Supermarket entry and the survival of small stores. Rev. Ind. Organ. 2014, 44, 73–93.
47. Dickson, P.R.; Sawyer, A.G. The price knowledge and search of supermarket shoppers. J. Mark. 1990, 54, 42–53.
48. Lal, R.; Rao, R. Supermarket competition: The case of everyday low pricing. Mark. Sci. 1997, 16, 60–80.
49. Levy, D.; Bergen, M.; Dutta, S.; Venable, R. The magnitude of menu costs: Direct evidence from large US supermarket chains. Q. J. Econ. 1997, 112, 791–825.
50. Russo, J.E.; Staelin, R.; Nolan, C.A.; Russell, G.J.; Metcalf, B.L. Nutrition information in the supermarket. J. Consum. Res. 1986, 13, 48–70.
51. Eisenhauer, E. In poor health: Supermarket redlining and urban nutrition. GeoJournal 2001, 53, 125–133.
52. Sutherland, L.A.; Kaley, L.A.; Fischer, L. Guiding stars: The effect of a nutrition navigation program on consumer purchases at the supermarket. Am. J. Clin. Nutr. 2010, 91, 1090S–1094S. [PubMed]
53. Smith, A.D. Retail-based loyalty card programmes and CRM concepts: An empirical study. Int. J. Innov. Learn. 2010, 7, 303–330.
54. Duffy, R.; Fearne, A.; Hornibrook, S.; Hutchinson, K.; Reid, A. Engaging suppliers in CRM: The role of justice in buyer–supplier relationships. Int. J. Inf. Manag. 2013, 33, 20–27.
55. Tolich, M.B. Alienating and Liberating Emotions at Work Supermarket Clerks' Performance of Customer Service. J. Contemp. Ethnogr. 1993, 22, 361–381.
56. Sirohi, N.; McLaughlin, E.W.; Wittink, D.R. A model of consumer perceptions and store loyalty intentions for a supermarket retailer. J. Retail. 1998, 74, 223–245.
57. Disney, J. Customer satisfaction and loyalty: The critical elements of service quality. Total Qual. Manag. 1999, 10, 491–497.
58. Bolton, R.N.; Kannan, P.K.; Bramlett, M.D. Implications of loyalty program membership and service experiences for customer retention and value. J. Acad. Mark. Sci. 2000, 28, 95–108.
59. Vazquez, R.; Rodríguez-Del Bosque, I.A.; Ma Díaz, A.; Ruiz, A.V. Service quality in super-market retailing: Identifying critical service experiences. J. Retail. Consum. Serv. 2001, 8, 1–14.
60. Sands, S.F.; Sands, J.A. Recording brain waves at the supermarket: What can we learn from a shopper's brain? Pulse IEEE 2012, 3, 34–37.
61. Andersson, P.K.; Kristensson, P.; Wästlund, E.; Gustafsson, A. Let the music play or not: The influence of background music on consumer behavior. J. Retail. Consum. Serv. 2012, 19, 553–560.
62. Morrison, M.; Gan, S.; Dubelaar, C.; Oppewal, H. In-store music and aroma influences on shopper behavior and satisfaction. J. Bus. Res. 2011, 64, 558–564.
63. Harris, J.L.; Schwartz, M.B.; Brownell, K.D. Marketing foods to children and adolescents: Licensed characters and other promotions on packaged foods in the supermarket. Public Health Nutri. 2010, 13, 409–417.
64. Drewnowski, A.; Aggarwal, A.; Hurvitz, P.M.; Monsivais, P.; Moudon, A.V. Obesity and supermarket access: Proximity or price? Am. J. Public Health 2012, 102, e74–e80. [PubMed]
65. Sommer, R.; Herrick, J.; Sommer, T.R. The behavioral ecology of supermarkets and farm-ers' markets. J. Environ. Psychol. 1981, 1, 13–19.

66. Jones, P.; Comfort, D.; Hillier, D. A case study of local food and its routes to market in the UK. Br. Food J. 2004, 106, 328–335.

67. FARMA. What is a farmers' market? Available online: Http://www.localfoods.org.uk/info/10-farmers-markets-faq/26-what-is-a-certified-farmers-market (accessed on 1 December 2013).

68. Bentley, G.; Hallsworth, A.G.; Bryan, A. The Countryside in the City-Situating a Farmers' Market in Birmingham. Local Econ. 2003, 18, 109–120.

69. McEachern, M.G.; Warnaby, G.; Carrigan, M.; Szmigin, I. Thinking locally, acting locally? Conscious consumers and farmers' markets. J. Mark. Manag. 2010, 26, 395–412.

70. Guthrie, J.; Guthrie, A.; Lawson, R.; Cameron, A. Farmers' markets: The small business counter-revolution in food production and retailing. Br. Food J. 2006, 108, 560–573.

71. Spiller, K. It tastes better because consumer understandings of UK farmers' market food. Appetite 2012, 59, 100–107. [PubMed]

72. La Trobe, H.L.; Acott, T.G. Localising the global food system. Int. J. Sustain. Dev. World Ecol. 2000, 7, 309–320.

73. Brooker, J.R.; Eastwood, D.B.; Gray, M.D. Direct marketing in the 1990's: Tennessee's new farmers' markets. J. Food Distrib. Res. 1993, 24, 1–6.

74. Gerbasi, G.T. Athens farmers' market: Evolving dynamics and hidden benefits to a southeast Ohio rural community. FOCUS Geogr. 2006, 49, 127–138.

75. Wallgren, C. Local or global food markets: A comparison of energy use for transport. Local Environ. 2006, 11, 233–251.

76. Svenfelt, Å.; Carlsson-Kanyama, A. Farmers' markets—linking food consumption and the ecology of food production? Local Environ. 2010, 15, 453–465.

77. Åsebø, K.; Jervell, A.M.; Lieblein, G.; Svennerud, M.; Francis, C. Farmer and consumer attitudes at farmers' markets in Norway. J. Sustain. Agric. 2007, 30, 67–93.

78. Jervell, A.M.; Borgen, S.O. New marketing channels for food quality products in Norway. Food Econ.-Acta Agric. Scand. Section C 2004, 1, 108–118.

79. Storstad, O.; Bjørkhaug, H. Foundations of production and consumption of organic food in Norway: Common attitudes among farmers and consumers? Agric. Hum. Values 2003, 20, 151–163.

80. Latacz-Lohmann, U.; Foster, C. From "niche" to "mainstream"—strategies for marketing organic food in Germany and the UK. Br. Food J. 1997, 99, 275–282.

81. Wüstenhagen, R.; Bilharz, M. Green energy market development in Germany: Effective public policy and emerging customer demand. Energy policy 2006, 34, 1681–1696.

82. Vecchio, R. European and United States farmers' markets: Similarities, differences and potential developments. In Proceedings of the 113th EAAE Seminar "A Resilient European food Industry and Food Chain in a Challenging World," Crete, Greece, 3–6 September 2009.

83. Evans, N.; Morris, C.; Winter, M. Conceptualizing agriculture: A critique of post-productivism as the new orthodoxy. Prog. Hum. Geogr. 2002, 26, 313–332.

84. Parrott, N.; Wilson, N.; Murdoch, J. Spatializing quality: Regional protection and the alternative geography of food. Eur. Urban Reg. Stud. 2002, 9, 241–261.

85. Sonnino, R.; Marsden, T. Beyond the divide: Rethinking relationships between alternative and conventional food networks in Europe. J. Econ. Geogr. 2006, 6, 181–199.

86. Rizov, M.; Mathijs, E. Farm survival and growth in transition economies: Theory and empirical evidence from Hungary. Post-Communist Econ. 2003, 15, 227–242.

87. Blaas, G. Individual farmers in Slovakia: Typology, recruitment patterns and sources of livelihood. Sociológia 2003, 35, 557–578.
88. Spilková, J.; Perlín, R. Farmers' markets in Czechia: Risks and possibilities. J. Rural Stud. 2013, 32, 220–229.
89. Alkon, A. Paradise or pavement: The social constructions of the environment in two urban farmers' markets and their implications for environmental justice and sustainability. Local Environ. 2008, 13, 271–289.
90. Tregear, A. Origins of taste: Marketing and consumption of regional foods in the UK. In Proceedings of the ESRC Seminar on Global Consumption in a Global Context, Cardiff, UK, 29 November 2005.
91. Brown, A. Farmers' market research 1940–2000: An inventory and review. Am. J. Altern. Agric. 2002, 17, 167–176.
92. Guthman, J. "If they only knew": Color blindness and universalism in California alternative food institutions. Prof. Geogr. 2008, 60, 387–397.
93. Moore, O. Understanding post organic fresh fruit and vegetable consumers at participatory farmers' markets in Ireland: Reflexivity, trust and social movements. Int. J. Consum. Stud. 2006, 30, 416–426.
94. Zepeda, L. Which little piggy goes to market? Characteristics of US farmers' market shoppers. Int. J. Consum. Stud. 2009, 33, 250–257.
95. Feagan, R.; Morris, D.; Krug, K. Niagara region farmers' markets: Local food systems and sustainability considerations. Local Environ. 2004, 9, 235–254.
96. Jones, P.; Hill, C.C.; Hiller, D. Case study: Retailing organic food products. Br. Food J. 2001, 103, 358–365.
97. European Commission. Available online: http://ec.europa.eu/agriculture/organic/home_en (accessed on 25 January 2013).
98. Lampkin, N.; Foster, C.; Padel, S. The Policy and Regulatory Environment for Organic Farming in Europe: Country Reports; Universität Hohenheim: Stuttgart, Gernman, 1999.
99. Lockie, S.; Lyons, K.; Lawrence, G. Going Organic: Mobilizing Networks for Environmentally Responsible Food Production; CAB International: Wallingford, UK, 2006.
100. Hamzaoui-Essoussi, L.; Zahaf, M. Decision making process of community organic food consumers: An exploratory study. J. Consum. Mark. 2008, 25, 95–104.
101. Bernet, T.; Kazazi, I.S. Organic Agriculture in Albania, Sector Study 2011. Available online: Http://orgprints.org/22709/1/berner-and-kazazi-2012-sectorstudy2011-albania.pdf (accessed on 18 October 2014).
102. Seppanen, L.; Helenius, J. Do inspection practices in organic agriculture serve organic values?—A case study from Finland. Agric. Hum. Values 2004, 21, 1–13.
103. McDonagh, P.; Prothero, A. Green Management: A Reader; International Thomson Business Press: London, UK, 1997.
104. Schifferstein, H.N.J.; Oude Ophuis, P.A.M. Health-related determinants of organic food consumption in the Netherlands. Food Qual. Prefer. 1998, 9, 119–133.
105. Lilliston, B.; Cummins, R. Organic vs "organic": The corruption of a label. Ecologist 1998, 28, 195–200.
106. Archer, G.P.; Sanchez, J.G.; Vignali, G.; Chaillot, A. Latent consumers attitude to farmers' markets in North West England. Br. Food J. 2003, 105, 487–497.
107. Vanhonacker, F.; Verbeke, W.; Guerrero, L.; Claret, A.; Contel, M.; Scalvedi, L.; Gutkowska, K.; Żakowska-Biemans, S.; Sulmont-Rossé, C.; Raude, J.; et al. How European consumers

define the concept of traditional food: Evidence from a survey in six countries. Agribusiness 2010, 26, 453–476.

108. Sparks, P.; Shepherd, R. Self-Identity and the Theory of Planned Behavior: Assesing the Role of Identification with "Green Consumerism." Soc. Psychol. Q. 1992, 55, 388–399.

109. Worcester, R.M. Ethical Consumerism Research; The Co-operative Bank: London, UK, 2000.

110. Makatouni, A. What motivates consumers to buy organic food in the UK? Results from a qualitative study. Br. Food J. 2002, 104, 345–352.

111. Huang, C.L. Consumer preferences and attitudes towards organically grown produce. Eur. Rev. Agric. Econ. 1996, 23, 331–342.

112. Grankvist, G.; Biel, A. The importance of beliefs and purchase criteria in the choice of eco-labeled food products. J. Environ. Psychol. 2001, 21, 405–410.

113. Fotopoulos, C.; Krystallis, A. Purchasing motives and profile of the Greek organic consumer: A countrywide survey. Br. Food J. 2002, 104, 730–765.

114. Lockie, S.K.L.; Lawrence, G.; Grice, J. Choosing organics: A path analysis of factors underlying the selection of organic food among Australian consumers. Appetite 2004, 43, 135–146. [PubMed]

115. Bruhn, C.M.; Diaz-Knauf, K.; Feldman, N.; Harwood, J.A.N.; Ho, G.; Ivans, E.; Kubin, L. Consumer food safety concerns and interest in pesticide-related information. J. Food Saf. 1991, 12, 253–262.

116. Lea, E.; Worsley, A. Australian consumers' food-related environmental beliefs and behaviours. Appetite 2008, 50, 207–214. [PubMed]

117. Hamzaoui-Essoussi, L.; Zahaf, M. Exploring the decision-making process of Canadian organic food consumers: Motivations and trust issues. Qual. Mark. Res. Int. J. 2009, 12, 443–459.

118. Drioucech, N.; XDulja, X.; Capone, R.; Dernini, S.; El, H. Albanian Consumer Attitude and Behaviour toward Ethical Values of Agro-Food Products. In Proceedings of the Fourth International Scientific Symposium "Agrosym 2013," Jahorina, Bosnia and Herzegovina, East Sarajevo, Republika Srpska, 3–6 October 2013; pp. 706–712.

119. Porjes, S. Fresh and Local Food in the US Packaged Facts, Rockville. In Green Technologies in Food Production and Processing; Springer Science & Business Media: London, UK, 2012.

120. Aarset, B.; Beckmann, S.; Bigne, E.; Beveridge, M.; Bjorndal, T.; Bunting, J.; McDonagh, P. The European consumers' understanding and perceptions of the "organic" food regime: The case of aquaculture. Br. Food J. 2004, 106, 93–105.

121. Schulman, G.I.; Worrall, C. Salience patterns, source credibility, and the sleeper effect. Public Opin. Q. 1970, 34, 371–382.

122. Seyfang, G. Avoiding Asda? Exploring consumer motivations in local organic food networks. Local Environ. 2008, 13, 187–201.

123. Siti, A.; Binti, N.B. Organic food: A study on demographic characteristics and factors influencing purchase intentions among consumers in Klang Valley, Malaysia. Int. J. Bus. Manag. 2010, 5, 105–118.

124. Chinnici, G.; D'Amico, M.; Pecorino, B. A multivariate statistical analysis on the consumers of organic products. Br. Food J. 2002, 104, 187–199.

125. Young, W. The True Cost of Supermarkets: Supermarket Shopping; Vision Paperbacks: London, UK, 2004.

126. Oosterveer, P.; Guivant, J.S.; Spaargaren, G. Shopping for Green Food in Globalizing Supermarkets: Sustainability at the Consumption Junction; Sage Publication: Thousand Oaks, CA, USA, 2007; pp. 411–428.
127. Shields, P.M.; Rangarajan, N. A Playbook for Research Methods: Integrating Conceptual Frameworks and Project Management; New Forums Press: Stillwater, OK, USA, 2013.
128. Chisnall, P. Marketing Research, 7th ed.; McGraw Hill Education: New York, NY, USA, 2005.
129. Deirdre, S.; Clarke, I. Belief formation in ethical consumer groups: An exploratory study. Mark. Intell. Plan. 1999, 17, 109–120.
130. Hammitt, J.K. Risk Perceptions and Food Choice: An Exploratory Analysis of Organic-Versus Conventional-Produce Buyers. Risk Anal. 1990, 10, 367–374. [PubMed]
131. Strauss, A.; Corbin, J. Basics of Qualitative Research, 3rd ed.; Sage Publication: Thousand Oaks, CA, USA, 2008.
132. Bryman, A.; Bell, E. Business Research Methods; Oxford University Press: Oxford, UK, 2003.
133. Kvale, S. Interviews: An introduction to Qualitative Research Interviewing; Sage Publication: Thousand Oaks, CA, USA, 1996.
134. Skreli, E.; Imami, D.; Canavari, M.; Chan-Halbrendt, C.; Zhllima, E.; Pire, E. Consumers' preferences for organic food applying conjoint analysis—The case of tomato in Albania. Available online: Http://ifama2014.com/Programme/2014_Symposium_e_-1-.pdf (accessed on 12 October 2014).
135. Arnould, E.; Thompson, C. Consumer culture theory (and we really mean theoretics): Dilemmas and opportunities posed by an academic branding strategy. Res. Consum. Behav. 2007, 11, 3–22.

CHAPTER 3

Analysis of Production and Consumption of Organic Products in South Africa

Maggie Kisaka-Lwayo and Ajuruchukwu Obi

3.1 INTRODUCTION

Food and nutritional security remain an issue of global concern especially in developing countries. The practice of organic agriculture has been identified as a pathway to sustainable development and enhancing food security. Arguably, the most sustainable choice for agricultural development and food security is to increase total farm productivity in situ, in developing countries particularly sub-Saharan Africa. Attention must focus on the following: (i) the extent to which farmers can improve food production and raise incomes with low-cost, locally available technologies and inputs (this is particularly important at times of very high fuel and agro-chemical prices); (ii) whether they can do this without causing further environmental damage; and (iii) the extent of farmers' ability to access markets [1]. Organic farming is one of the sustainable approaches to farming that can contribute to food and nutritional security [2]. Driven by increasing demand globally, organic agriculture has grown rapidly in the past decade [3]. Policy makers at the primary end of the food chains must wrestle with the dual objective of reducing poverty and increasing the flow of ecosystem services from rural areas occupied by small-scale farmers and/or family farms [4].

Expectedly, a paradigm shift towards this realization of organic agriculture's role in food and nutritional security is emerging [5]. The United Nations Environmental Programme-United Nations Conference on Trade and Development, UNEP-UNCTAD [6] indicates that organic agriculture offers developing countries a wide range of economic, environmental, social and cultural benefits. On the development side, organic production is particularly well-suited for smallholder farmers, who make up the majority of the worlds' poor. Resource poor farmers are less dependent on external resources, experience higher yields on their farms and enjoy enhanced food security [7]. Organic agriculture in developing countries builds on and keeps alive their rich heritage of traditional knowledge and traditional land races. It has been observed to strengthen communities and give youth incentive to keep farming, thus reducing rural-urban migration. Farmers and their families and employees are no longer exposed to hazardous agro-chemicals, which is one of the leading causes of occupational injury and death in the world [7].

As organic production increases, so does the interest in organic market dynamics and studies are being carried out in order to analyse the future potential for organic agriculture. Figure 3.1 shows the global markets for certified organic products. In 2009, the global market for certified organic food and drink was estimated to be 54. 9 billion US dollars [8]. This represents a 37% growth from 2006 sales estimated at 40. 2billionUS dollars and a 207% increase from year 2000 sales estimated at17. 9 billion US dollars. In Africa, most of the organic farms are small family smallholdings [9] and certified organic production is mostly geared to products destined for export beyond Africa's shores. However, local markets for certified organic products are growing, especially in Egypt, South Africa, Uganda and Kenya [10]. Figure 3.2 shows the ten countries in Africa with the largest proportion of land allocated to organic agriculture. South Africa has the third largest area under organic farming with 50, 000 hectares (ha), trailing Tunisia which has the largest area of 154, 793ha and Uganda with 88, 439ha [11]. Approximately 20% of the total area under certified organic farming in Africa is in South Africa, with 250 certified commercial farms [12]. With a few exceptions, notably Uganda, most African countries do not have data collection systems for organic farming and certified organic farming is relatively underdeveloped, even in comparison to other low-income continents. Some expert opinions suggest that this is due to lack of awareness, low-income levels, lack of local organic standards and other infrastructure for local market certification [13].

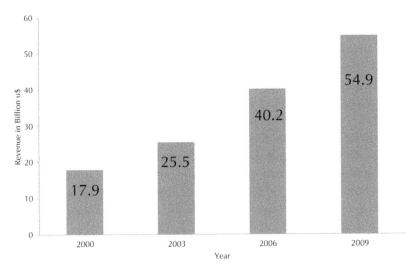

FIGURE 3.1   Development of the global market for organic products

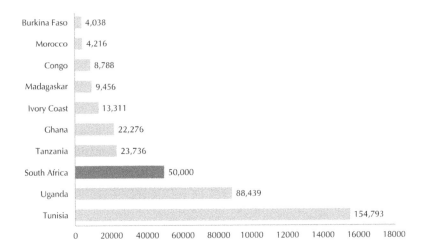

FIGURE 3.2   The ten countries in Africa with the most organic agricultural land in hectares

3.2   ORGANIC AGRICULTURE IN SOUTH AFRICA

In 1999, only 35 farms were certified in South Africa, whereas in 2000 this number had increased to approximately 150 [15]. GROLINK [16] estimates that 240 farms with a total area of 43 620 ha (including pastures and in-conversion land) were certified in 2002. Certified organic produce in South Africa started with mangoes, avocadoes, herbs, spices, rooibos tea and vegetables [17]. This has now expanded to include a much wider range of products. Organic wines, olive oil and dairy products are now being produced [18]. The Organic Agricultural Association of South Africa (OAASA) estimates that there are approximately 100 non-certified farmers, farming about 1000 hectares, following organic principles, who market informally through local villages or farmers markets (ibid). In the latter case, no differentiation is made between organic and non-organic produce.

South Africa has had an organic farming movement dating back many years, although it has grown in "fits and starts" [19]. Organic approaches have to make a trade off between market oriented commercial production and increasing the productive capacity of marginalized communities [20]. The growth of the organic industry has resulted in organic farming being practised in the Western Cape, KwaZulu-Natal, Eastern Cape, Northern Cape and Gauteng Province (Table 3.1). As discussed by [21] and [22] changing consumer preferences towards more health and environmental awareness has led to an increase in the demand for products produced using sustainable production methods. GROLINK [16] states that South Africa has in contrast with other Sub-Saharan countries, a substantial domestic market for organic products. This is an indication that the potential for organic farming in South Africa is not only based on access to the export market in Europe and the USA but also on the local demand. The domestic market is robust with two domestic retailers (Woolworth and Pick 'n' Pay) selling reasonable amounts of organic produce and both are now starting to insist on certification for this produce as well as farmers markets attracting large number of buyers.

One approach taken to improve smallholder access to organic markets has been the formation of certified organic groups using guidelines developed by the International Federation of Organic Agriculture Movement (IFOAM) and enforced by certification agencies such as Ecocert/AFRISCO (African Farmers Certified Organic) in the case of South Africa [23]. Under the group certification system, organic farmers can either grow and market

TABLE 3.1: Summary statistics of sampled farmers in KwaZulu-Natal (n=200)

Variable	Fully-certified organic		Partially-certified organic		Non-organic	
	Mean	Std. Dev	Mean	Std. Dev	Mean	Std. Dev
Age (years)	52.60	1.90	48.60	1.41	52.70	2.11
Gender (1=female)	0.82	0.05	0.71	0.05	0.84	0.05
Education (years)	4.94	4.24	4.37	4.49	3.38	0.61
Household size	9.49	5.23	7.72	3.68	6.60	3.46
Land size (hectares)	0.59	1.22	0.71	1.16	0.67	1.43
Input costs (rand/year)	812.90	884.90	309.30	343.40	318.20	302.90
Proportion of income from farming	0.62	0.79	0.38	1.04	0.39	0.63
Farm income (rands/year)	973.17	1074.51	417.26	271.50	400.53	429.53
Location	2.56	0.60	1.91	0.54	4.00	0.00
Arrow Pratt Risk Aversion coefficient	0.55	0.29	0.58	0.31	0.76	0.29
Land rights (0 = no)	1.98	0.14	1.75	0.56	1.93	0.33
Chicken ownership	15.29	13.16	9.25	8.69	6.40	6.62
Asset ownership (index)	0.98	0.60	0.56	0.59	0.67	0.75

their produce collectively or produce individually but market collectively. This ensures that smallholder farmers especially in developing countries are not marginalised and unduly excluded from the organic sector due to factors beyond their control. Several organic farming groups have emerged in South Africa in the last decade notably Ezemvelo Farmers Organization (EFO), Vukuzakhe Organic Farmers Organization (VOFO), Ikusasalethu Trust and Makhuluseni Organic Farmers Organisation.

The question of how to face the growing problem of food insecurity in Africa becomes more and more important, especially due to the steadily increasing world population and the changing consumption pattern. According to [24], while organically produced food seems not to be able to feed the World's Population, there are strong evidences that organic agriculture might help to alleviate the number of people suffering from hunger especially in developing countries. Given the strong negative externalities of conventional agriculture, the diversification of production as a basic principal of organic agriculture can contribute to the improvement of food security [25] which may improve the nutritional level in rural communities. The expanding global market for organic products [26, 27] and the possibilities for smallholder farmers in developing countries to access markets [24] can have very positive effects on the rural economies, triggering rural development. The increasing awareness of what people consume also has positive effects on organic agriculture as an alternative option for agricultural production. Organic agriculture may thus be an option in some areas to strongly support rural development.

Against this background, the objective of this paper is to provide, through an exploratory analysis of data from farm and households surveys, empirical insights into determinants of organic farming adoption, differentiating between fully certified organic, partially certified organic and non-organic farmers; eliciting farmers risk preferences and management strategies and; exploring consumer awareness, perceptions and consumption decisions. By exploring a combination of adoption relevant factors in the context of real and important land management choices, the paper provides an empirical contribution to the adoption literature and provides valuable pointers for the design of effective and efficient public policy for on-farm conservation activities. Similarly, achieving awareness and understanding the linkage between awareness and purchasing organics is fundamental to impacting the demand for organically grown products. Consumer awareness of organic foods is the first step in developing demand for organic. Section 3.3 describes the materials and methods, outlining the study areas and study methodology. Section 3.4 presents the results and discussion. Finally, Section 3.5 provides concluding remarks.

3.3   MATERIALS AND METHODS

The study was carried out in the two provinces of KwaZulu-Natal and the Eastern Cape Provinces in South Africa (Figure 3.3). The selected study areas are in the rural Umbumbulu Magisterial District in KwaZulu-Natal Province and the OR Tambo and Amatole District Municipalities in the Eastern Cape Province.

The Umbumbulu area is one of the former homelands of the KwaZulu-Natal. The province has the largest concentration of people who are relatively poor, and social indicators point to below average levels of social development. According to the mid-year populations estimates by Statistics South Africa [28], the province has a population of 10. 6 million people,

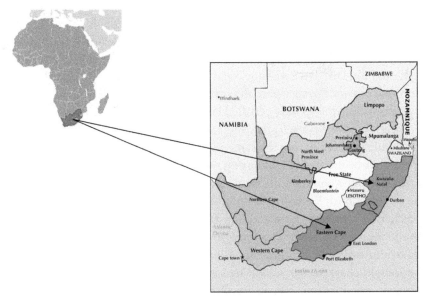

FIGURE 3.3   Map of study area

67 percent of whom reside in communal areas of the former KwaZulu-Natal homeland [28]. The OR Tambo District Municipality is the second poorest municipality in the Eastern Cape Province with some areas having poverty levels of as high as 82 % [29]. About 67% of the households within the district have income levels that range between R0 and R6, 000. The District Municipality has the second highest population of all the districts with more than 1, 504, 411 inhabitants [29]. For a mostly rural district it also has a high population density of 90 people per square kilometre. The Amatole District Municipality is named after the legendary Amatole Mountains and is the most diverse District Municipality in the Province. Two-thirds of the District is made up of ex-homeland areas. The District has a moderate Human Development Index of 0. 52 with over 1, 635, 433 inhabitants [30], and a moderately high population density of 78 people per square kilometre. The population is mainly African with some whites and coloureds. Amatole District Municipality has the second highest economy in the province.

The Eastern Cape Province is bordering KwaZulu-Natal with similarities in the socio-economic status and rurality of the two Provinces. Both Provinces' economic dependence is on agriculture with huge potential for organic agriculture development. The Eastern Cape is also a major consumer of produce from KwaZulu-Natal. A total of 400 respondents were interviewed, representing 200 farmer respondents from KwaZulu-Natal and 200 consumer respondents from Eastern Cape Provinces. The survey farmers in Umbumbulu District, KwaZulu-Natal were stratified into three groups: fully certified organic farmers, partially certified organic farmers and non-organic farmers. While the 48 fully certified farmers and 103 partially certified farmers were purposively selected, the sample of 49 non-organic farmers was randomly selected within the same region from a sample frame constructed from each of the five neighbouring wards. The survey was conducted by a team of trained enumerators from the study area. These enumerators had to be fluent in both English and Zulu. A questionnaire was used to record all household activities (farm and non-farm), enterprise types, crop areas and production levels, inputs, expenditures and sales for the past season. The questionnaires also captured socio-economic and institution data such as household characteristics, land size and tenure arrangements, farm characteristics and investment in assets. Other questions related to farmers' management capacity and demographic characteristics such as the supply of on-farm family labour and education status.

The farmers' risk attitude was elicited using the experimental gambling approach as outlined by [31]. Here, the study farmers were presented with a series of choices among sets of alternative prospects (gambles) that do not involve real money payments. Respondents were required to make a simple choice among eight gambles whose outcomes were determined by a flip of a coin. The experimental approach remedies some of the more serious measurement flaws of the direct elicitation utility (DEU) interview method reporting that evidence on risk aversion using direct elicitation utility through pure interviews is unreliable, nonreplicable and misleading even if one is interested only in a distribution of risk aversion rather than reliable individual measurements [31, 32]. The farmers were further asked in the field survey to give their perceptions of the main sources of risk that affect their farming activity by ranking a set of 20 potential sources of risk on likert-type scales ranging from 1 (no problem) to 3 (severe problem). These sources of risk were developed from findings of the research survey and from past research on the sources of risk in agriculture, challenges that smallholder farmers face in trying to access formal supply chains. The farmers were also requested to score any other sources of risk(s) that they wanted to add to the list of hypothesized sources of risk. These sources of risk are ranked from 1 (being the most important source of risk) to 20(being the least important source of risk ones). The ranking was done by averaging the scores on each source of risk and assigning a rank accordingly.

The study area in the Eastern Cape was stratified into the OR Tambo District Municipality and the Amatole District Municipality representing a broad spectrum of consumers across the Province. The stratified study areas were further clustered into rural, peri-urban and urban areas. The respondents were selected by simple random sampling to avoid bias. A total of 100 consumers were selected from OR Tambo District Municipality and represented by a selection of 30 respondents from peri-urban location, 40 respondents from urban suburbs and 30 respondents from rural areas. In the Amatole District Municipality, 100 consumers selected and interviewed included 30 respondents from rural Cata, 40 urban respondents from the East London Suburbs and lastly 30 respondents drawn from the peri urban area of Kwezana and Tsathu villages. A structured questionnaire was used that covered the respondent's socio- economic and demographic background, consumer knowledge and awareness of organic products, perceptions, attitudes as well as consumption decisions.

The ordered probit model was used to identify the determinants of farmers' decision to participate in organic farming. The dependent variable is the farmer's organic farming status and was placed in three ordered categories in the survey. The model is estimated as:

$$\text{Organic farming status} = f \begin{pmatrix} age, gender, education, household\ size, \\ farm\ size, farm\ income, off\ farm\ income, \\ input\ costs, land\ tenure, location, \\ land\ tenure, livestock, chicken \\ ownership, risk\ attitudes\ and\ assets \end{pmatrix} \quad (1)$$

The organic farming status is modelled using the ordered probit model with the model outcomes:

$S_i=3$ (fully-certified organic),
$S_i=2$ (partially-certified organic farming) and
$S_i=1$ (non-organic farmers).

The farmer's decision on their organic farming status is unobserved and is denoted by the latent variable s_i^*. The latent equation below models how s_i^* varies with personal characteristics and is represented as:

$$s_i^* = X_i'\alpha + \varepsilon_i \qquad (2)$$

Where:

- the latent variable s_i^* measures the difference in utility derived by individual i from either being fully-certified organic, partially-certified organic or non-organic.
- (i=1, 2, 3.............. n) n represents the total number of respondents. Each individual i belongs to one of the three groups.
- X_i is a vector of exogenous variables,
- α is a conformable parameter vector, and
- the error term ε_i is independent and identically distributed as standard normal, that is $\varepsilon_i \sim$ NID (0, 1).

The observed variable (S_i) relates to the latent variable (s_i^*) such that

$$S_i = \begin{array}{ll} 1 & \text{if } s_i^* \leq 0 \\ 2 & \text{if } 0 < s_i^* > \gamma \\ 3 & \text{if } s_i^* > \gamma \end{array} \qquad (3)$$

Taking the value of 3 if the individual was fully certified organic and 1 if the individual was non-organic. The implied probabilities are obtained as:

$$Pr\{S_i = 1 \mid X_i\} = Pr\{s_i^* \leq 0 \mid X_i\} = \Phi(-X_i'\alpha)$$
$$Pr\{S_i = 3 \mid X_i\} = Pr\{s_i^* > \gamma \mid X_i\} = 1 - \Phi(\gamma - X_i'\alpha)$$
and
$$Pr\{S_i = 2 \mid X_i\} = \Phi\{\gamma - X_i'\alpha\} - \Phi(-X_i'\alpha) \qquad (4)$$

Where γ is the unknown parameter that is estimated jointly with α. Estimation is based upon the maximum likelihood where the above probabilities enter the likelihood function. The interpretation of the α coefficients is in terms of the underlying latent variable model in equation 11.

The probability of the farmer being fully certified organic can be written as

$$Pr(S_i = 1) = \Phi(X_i'\alpha_1) \qquad (5)$$

Where $\Phi()$ is the cumulative distribution function (cdf) of the standard normal [33].

A measure of goodness of fit can be obtained by calculating

$$\rho^2 = 1 - \left[\ln L_b / \ln L_o \right] \tag{6}$$

Where $\ln L_b$ is the log likelihood at convergence and $\ln L_o$ is the log likelihood computed at zero.

This measure is bounded by zero and one. If all model coefficients are zero, then the measure is zero. Although ρ^2 cannot equal one, a value close to one indicates a very good fit. As the model fit improves, ρ^2 increases. However the ρ^2 values between zero and one do not have a natural interpretation [34]. Another similar informal goodness of fit measure that corrects for the number of parameters estimated is

$$\rho^2 bar = 1 - \left[\ln L_b K / \ln L_o \right] \tag{7}$$

Where K is the number of parameter estimates in the model (degrees of freedom)

For the experimental gambling approach, the utility function with Constant Partial Risk Aversion (CPRA) is used to get a unique measure of partial risk aversion coefficient for each game level. This depicted as the equation below:

$$U = (1 - S)c^{(1-S)} \tag{8}$$

Where
S=coefficient of risk aversion, and
c=certainty equivalent of a prospect.

The Herfindahl index (DHI) is used to calculate enterprise diversification and represent the specialization variable. Although, this index is mainly used in the marketing industry to analyze market concentration, it has also been used to represent crop diversification [35, 36]. Herfindahl index (DHI) is the sum of square of the proportion of individual activities in a portfolio. With an increase in diversification, the sum of square of the proportion of activities decreases, so also the indices. In this way, it is an inverse measure of diversification, since the Herfindahl index decreases with an increase in diversification. The Herfindahl index is bound by zero (complete diversification) to one (complete specialization).

$$\text{Herfindhal index } (\text{DHI}) = \sum_{i=1}^{N} s_i^2 \tag{9}$$

Where
N=number of enterprises and
s_i =value share of each i-th farm enterprise in the farm's output. Equation 3.9a
is the proportion of the i-th activity in acreage / income.

3.4   RESULTS AND DISCUSSIONS

3.4.1   DETERMINANTS OF ADOPTION OF ORGANIC FARMING

The summary statistics in Table 3.1 show that the average age of the farmers was over 50 years with younger people migrating to urban centres in
search of better jobs. In the study area, most of the men are engaged in wage
employment at the neighbouring sugarcane farms or as migrant workers in
the cities of Durban, Johannesburg. Hence over 70% of the farmers were
female. Education levels are low and are consistent with most rural farming communities in South Africa, where formal education opportunities are
limited. Household sizes were large with family labour playing a major role
in tilling the land. Small farm sizes averaging 0.59 hectares for fully certified
organic farmers, 0. 67 hectares for non-organic farmers and 0. 71 hectares
for partially certified farming was common in KwaZulu-Natal.

The main sources of income were farm and off farm employment, the
latter constituting wages or salary income and remittances. Farm income
was highest for fully certified organic farmers. This is an indication that
the adoption of fully certified organic farming and its commercialization
has brought economic benefits to these otherwise poor rural households
and is an important contributor to household income. The proportion of
income from farming was highest among the fully certified organic farmers.
While the average farmer was classified as risk averse, non-organic farmers
were more risk averse than their organic counterparts. Risk-averse farmers are reluctant to invest in innovations of which they have little firsthand
experience. Despite the tenure system being communal, farmers felt they
had tenure rights through the permission to occupy with allocation done
by the traditional chief of the tribe (*inkosi*) and his headman (*induna*). On
average the farmers acknowledged that the household had rights to exercise
on its own cropland the building of structures, planting trees and bequeath-
ing to family members or leasing out. Fully certified farmers had more assets
than their non-organic counterparts as well as chicken and livestock.

The ordered probit model results are presented in Table 3.2. The model successfully estimated the significant variables associated with the farmer's adoption decisions. The Huber/White/ sandwich variances estimator was used to correct for heteroscedasticity. The explanatory variables collectively influence the farmer's decision to be a certified organic with the chi- square value significant at one percent. The following variables were found to be significant determinants in the organic farming adoption decision by smallholder farmers in the study area: age, household size, land size, locational setting of the farmer depicted by the sub-wards Ogagwini, Ezigani, and Hwayi, farmer's risk attitude, livestock ownership (chicken and goat ownership), land tenure security as depicted by the rights the farmer can exercise on his/her own cropland to build structures and asset ownership.

TABLE 3.2: Adoption of organic farming among smallholder farmers: Ordered probit model results

Variables	Parameter	Robust std error	P-values
Age	0.0194072	0.0079204	0.014***
Gender	0.3796234	0.2707705	0.161
Household size	0.0504668	0.0271520	0.063*
Land size	-0.2352607	0.1083583	0.030**
Off Farm Income	-0.0001223	0.0001129	0.279
Location (sub-ward)			
Location (1= ogagwini)	2.894311	0.6380815	0.000***
Location (1=ezigani)	4.191274	0.7234394	0.000***
Location (1=hwayi)	5.158803	0.8495047	0.000***
Risk attitudes	-0.759508	0.3773067	0.044**
Fertility (Manure)			
Chicken ownership	0.0424046	0.0148472	0.004***
Cattle ownership	-0.0418692	0.0431078	0.331
Goat ownership	-0.1005212	0.0569375	0.077*
Land tenure rights			
Land tenure (1= build structures)	0.4803418	0.2372247	0.043**
Land tenure (1= plant trees)	0.0235946	0.3023182	0.938
Land tenure (1= bequeath)	0.1335225	0.2619669	0.610
Land tenure (1= lease out)	-0.3840883	0.2593139	0.139
Land tenure (1 = sell land)	0.0829177	0.2978485	0.781
Asset ownership	0.5853967	0.205389	0.004***

*Significance levels: *** $p<0.01$: ** $p<0.05$: * $p<0.1$*

(Source: Field Data)

The study established that older female farmers with large household sizes were more likely to be certified-organic. Similarly, farmers who reside in the sub-wards Ogagwini, Ezigani, and Hwayi were more likely to be certified organic. This suggests the presence of local synergies in adoption which raises the question about the extent to which ignoring these influences biases policy conclusions. The negative correlation between land size and adoption implies that smaller farms appear to have greater propensity for adoption of certified organic farming. This finding is supported by several studies reviewed in the literature that allude to the fact that organic farms tend to be smaller than conventional farms. The significance of livestock is explained by the importance of manure for organic farming. The study also found that older farmers tend to be adopters supporting findings by [37]. The propensity to adopt was also positively influenced by asset index which is a proxy for wealth.

3.4.2 RISK AVERSION AND RISK MANAGEMENT STRATEGIES

The distribution of risk aversion preferences for each prospect for the fully certified organic, partially certified organic and non-organic crop farmers are presented in Table 3.3. The distribution of responses was spread across all classes of risk aversion for the pooled data. It can be noted that on average, the majority of the respondents revealed their preference for prospects representing intermediate and moderate risk aversion alternatives across the three farmer groups. Table 3.3 further shows that non-organic farmers were the most risk averse being classified as extremely risk averse at 20.4%, compared to fully and partially certified organic farmers at 7.3% and 4.2%, respectively. This explains their non-adoption of certified organic farming, despite its introduction in the area since the year 2000. On the other hand, the fully certified organic farmers were the least risk averse, being classified as neutral to risk preferring at 9.1% compared to 7.3% and 4.1% for the partially certified and non-organic farmers respectively. These results conform to a priori expectations regarding the risk preference patterns of smallholder farmers.

According to Figure 3.4, the non-organic farmers constituted 55.6% of respondents within the extreme risk aversion class compared to 22.2% for fully certified organic and 22.2% for partially certified organic farmers. This is a confirmation of previous findings in this study that explains the non-adoption of certified organic farming by the non-organic farmers.

TABLE 3.3: Distribution of smallholder farmers according to risk preference patterns in KwaZulu-Natal

Farmer group	Risk aversion classification					
	Extreme	Severe	Intermediate	Moderate	Slight to neutral	Neutral to preferring
Fully certified organic (n = 48)	7.30	5.50	30.90	40.00	7.30	9.10
Partially certified organic (n = 95)	4.20	8.30	44.80	29.20	5.20	7.30
Non-organic (n= 46)	20.40	8.20	30.60	30.60	0.00	4.10
Pooled data (n = 189)	9.00	7.50	37.50	32.50	4.50	7.00

Source: Field data

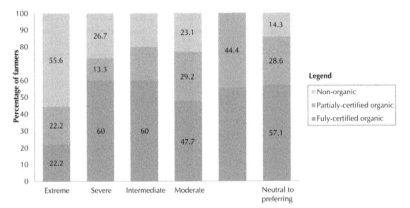

FIGURE 3.4 Frequency distribution within risk aversion classes across the farmer groups

In the risk neutral to preferring category, the non-organic farmers constitute only 14.3%. Fully certified organic farmers constituted 57.1% and partially certified organic farmers constituted 28.6%.

A comparison of the results from the South African study, which applied the general experimental method, with similar studies using the same methodology was for farming communities in the Côte d'Ivoire [38], Ethiopia [39], Zambia [40], Philippines [41] and India [31], shows similarities in the findings of the studies done in India, Philippines, Zambia and Côte d'Ivoire, India

[31], where the majority of the respondents are classified as intermediate to moderate risk aversion (Table 3.4). Similarly, these results suggest that farm households in South Africa are less risk averse than in Ethiopia, Zambia and Côte d'Ivoire but are much more risk averse than in India and Philippines.

Farmers identified their sources of risk and significance in terms of the potential impact to their farming activity as presented in Table 3.5. The fully certified organic farmers cited in order of priority, uncertain climate (mean 2.96), lack of cash and credit to finance inputs (mean 2.78) and tractor unavailability when needed (mean 2.76). These risk sources have a direct bearing on production of organic produce. Climatic conditions are beyond the farmers' control, and the top ranking probably reflects the farmers' concerns about the effects of recent drought in Umbumbulu district. These impact negatively on crop yield. Due to communal land ownership

TABLE 3.4: Percentage distribution of revealed risk preferences in five experimental studies

Studies	Extreme to severe risk aversion	Intermediate to moderate risk aversion	Risk-neutral to risk preferring	Number of responses
India [31]				
50 rupee	8.4	**82.2**	9.4	107
500 rupee	16.5	**82.6**	0.9	115
Philippines [41]				
50peso	10.2	**73.5**	16.3	49
500peso	8.1	**77.6**	14.3	49
Zambia [40]				
1000kw	29.1	**46.4**	24.5	423
10000kw	36.7	**52.5**	11	137
Ethiopia [39]				
5bir	**45.4**	33.6	21	262
15bir	**55.7**	27.5	16.8	262
Côte d'Ivoire [38]				
1000FCFA	32.8	**53.9**	13.3	362
5000FCFA	**46.1**	45.9	8	362
*South Africa [42]				
400Rands	16.5	**70**	11.5	196

*Source: Field work

TABLE 3.5: Identification and ranking of risk sources by farmers

Constraint	Fully-certified organic			Partially-certified organic			Non-organic		
	Mean	Std.Dev.	Rank	Mean	Std.Dev.	Rank	Mean	Std.Dev.	Rank
Livestock damage	2.56	0.774	7	2.82	0.448	4	2.8	0.539	2
Uncertain climate	2.96	0.189	1	2.83	0.409	3	2.82	0.486	1
Uncertain prices for products sold to pack house	2.21	0.793	13	2.13	0.591	16	-	-	-
Uncertain prices for products sold to other markets	1.94	0.811	17	2.02	0.595	18	2.17	0.761	10
Huge work load	2.58	0.599	6	2.32	0.688	12	2.53	0.649	4
Lack of cash and credit to finance inputs	2.78	0.567	2	2.58	0.615	6	2.78	0.468	3
Lack of information about producing organic crops	2.02	0.687	15	2.2	0.632	14	2.16	0.717	11
Lack of information about alternative markets	2.38	0.623	10	2.29	0.602	13	-	-	-
Lack of proper storage facilities	2.56	0.66	7	2.46	0.543	9	2.41	0.643	7
Lack of affordable transport for products	2.72	0.492	4	2.42	0.56	11	2.06	0.852	12
Lack of telephone to negotiate sales	2.69	0.509	5	2.55	0.633	8	2.22	0.771	8
Inputs not available at affordable prices	2.52	0.642	9	2.8	0.447	5	2.51	0.545	5
Tractor not available when needed	2.76	0.501	3	2.89	0.416	1	2.46	0.713	6
Cannot find manure for purchase	1.92	0.778	18	2.56	0.66	7	2.2	0.645	8
Cannot find labour to hire	1.73	0.764	20	1.76	0.816	20	-	-	-
Cannot access more cop land	1.95	0.753	16	1.98	0.805	19	1.92	0.764	13
Delay of payment of products sent to pack house	2.22	0.723	12	2.89	0.315	1	-	-	-

TABLE 3.5: Continued

Constraint	Fully-certified organic			Partially-certified organic			Non-organic		
	Mean	Std.Dev.	Rank	Mean	Std.Dev.	Rank	Mean	Std.Dev.	Rank
Lack of bargaining power over product prices at the pack house	2.16	0.672	14	2.2	0.704	14	-	-	-
Lack of information about consumer preferences for organic products	2.23	0.654	11	2.44	0.604	10	-	-	-
No reward system or incentive for smallholder producers	1.86	0.78	19	2.02	0.866	17	-	-	-

and strict conditions for credit, farmers have limited options to obtain credit from financial institutions. Among the sampled farmers only 21 farmers were able to access credit. Farmers in the study area lack collateral that is acceptable to banks. For example, banks required title deeds as proof of land ownership but the majority of black farmers in South Africa and especially in the former homelands still lacked this vital documentation. Tractor unavailability can be attributed to the fact that there is one tractor that has been allocated to the members of Ezemvelo Farmers Organisation. The tractor is leased out at a rental fees. This poses a challenge during the land preparation phase when the demand for its services is at peak.

Similarly, partially certified farmers also ranked tractor not being available when needed (mean 2.89) and uncertain climate (mean 2.83) as identified sources of risk (Table 3.5). The risk of delays in payment for products sent to pack house (mean 2.89) are attributed to various factors, among them the contractual obligation the agent has with the retailer which has a bearing on the duration of payment. Payment is only made to the farmer once the supply has been forwarded to the retailer and there is confirmation of the quantity of produce that has been rejected. The process flow delays payments to farmers. Non-organic farmers also cited uncertain climate (mean 2.82), livestock damage to crops (mean 2.80) and lack of cash and credit to finance farm inputs (mean 2.78). The livestock damage is a result of lack of fencing around the crops planted.

The most important traditional risk management strategies used by the farmers were identified as crop diversification, precautionary savings and participating in social network. The overall Herfindahl index of crop diversification is estimated at 0.61 which indicates that the cropping system is relatively diverse (Table 3.6). These results confirm previous findings by [43] who obtained an estimated DHI of 0. 49–0.69 among smallholder farmers in three regions in Bangladesh. As shown in Table 3.6, non-organic farmers practiced more crop diversification with a DH index of 0. 23 compared to organic farmers with a DHI of 0. 72. These results are consistent with previous findings in this study measuring farmers risk attitudes and presented in Figure 3.6, that established that smallholder farmers in the study area tend to diversify due to their risk-averse nature and that non-organic farmers are more risk averse than organic farmers.

According to Table 3.6, a total of 69.1% of fully certified farmers practised crop diversification compared to 96.8% of the non-organic farmers. A total of 81.2% of the partially certified farmers practised crop diversification. The

TABLE 3.6: Risk management strategies used by the different farmer groups

No.	Risk management strategy	Fully-certified organic	Partially-certified organic	Non-organic
		n = 48	n = 103	n = 49
1	Enterprise diversification index (DH)	0.72	0.89	0.23
2	Practice crop diversification (% of respondents)	69.10	81.20	96.80
3	Savings bank account (% of respondents)	60.90	48.90	46.80
4	Current level of savings (% of respondents)			
	less than R500	27.27	37.84	35.29
	R501 -R1000	45.45	29.73	41.18
	R1001 -R5000	21.21	29.73	17.65
	More than 5000	6.07	2.70	5.88
5	Social networks (% of respondents)			
	Membership of EFO	100.00	100.00	10.00
	Others (burial clubs, *stockvels*)	33.00	25.00	25.00

common crops grown by the organic farmers are amadumbe, potatoes, sweet potatoes and green beans while non-organic farmers grew amadumbe, potatoes, sweet potatoes, green beans, maize, sugarcane, bananas, chillies and peas.

Precautionary saving occurs in response to risk and uncertainty [44]. The smallholder farmers' precautionary motive was to delay/minimise consumption and save in the current period due to their lack of crop insurance markets. According to [45], the quantitative significance of precautionary saving depends on how much risk consumers face. Whereas 60. 9% of the fully certified farmers had savings bank accounts, only 46.8% non-organic farmers had bank accounts. The current level of saving in the study area was low with savings ranging from less than R500 to over R5000 per month. The level of savings was low across all groups. Among the fully certified organic group, most of the respondents (45.45%) saved between R1000–R5001 whereas most of the partially certified farmers (37.84%) saved less than R500 per month. Most of the non-organic farmers (41.18%) saved between R501–R1000 per month. Across all groups, however the level of saving greater than R5000 was minimal.

The farmers also engage in social networks as a risk sharing strategy. There were two main categories of social networks that the farmers engaged in. These are farmers association and other social networks most notably burial clubs and stockvels. The farmers association is used as a vehicle by the organic farmers to gain access to markets for their organic produce while the burial clubs and stockvels are sources of access to credit and/or loans. In the latter instance, farmers do not have to produce collateral. The burial clubs and stockvels are common in most rural areas and are a source of mitigating liquidity and financial risk where possible.

3.4.3 CONSUMER AWARENESS, PERCEPTIONS AND CONSUMPTION DECISIONS

The summary statistics of consumers presented in Table 3.7 showed that the majority of the consumers were females within 25–34 age category. Previous studies for example [46] found that women were the predominant purchasers of organic food and responsible for household consumption. The younger generation consumers represent an important target group in the advancement of consumer demand for organic products. The level of education was generally low especially among rural consumers. The unemployment rates in the former homelands demonstrate a substantial skewering of the demographic profile of the district and high dependency rates of those not economically and productively active. It also reflects the levels of out migration of economically active population from the province to other parts of South Africa. Unemployment was also lower in urban areas than rural areas. The income distribution of the respondents is especially concentrated in the R1000–R5000/month category. However the majority of the respondents within this category were in the rural areas. This can be attributed to limited economic activity in rural areas. The household size was within the provincial estimate of 4-5 persons per household [47] with rural households having higher numbers. Majority of the respondents had children under the age of 18 years in the household. The average distances to the nearest shops were estimated at between 6–9kms. In the urban areas however this was reduced to 1.38kms.

There is a general understanding of term 'organic foods' among consumers. Consumers defined organic foods as healthy and nutritious, associated with traditional and or indigenous methods of production and free from chemicals. There were low levels of awareness about local standards

TABLE 3.7: Summary statistics of consumers in the Eastern Cape Province

		Former Homelands		Locality		
		OR Tambo DM	Amatole DM	Rural	Peri-urban	Urban
Variable		**n = 100**	**n = 100**	**n = 30**	**n = 30**	**n = 40**
Gender	Male	43	34	28	40	44
	Female	57	66	72	60	56
Age in years	18-24	17	13	18	16	12
	25-34	29	33	12	34	26
	35-44	27	16	14	23	41
	45-55	20	17	21	19	16
	>55	7	21	35	8	5
Education Level	None	4	9.7	16.1	6.5	1.2
	Primary	21	29.1	46.4	32.3	5.9
	High school	39	39.8	37.5	48.4	34.1
	Tertiary	36	21.4	0	12.9	58.8
Employment Status	Unemployed	29.4	31	52.6	48.3	2.4
	Student	9.8	4	5.3	5	9.4
	Housewife/man	10.8	8	19.3	10	2.4
	Retired	5.9	1	8.8	1.7	1.2
	Working part-time	14.7	11	8.7	18.3	11.8
	Working full time	29.4	45	5.3	16.7	72.8
Income Level	<1000	10	12.5	0	4.8	23.5
	1001 -5000	16	5.8	0	0	25.9
	5001-10 000	20	17.3	5.3	17.7	28.2
	10 001 - 15 000	30	49	66.7	46.8	16.5
	>15 000	24	15.4	28.1	30.6	5.9
Household size in number		5.2	4.33	5.18	4.98	4.31
Children < 18 years		79	55.8	71.9	71	61.2
Distance in Kms		6.71	9.63	12.67	9.32	1.38

for organic products, the identification of organic products using an organic logo, existence of a national organic movement and/or the presence of an organic certification body in South Africa. Therefore consumers could not readily identify certified organic against non-certified organic products.

Notwithstanding, consumers argued that there was a need for certification and verification of organic products and hence are unable to make informed decisions on the organic status of products in the market.

Trust of organic labels can be increased once more information is available to consumers on the various organic labels, their meaning and on the difference between certified and non certified products in the shelves. In the absence of this information, producers and likewise consumers may not get value for money. Certification and labelling is essentially in regulating and facilitating the sale of organic products to consumers. The perception of the high price of organic products is a deterrent to the purchase of organic products and hence the growth of organic industry especially for the emerging organic market of South Africa. To increase the consumption of organic products, it will be important to motivate new consumer segments to buy organic food. Hence trust is a crucial aspect when consumers decide whether to buy or not to buy organic products [48].

Trust is a 'credence attribute' which is not directly observable by consumers. Enhancing consumers trust about the labels of organic products can be achieved through among others, effective communication strategies on the traceability of organic products and ensuring compliance and adherence by retailers selling organic products to the certification standards and availability of information on the organic status of products. Some of the reasons advanced in the study to increase consumers' trust for organic products are to:

- purchase from specific shops that sell organic
- check for organic certification label
- practice own organic farming

In South Africa, food retailers have the largest share of the organic industry [49]. Similarly, most products are sold through the export market due to the higher revenue from exports. Irwin [50] says that South Africa has a favourable position for expansion in the domestic market as a result of the following developments in the organic sector over the past few years:

- establishment of separate organic section in major retail stores
- national regulation/standards for organic products
- establishment of South Africa organic certification bodies
- formation of South African organic associations.

Food purchasing is an important part of food behaviours. In this study the apportioning, explicitly or tacitly, of the responsibility of household food shopping depends on a number of factors as food purchasing is an important part of food behaviours. This responsibility was closely shared among various members of the household with majority of the consumers being responsible for the decision making of organic food demand and purchase. The general finding in the study was that most consumers shop in supermarkets, grocery stores and spaza (kiosks) shops. The majority of consumers who shop in supermarkets reported that local shops do not provide the services people demand and that food choice and quality are limited. This is coupled with discount promotions common with supermarkets and variety of products. The findings from this study are consistent with findings from the Food Safety Agency [51] that state that a vast majority (92%) of consumers continue to use supermarkets for most of their food shopping. However, local shops play an important role in 'top-up' shopping, being used by 75% consumers for some of their food purchases.

Commonly consumed organic products included fresh vegetables, fresh fruits, meat/meat products and milk/ milk products. However, the general trend in Figure 3.5 and Figure 3.6 shows that there are marked increases in the future demand of all organic products. This augurs well for the growth of the organic industry in the Eastern Cape and in South Africa in General. The findings of this study are consistent with [52] who stated that a study by Pick-n-Pay, one of the major national retail supermarket chains and supporter of the development of the retail organic market in South Africa, on the performance and trends of fresh organic produce showed that fresh produce completely dominated the sales.

This is an indication that the consumption of organic products is closely related to consumer awareness and knowledge of organic products. Increasing awareness about organic products to consumers is important to spur its demand. Most of the consumers had consumed organic products in South Africa with non-consumers showing a general interest in organic products. Authors [53] state that consumer awareness of organic foods is the first step in developing demand for organic products. Yet, awareness does not necessarily equate with consumption. While organic refers to the way agricultural products are grown and processed [54], interest in consuming organic products may relate to food safety concerns where organic products may be a partial answer to recent food scares associated with production and handling (e. g. BSE, dioxins, Salmonella, etc.). Food safety issues have driven consumers to search for safer foods whose qualities and attributes are guaranteed [55]. The main reasons advanced for the consumption of organic products are that organics are healthy and nutritious, have a better appearance and taste, are affordable and are safe to consume. Identified hindrances to the consumption of organics are that they are expensive and not readily available. Price and affordability of organic products was ranked

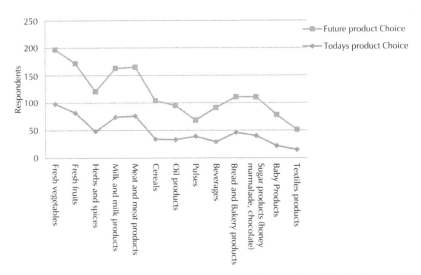

FIGURE 3.5   Demand difference between organic products today and in the future in OR Tambo District Municipality

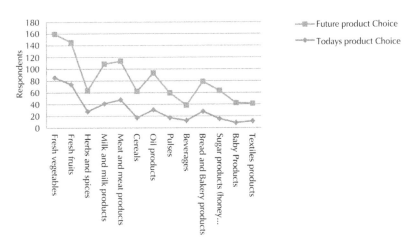

FIGURE 3.6   Demand difference between organic products today and in the future in Amatole District Municipality

as the most important consideration among all consumers when buying organic products in South Africa.

3.5   CONCLUSIONS

The global markets for organic products have grown rapidly over the past two decades [8]. Currently 32. 2 million ha are being managed organically world-wide by more than 1.2 million producers [11]. In Africa, South Africa has the third largest area (50, 000ha) under organic farming [11]. Organic production is particularly well-suited for smallholder farmers, who comprise the major-ity of the world's poor. The promotion of organic agriculture does not only constitute an important option for producers but also responds to consumers' desire for higher food quality and food production methods that are less dam-aging to the environment. The consumers' concerns for food safety, quality and nutrition are increasingly becoming important across the world, which has provided growing opportunities for organic foods in recent years. Expectedly, the demand for organic food is steadily increasing in the developing countries. The untapped potential markets for organic foods in the countries like South Africa need to be realised with organised interventions on various fronts, which require a better understanding of the consumers' preference for organic food. Therefore, an analysis of consumer's awareness of various aspects of organic products may be considered as important ground to build the markets for organic food in the initial phase of market development. Recent analysis [53] indicates that consumer awareness of organic foods is the first step in developing demand for organic products. By identifying independent vari-ables that explain the adoption of organic farming, the present study sought to contribute to policy formulation to promote adoption in South Africa and the rest of Africa. The identified sources of risk faced by smallholder farm-ers provide useful insights for policy makers, advisers, developers and sellers of risk management strategies. This information can yield substantial payouts in terms of the development of quality farm management and education pro-grams as well as the design of more effective government policies.

REFERENCES

1. Tilman D, Cassman KG, Matson PA, Naylor R, Polasky S. Agricultural Sustainability and Intensive Production Practices. Nature 2002; (418)671-677.
2. Food and Agricultural organization. FAO. Assessment of the World Food Security Situation, Report CFS2007/2. Rome; 2007.
3. Willer H, Rohwedder M, Wynen E. Organic Agriculture Worldwide: Current Statistics. In Willer H, Kilcher L (ed.): The World of Organic Agriculture. Statistics and Emerging Trends 2009. ITC Geneva; 2009.

4. Oelofse M, Høgh-Jensen H, de Abreu LS, de Almeida GF, Sultan T, Yu Hui Q, de- Neergaard A. Certified Organic Agriculture in China and Brazil: Market Accessibility and Outcomes Following Adoption. Ecological Economics 2010; (69) 1785-1793.
5. Byerlee D., Alex G. Organic farming: a contribution to sustainable poverty allevia- tion in developing countries? Forum for Environment and Development, Bonn. Report; 2005.
6. UNEP-UNCTAD. Sector Background Note – Organic Agriculture. http://www. un-ep-unctad.org/cbtf/events/geneva5/WordBackgroundnoteorganicagriculture_ 01102007.pdf (accessed 15 June 2010).
7. UNCTAD. Organic Agriculture: A Trade and Sustainable Development Opportunity for Developing Countries. In UNCTAD/ DITC/TED/2005/12. (ed.)UNCTAD Trade and Environment Review 2006. United Nations New York and Geneva; 2006.
8. Sahota A. The Global Market for Organic Food and Drink. In Willer H, Kilcher L. (Eds.) The World of Organic Agriculture. Statistics and Emerging Trends 2011. IFOAM Bonn and FiBL Frick; 2011.
9. Willer H, Yussefi M. The World of Organic Agriculture 2005. Statistics and Emerging Trends. IFOAM Publication, Tholey-Theley, Germany; 2006.
10. Parrott N, Ssekyewa C, Makunike C, Ntambi S M. Organic Farming in Africa. In Willer/ Yussefi (eds.) The World of Organic Agriculture. Statistics and Emerging Trends 2006. IFOAM, Bonn; 2006.
11. Willer H, Klicher L. The World of Organic Agriculture. Statistics and Emerging Trends 2009. IFOAM, Bonn, Germany and Research Institute of Organic Agriculture, FIBL, Frick, Switzerland; 2009.
12. Walaga C. Organic Agriculture in the Continents. In Yusseffi M, Willer H. (ed.) The World of Organic Agriculture: Statistics and Future Prospects; 2003.
13. Ssekyewa C. Organic agriculture research in Uganda. Paper presented at the International Society for Organic Agriculture Research (ISOFAR) Scientific Conference, 20th-23rd September 2005, Adelaide, Australia; 2005.
14. FiBL/IFOAM: Key Results from Survey on Organic Agriculture Worldwide. www. fibl. org (accessed 9 September 2009).
15. Moffet J. Principles of organic farming. Paper contributed at the 1st short course in organic farming, 25th-27th October, Spier Institute and Ellensburg Agricultural Col- lege; 2001.
16. GROLINK. Feasibility study for the establishment of certification bodies for organic agriculture in Eastern and Southern Africa, Report commissioned by Sida/INEC, Höje; 2002.
17. International Trade Centre. ITC. Organic Food and Beverages: World Supply and Major European Markets. www.intracen.org/mds/sectors/organic/welcome.htm (accessed 10 January 2010).
18. Scialabba NE, Hattam C. Organic Agriculture, Environment and Food Security. http://www.fao.org/docrep/005/y4137e/y4137e00.HTM (accessed 2 July 2011).
19. Arnold G. The Resources of the Third World. Cassell, London; 1997.
20. Millstone E, Lang T. The Atlas of Food: Who Eats What, Where and Why? Earth Scan, London; 2002.
21. Mahlanza B, Mendes E, Vink N. Comparative Advantage of Organic Wheat Production in the Western Cape. Agrekon 2003; (42) 144-162.

22. Troskie D P. Factors influencing organic production: An economic perspective. Paper presented at 12 Annual Interdisciplinary Symposium, 12 September, ARC-PPRI, Stellenbosch; 2001.
23. International Federation of Organic Agriculture Movements. IFOAM. IFOAM's Position on Smallholder Group Certification for Organic Production and Processing. http://www.ifoam.org/press/positions/Small_holder_group_certification. html(accessed 17 April 2011).
24. Naegeli F, Torrico JC. The Potential of Organic Agriculture to Improve Food Security. CienciAgro 2009; (1) 144-151.
25. Zundel C, Kilcher L. Organic agriculture and food availability. Issues Paper, International Conference on Organic Agriculture and Food Security, 3-5 May, Rome; 2007
26. Connor D J. Organic Agriculture Cannot Feed the World. Field Crops Research 2008 (106)187-190.
27. Badgley C, Moghtader J, Quintero E, Zakem E, Chappell MJ, Avilés-Vázquez K, Samulon A, Perfecto I. Organic Agriculture and the Global Food Supply. Renewable Agricultural Food Systems 2007 (22): 86–108.
28. Statistics South Africa. STATSSA. Mid-year Population Estimates: Statistical Release P0302. http: //www. statssa. gov. za/ publications/P0302/P03022010. pdf(accessed 12 July 2011).
29. OR Tambo District Municipality Integrated Development Plan. ORTIDP. OR Tambo District Municipality Integrated Development Plan. http: //www. ortambodm. org. za/ files/PDF/pplan2012to2013. pdf (accessed 5 October 2011).
30. Community Survey. CS. Statistical Release P0301. 1. Basic Results for Municipalities. http://www.statssa.gov.za/publications/p03011/p030112007.pdf(accessed 5 October 2011).
31. Binswanger HP. Attitudes towards Risk: Experimental Measurement in Rural India. American Journal of Agricultural Economics 1980(62)395-407.
32. Young DL. Risk Preferences of Agricultural Producers: Their Use in Extension and Research. American Journal of Agricultural Economics 1979 (61) 1063-70.
33. Verbeek M. A Guide to Modern Econometrics. John Wiley and Sons Ltd; 2008
34. Greene W H. Econometric Analysis. 5th ed. Prentice Hall, New Jersey; 2003.
35. Llewelyn RV, Williams JR. Nonparametric Analysis of Technical, Pure Technical and Scale Efficiencies for Food Crop Production in East Java, Indonesia. Agricultural Economics 1996 (15) 113-126.
36. Bradshaw B. Questioning Crop Diversification as Response to Agricultural Deregulation in Saskatchewan, Canada. Journal of Rural Studies 2004 (20)35-48.
37. Feng XHE, Chenqi D. Adoption and diffusion of sustainable agricultural technology: an econometric analysis. Report. Chongqing Institute of Technology China; 2010.
38. Kouamé EB H. Risk, risk aversion and choice of risk management strategies by cocoa farmers in Western Côte d'Ivoire. University of Cocody-AERC Collaborative PhD Programme. Abidjan-Côte d'Ivoire; 2010.
39. Yesuf M. Risk aversion in low income countries: experimental evidence from Ethiopia. Discussion paper 715. IFPRI, Washington D. C; (2007).
40. Wik M, Holden ST, Taylor E. Risk, market imperfections and peasant adaption: evidence from Northern Zambia. Discussion paper no. 28/1998, Department of Economics and Social Sciences, Agricultural University of Norway; 1998.

41. Sillers DA. Measuring risk preference of rice farmers in Nueva Ecija, Philippines: An experimental approach. PhD dissertation, Yale University, USA; 1980.
42. Lwayo M. Risks preference and consumption decisions in organic production: The case of KwaZulu-Natal and Eastern Cape Provinces. PhD thesis, University of Fort Hare, South Africa; 2012.
43. Rahman S. Whether Crop Diversification is a Desired Strategy for Agricultural Growth in Bangladesh? Food Policy 2009 (34)340-349.
44. Feigenbaum J. Precautionary Saving or Denied Dis-saving. Economic Modelling 2011 (28)1559-1572.
45. Cunha F, Heckman J, Navarro S. Separating Uncertainty from Heterogeneity in Life Cycle Earnings. Oxford Economic Papers 2005(57) 191-261.
46. Mutlu N. Consumer attitude and behaviour towards organic food: Cross-cultural study of Turkey and Germany. Master's thesis, University of Hohenheim Germany; 2007.
47. PROVIDE. A profile of the Eastern Cape Province: Demographics, poverty, income, inequality and unemployment from 2000 till 2007. Background paper series: 1(2). Ellensburg. Cape Town; 2009.
48. Zanoli R. The European consumer and organic food. School of Management and Business, Wales; 2004.
49. Botha L, Van Schalkwyk H D. Concentration in the South African food retailing industry: proceedings of the 44th annual AEASA conference, September 20–22, Grahamstown, South Africa; 2006.
50. Irwin BL. Small-scale previously disadvantaged producers in the South African organic market: Adoption model and institutional approach. Masters' dissertation. Michigan State University, USA; 2002.
51. Food Safety Agency. FSA: Consumer Attitudes. www.food.gov.uk/multimedia/pdfs/casuk05.pdf. (accessed 5 October 2011).
52. Vermeulen H, Bienabe E. What about the food 'quality turn' in South Africa? Focus on the organic movement development. Paper presented at the 105th EAAE Seminar International Marketing and International Trade of Quality Food Products, March 8-10, Bologna, Italy; 2007.
53. Briz T, Ward RW. Consumer Awareness of Organic Products in Spain: An Application of Multinomial Logit Models. Food Policy 2009 (34) 295-304.
54. Organic Trade Association. OTA. Manufacturer Survey. www.ota.com (accessed 10 October 2011).
55. Zalewski RI, Skawiska E. Food safety: commodity science point of view. Paper presentation at the International Association of Agricultural Economists Conference, August 12–18, Gold Coast, Australia; 2006.

PART 3
Localizing the Food System

Empowering the Citizen-Consumer: Re-Regulating Consumer Information to Support the Transition to Sustainable and Health Promoting Food Systems in Canada

Rod Macrae, Michelle Szabo, Kalli Anderson, Fiona Louden, and Sandi Trillo

4.1 INTRODUCTION

Both health and sustainability are stated public policy objectives, but Canada's food information rules and practices may not be optimal to support their achievement. Lacking a joined up national food policy [2], and thus clear purpose for public information about food, the information presented is frequently left to the marketers of product. No one has responsibility for determining the overall coherence of consumer food messages with these stated policy objectives. Individual firms provide information that shows their products to best advantage. As a result, consumers receive information that is incomplete, and which may contradict the information provided by another firm or government agency. Individual consumers do not have the resources to determine with any ease the accuracy or completeness of any firm's messages, particularly when faced with the size of food industry advertising budgets. Equally problematic, the national government has recently decided it will "no longer verify nutrition claims on food labels, and will instead set up a website where consumers can take their concerns directly to food producers" [3]. Partly in response to these information and monitoring gaps, for some time now third parties have been filling the consumer food information void, providing endorsements and health and eco-labels that can affect consumer purchasing behaviour.

Government rules confound this problem because there is limited coherence between the parts and levels of government that have responsibility for advertising rules, labelling and grading systems. The healthy eating messages of health departments are often competing with contradictory messages permitted by the regulatory framework of other arms of government. Investments in programs that successfully promote environmental stewardship in agriculture are undercut in the market because consumer information rules do not permit consumers to identify many of these products and support those stewardship efforts with their purchases and eating.

It has been known for some time that diet is a significant risk factor in at least 60% of diseases [4]. Many chronic diseases and conditions, including cardiovascular disease, hypertension and stress, cancer, diabetes, low birth weight infants (and its associated problems), anaemia, and some infections in children now pose major public health challenges. All of these chronic diseases and conditions are related to nutrition. They affect both the food-rich (those with sufficient income to acquire whatever foods they desire) and the food-poor (those experiencing food insecurity). Very significant percentages of the Canadian population are at risk of these diseases because they do not eat in a manner optimal for health. Of course, healthy eating is a complex undertaking, but information about food is certainly a significant factor.

Canadians collectively pay, through publicly funded health insurance, for the costs of individuals' food choices or hunger. The food system has limited responsibility for the social consequences of consumption of its products. The efforts of ministries of health to promote healthy eating are frequently compromised by agribusiness expenditures encouraging unhealthy eating patterns.

Equally, food firms bear little responsibility for the negative environmental impacts of their products. Canada faces significant environmental challenges related to greenhouse gas emissions, energy inefficiencies, nitrogen and phosphorous contamination of waterways, biodiversity losses, and excessive food waste [5,6]. It lags behind most OECD countries in agri-environmental performance [7]. In the industrial world, most of these environmental costs are externalized, absorbed in the environment or remediated by the state, using taxpayer money [8].

Confronted with a myriad of food system problems, governments are failing on many fronts related to health promotion and sustainability, including changing food information rules to enlist eaters in a public effort to create change. This hands-off approach to consumer information is particularly problematic against the backdrop of a growing movement of "citizen-consumers," eaters who bring a set of values to their shopping and eating decisions that go beyond individual concerns.

Other countries have recognized this problem and taken actions to solve it, Norway and the Netherlands representing two contrasting approaches. In the 1970s, Norway aligned food production and nutrition information with their nutrition policy objectives, to motivate better dietary habits and to develop skills for making more informed food choices. The government recognized that "present marketing practices are in relatively large disaccord with the nutritional objectives. The factors which today regulate sales are only to a small degree dictated by nutritional considerations" [9]. Following a different strategy, because they were unwilling to extensively regulate environmental changes in the agricultural sector, the Dutch government has actively facilitated eco-label programming by NGOs, farm associations and firms to allow consumers to be better informed about the sustainability attributes of food products. Whether this is the most effective approach is subject to some debate [10], but it highlights how environmental information can potentially drive supply chain changes with well-designed and profound programming.

The challenge is to redesign consumer information systems so that they help governments achieve their national nutritional, health and environmental objectives for the population. But given shifts underway that reduce state authority over the food system, limited mechanisms of governance, and reduced use of potentially effective policy instruments [11], the limited approaches of earlier periods must be modified. We explore the possibilities for change in this paper, focusing particularly on information conveyed in the supermarket, where most Canadians acquire the bulk of their food for consumption in the home.

4.2　THE SHIFTING TERRAIN OF FOOD INFORMATION REGULATION

According to classical market theory, consumers are presumed to be acting rationally when they make purchases, acting in their self-interest with full awareness. In order to act rationally, they need all the relevant information. Having all the relevant information allows the market to send clear signals to buyers and sellers.

When considering consumer food information needs, regulators have focussed primarily on price, quality and convenience. These parameters have been fairly narrowly defined. For example, food quality has been defined primarily by the safety of the product and, particularly with fresh foods, its cosmetic appearance. Convenience is about the speed and simplicity of preparation.

For their part, the majority of food firms are largely silent on the social, environmental and health impacts of food production, processing and distribution. For example, how a product's nutritional profile might have been affected by agricultural, storage and distribution practices has not traditionally been seen as relevant to consumers. According to market theory, this absence of full information helps to create a dysfunctional food marketplace in which partial and contradictory signals are sent to both producers and consumers. In turn, these distorted signals mean that resources in the food system are improperly allocated, particularly, those that help to ensure health, environmental sustainability and equitable access.

A traditional positive role of government is to shape, monitor and correct deficiencies in the marketplace. Regulation is one of several tools used by government to carry out this role. In particular, regulation serves to influence the actions of market players, define products and processes, determine what is allowed in the market under what conditions, and provide penalties for non-compliance.

However, policy makers have failed to implement strategies based on how more fully informed consumers can help achieve public policy objectives (e.g., improved health and sustainability). The policy tradition is to manage food supply, rather than actively managing demand [2]. The traditional market view is that prices convey accurate and sufficient information on costs and value that permit consumers to act rationally. In reality, the very conditions required to convey accurate information do not normally exist in modern markets [12]. Economists have long recognized that information asymmetry is a chronic problem in food markets, related to issues of limited competition, the absence of accurate prices and incomplete state regulation. Many have devised theory to explain firm and consumer behaviour in asymmetrical environments. Unfortunately, while this evolving theory explains behaviour, it has not advanced policy changes in Canada that have substantially improved the consumer information environment. While asymmetric information is the norm, the regulatory system does not fully recognize this, neither by regulating to fully correct asymmetry nor by actually normalizing it in regulatory approaches.

As well, this approach to regulation assumes that businesses have no broader social obligations, aside from those related to food safety and product promotion regulations. Yet historically, those obligations arose from public demand for regulation. Health and environmental concerns can be viewed as a contemporary equivalent.

Given this narrow view of business obligations, it is unlikely that many food firms will provide, without state oversight, the information that consumers require to make informed decisions. In 1998 interviews conducted for the TFPC (Toronto Food Policy Council) [1], food marketers stated that they had the interests of consumers at heart. It seems though that most marketers were caught up with achieving sales and brand share targets, keeping within budget and satisfying the demands of more senior management. Most marketers had not spent many hours critically thinking about what it really means to satisfy consumer needs. In fact, many of those interviewed were not familiar at that time with issues of increasing concern to many consumers, including gene manipulation, antibiotic and hormone usage, pesticides, and farm worker rights. It seems, then, that there was a gap between intention and practice. Also, interviewees indicated they were not really willing to pay for consumer education. Particularly with regard to nutritional and environmental concerns, they hoped that the media and government would do most of that work for them.

Ultimately, though, it is the marketers who determine which products consumers have to choose from, and, within existing guidelines, what information about them is conveyed in packaging and labeling. If there is no obvious advantage (as measured through potential increased brand sales and brand market share) to providing the information, then there is no incentive to do something that could ultimately mean increased product costs and lower profits. The degree of incentive changes if a competitor changes tactics and provides different information to consumers. Then a marketer may react to keep the playing field level.

So a contradiction emerges. Business may want informed consumers if it suits their marketing approach, but is not sufficiently informed and motivated to provide them information around broader social objectives. As well, firms have known for some time that consumers are confused about the information provided [1]. But in the absence of clear information, consumers may not be able to articulate what they want. In many cases, not enough consumers are vocalizing concerns in a manner that reaches and resonates with food market-ers. Hence, marketers may not react in any significant way with new products or product modifications. All in all, vocalized consumer needs/demands that allow managers to meet product target profit margins with products differenti-ated from those of the competition drive much of what marketers do regarding the products they market.

There has, however, been an explosion of third party health and environmen-tal organizations involved in consumer information provision, in part because government and business approaches are viewed as inadequate. According to Busch [13]:

> To be sure the role of the State has changed and perhaps diminished. Nation-states are now far less likely to regulate directly and far more likely to delegate regulatory authority to other organizations. Moreover, the opening of the world economy has restricted the ability of nation-states to intervene in markets with-out significant and often negative consequences.

From organizations involved in disease prevention, to food production cer-tifiers, these organizations have evolved what are often called private systems of regulation. These organizations are typically non-governmental, or represent associations of businesses. Such approaches allow farmers and firms to cre-ate a niche for themselves in food markets, and to ensure the authenticity of their claims, many producers and processors are going to third party certifiers that require changes in production, including low chemical use, the humane treatment of animals, preservation of biodiversity and wildlife, and just labour practices.

Howard and Allen ([14], p. 439) claim that these third party eco-labels serve "three primary functions":

- "They provide consumers with information about product characteristics that are not immediately apparent or verifiable by consumers themselves."
- They may be a possible instrument "for implementing public policy objec-tives, such as reducing the use of pesticides."
- They are able to "increase producer revenues, either through facilitating a price premium for growers or by providing a market niche for increased sales."

There are some efforts to reconcile state intervention with these new private regulatory initiatives. Sometimes referred to as regulatory reconfiguration [15], such new governance regimes embrace a wide range of coordinated and integrated instruments (including some traditional command and control regulations), well matched to the desired effect, and implemented by an equally wide range of state and non-state actors to have the best chance of success in the long run. The proposals that follow are rooted in such an approach.

4.3   A BRIEF OVERVIEW OF THE EXISTING REGULATORY SYSTEM

In the Canadian system, responsibilities for consumer information are divided amongst different acts, levels of government, and different units within government departments [16]. The most important pieces of federal legislation are the Food and Drugs Act and Regulations (FDA), and the Consumer Packaging and Labelling Act and Regulations (CPLA). Other legislation, including the Canadian Agricultural Products Act (CAPA), the Meat Inspection Act (MIA), the Fish Inspection Act (FIA), are also relevant in some cases. In addition, the federal Broadcasting Act and Regulations have an influence over food commercial messages. This act is the responsibility of the Canadian Radio-television and Telecommunications Commission (CRTC), with regard to the application of regulations and policy rulings. Other pieces of legislation with limited, but sometimes important, bearing on food include the Competition Act and the Trade-marks Act, administered by Industry Canada. In many provinces, there are also provincial rules pertaining to food labeling (e.g., in Ontario, the Farm Products Grades and Sales Act, Regulation 387). The Ontario government is also involved in grading, meat inspection, nutrition and food safety matters. Municipalities in many provinces have some responsibility for implementing provincial legislation regarding nutrition and food safety programs, as they relate to public health, and consequently have some consumer information duties.

The Canadian Food Inspection Agency (CFIA) and Health Canada (HC) share responsibility for regulating food labeling, based on the authorities provided primarily by the Food and Drugs Act (FDA) and Consumer Packaging and Labeling Act (CPLA). Health Canada sets food labeling policies dealing with health and safety matters and CFIA is responsible for non-health and safety food labeling regulations and policies, as well as investigating complaints, encouraging compliance and consumer protection.

Regarding the development of food labeling regulations, the prime concern for CFIA is that no information materials be "false," "misleading" or "deceptive" as stated in the Canadian Food Labeling Regulations. This situation exists because food safety legislation in Canada has been built on an anti-adulteration platform [17,18].

According to the CFIA, a label "serves three primary functions": "to provide basic product information"; to provide "health, safety and nutrition information"; and to provide a "vehicle for food marketing, promotion, and advertising" [16]. The policies pertaining to food labeling and packaging, are developed in order to:

- "Protect health and safety.
- Prevent product misrepresentation and fraud.
- Promote an informed food choice, by providing consumers with reliable and comparable information that reflects current food technology and nutrition recommendations and that can be easily understood.
- Support marketplace equity and fair competition.
- Respect obligations under international and federal provincial trade agreements.
- Do not entail costs of implementation that outweigh benefits to society."

Given the complexity of the current system, the jurisdictional divisions, and the gaps in coverage, the question remains whether the current regulatory environment does promote informed food choice, especially with the emergence of a broader range of consumer concerns. We next address this emerging phenomenon of the citizen-consumer.

4.4   THE CITIZEN-CONSUMER: A NEW FRAMEWORK FOR CONSUMER INFORMATION SYSTEMS

Historically, attention has focussed on perceived consumer concerns about food price, quality (usually defined by cosmetic appearance), convenience and safety. Now there is increasing evidence that consumer interests are more diverse and complex, challenging the traditional way in which companies have both informed consumers and merchandised food products.

For years, the food industry has publicly explained its behaviour in the marketplace by claiming it was responding to what consumers wanted. Mass produced and inexpensive food, convenience, packaging and extensive product variety have been explained as responses to market signals. Surveys of consumer attitudes have historically reinforced this view [1].

But the consumer marketplace is less homogeneous than earlier times. Consumers have been rebelling against mass produced foods for some time. Smart processors and retailers have diversified their product offerings, in the hope of capturing these new market segments. To do so, they have changed promotional strategies, and have invested in sophisticated market survey instruments. Some consumers, once offered new kinds of choices, have responded and changed their purchasing patterns. All these developments confirm the interactive and dynamic interconnections between product availability, consumer information and desires. It is increasingly clear that consumer demand is a product of individual and collective wants and needs, access and availability, and the type and manner of information provided.

Welsh and MacRae [19] suggested that if the population is to be engaged in food buying that supports sustainability and food security, and not just traditional conceptions of price, quality, convenience and safety, then the concept of 'consumer' is far too limited because it focuses primarily on the ability to buy (or reject) products and services. In contrast, the language of 'citizen' implies membership in a society, with both rights and responsibilities beyond those of consuming goods and services. Similarly, society is more than simply a marketplace.

It may now be that many consumers are moving into what Gabriel and Lang [20] have called the fourth wave of consumer activism. In their analysis, they separate the emergence of consumer movements into four waves. The first wave of co-operative consumers was sparked by the Co-operative Movement, which began in England in 1844 and which was based on the principle of self-help by the people without any distinctions between consumers and producers ([20], p. 41). The second wave consists of the 'value for money' movement, which came into its own during the 1930s in the United States ([20], p. 44). Consumer activists in the second wave were concerned about "the threat posed to consumers by increasing concentration and monopoly capital" and their aim was to "make the marketplace more efficient and to champion the interests of the consumer within it" through education and advocacy efforts ([20], pp. 44–45). The third wave, Naderism, also emerged in the United States. This movement originated with the publication of American lawyer, author and activist Ralph Nader's 1965 exposé of the American car industry, Unsafe at Any Speed. Gabriel and Lang argue that Naderism "brought a new punch to consumer politics" by tapping "a deep well of public unease about the power of large corporations vis-à-vis the individual customer" ([20], p. 47). Naderism places a lot of emphasis on debunking misinformation from large corporations and demanding that the state protect its citizens. The fourth wave of consumer activism consists of what Gabriel and Lang [20] term alternative consumerism, what others have called ethical consumerism [21] and what Micheletti [22] refers to as political consumerism. Political consumerism emerged in the 1970s and gained coherence as a movement in the 1990s [20]. The emergence of political consumerism coincides with the rise of the environmental movement in North America and Europe and one of the first streams of alternative consumerism was the green consumer movement which advocated for reduced consumption and/or the consumption of products that were more environmentally friendly. Gabriel and Lang [20] note that by the early twenty-first century, the green consumer movement had gone mainstream, and the reformist stream of green consumerism (which advocated consuming not less, but differently) had spawned a whole new category of green products and businesses.

A prominent feature of this fourth wave is what Micheletti [22] calls buycotting, a form of 'positive political consumerism' that involves buying particular goods and brand names over others based on claims made about the product. Buycotting is in contrast to 'negative political consumerism,' or boycotting, which involves refusing to buy specific products or brand names. Buycotting is generally dependent on labeling schemes set up by governmental or extra-governmental regulating bodies. Labeling schemes are important because they legitimize and govern the claims made about products marketed to political consumers. For consumers interested in the politics of products, labeling schemes help them to identify products that are (or at least claim to be) in line with their political values and concerns.

Out of these dynamics emerges the concept of the citizen-consumer [22,23,24], those actors who "combine the public role of citizens with the private role of consumers" ([22], p. 16). Unlike consumers whose motivations are based on purely private concerns, citizen-consumers are motivated (at least in part) by public concerns related to their identity as citizens.

Arguments regarding the effectiveness and viability of political consumerism are generally based on the concept of consumer sovereignty, which Dixon and Carsky define as "the power of consumers to determine, from among the offerings of producers of goods and services, what goods and services are and will be offered (produced) and/or created in the economic sphere of society" ([25], p. 29). Although the concept of consumer sovereignty is not new, dating back at least as far as Adam Smith, contemporary forms of 'voting with your dollar' rose to prominence in the late twentieth century alongside neo-liberal governance models that "actively promoted the idea of consumer choice in the market as a worthy complement to, and even substitute for the citizenship ideal of democratic participation" ([23], p. 246). Although contentious and with only piecemeal empirical data on the effectiveness of political consumerism [26], some of the contemporary advocates of political consumerism borrow from the logic of trickle-down economics [27] to argue that, in theory, the more people shop for products labeled as more ecologically friendly or socially just, the greater will be the market share of these products, and in turn, the more these products will be available and affordable to the general population [28].

However, Jacobsen and Dulsrud [29] found that consumers do not often have enough information about the politics beyond products to make informed political consumerist choices. This is the jumping off point for the rest of our analysis. How must consumer food information systems be redesigned to leverage the citizen-consumer phenomenon and advance government objectives around health and sustainability? Given that health and sustainability are collective good objectives, we follow Johnston's interpretation that draws a clear distinction between consumerism's "maximization of individual self-interest through commodity choice," and citizenship based on the prioritization of "the collective good," which "emphasizes responsibility to ensure the survival and wellbeing of others—human and non-human" ([29], p. 243). This is an important distinction because a significant assumption of the current system is that individual self-optimization generates wider societal value. While some actions may benefit both the individual and society [30], the view expressed here is that the state must facilitate collective good undertakings and must shape how individuals behave to achieve public objectives. Individual self-optimization is not sufficient to create the necessary aggregate effects. This is admittedly a challenging proposition for the state because the industrial and neo-liberal values that currently dominate food production and manufacturing have for some time been supported by state legislation, regulation and programming [31,32]. We highlight below certain health and environmental aspects of the consumer information system that could be modified in the current environment to allow this wider citizen-consumer notion of choice and societal value to be expressed.

A caveat—we recognize that improving information is only one of many changes to create a more sustainable and health promoting food system. Ultimately, our proposals will only be truly effective when integrated with other measures, as elaborated in a joined up food policy approach by MacRae [2]. Consistent with a joined up approach, we focus later on integrated solutions.

4.5   SHIFTING REGULATION TO LEVERAGE THE CITIZEN-CONSUMER PHENOMENON

4.5.1   EVOLUTIONARY CHANGE

Our objective is to identify changes to information systems that actively support healthy and sustainable food choices, as part of a programme of system change. Although as a society we understand reasonably well the kind of diet that will help produce a healthy population and sustainable food systems, governments and professionals have largely failed to provide the structures and resources to ensure that it happens. They have, instead, largely relied on an increasingly discredited approach to creating health and sustainability—individual behavioural change—without addressing significantly the deeper structural forces contributing to less desirable behaviours.

Given the current low level of support for significant change to consumer information systems in Canada, the transition will have to be a slow, evolutionary process requiring action by many different advocates, both within and outside of these systems. For this paper, key government documents were analyzed, and from these, we identified potential solutions, some currently in effect, others proposed, using an efficiency, substitution, redesign transition framework [33], key conceptual frames in health promotion [34], agroecology [35,36] and emerging ideas on regulatory pluralism [11].

The transition framework serves as both a guide to action, and an indicator of progress. Stage 1 strategies involve making minor changes to existing practices to help create an environment somewhat more conducive to the desired change. The changes would generally fit within current consumer information activities, and would be the fastest to implement. For example, changing the visual presentation on a packaging label would make existing information more accessible to many consumers. Second stage strategies focus on the replacement of one practice, characteristic or process by another, or the development of a parallel practice or process in opposition to one identified as inadequate. For example, a new system of providing nutrient value information on a food packaging label would replace what is currently provided. Finally, third stage strategies are based on the principles of healthy public policy, agroecology and the values of citizen-consumerism. These strategies are more joined up, take longer to implement and demand fundamental changes in the use of human and physical resources. This final, or redesign stage, is unlikely to be achieved, however, until the first two stages have been attempted. Ideally, strategies should be selected from the first 2 stages for their ability to inform analysts about redesign (the most underdeveloped stage at this point) and to contribute toward a smooth evolution to the redesign stage.

In the next few sections, we identify strategies to move us in this new direction, starting from the current dominant regulatory instruments for informing consumers: definitions of food terms, product labels, Point of Purchase (POP) information, and television advertising. Clearly, other consumer education vehicles are available and promoted, but these have not historically been regulated as significantly by the state, so we leave them for another discussion. As well, traditional instruments related to information do not represent the most profound interventions that the state can muster [37], but in the current political environment, it is unlikely that Canadian governments will engage in more direct and forceful regulatory action. Our focus is on currently regulated measures that could advance population health and sustainability in the food system. Our presumption is that the state will need to proactively intervene, sometimes with the help of other policy actors, to create regulatory change to optimise the value of the emerging "citizen-consumer" phenomenon. This is admittedly challenging, as the state does not typical respond to emerging phenomena in this way. A more common response is to wait until a significant mass of consumers (and some NGOs and businesses) are demanding changes before moving incrementally towards these demands. The proposals that follow are not a comprehensive set of required changes, but provide a flavour, within a transition framework, of the kinds of initiatives we believe can take advantage of the citizen-consumer phenomenon to drive change.

4.5.2   PROMOTING HEALTH

The efficiency and substitution stage strategies involve modifying existing instruments. The redesign strategy is to create an integrated and simple system providing more complete decision-making information.

4.5.2.1   MODIFYING PRODUCT LABELS, SHELF TALKERS AND ADVERTISING

4.5.2.1.1   Product Labels

As discussed earlier, the citizen consumer concept is predicated on the idea that eaters have purposes beyond self-maximizing consumption. Canada's consumer information system has historically been rooted in fraud prevention and the assumption that self-maximizing information makes markets work efficiently. Capitalist markets presume that individual self-maximizing behaviour is also optimal for social welfare; however, as discussed earlier, this is not always the case in food markets. But self-maximizing behaviour can have larger public benefits. Individuals closely following nutrition labels may improve their health, reduce morbidity and mortality and contribute to reductions in public health care costs. However, our interest here is more in how food information can progressively contribute more directly to public maximizing activities. We devote most of our remaining analysis to these limitations in the current system, rather than a more detailed critique of the self-maximizing challenges of current information rules.

In Canada, product labels usually provide information on the company, the name of the product, nutrition information, a list of ingredients, sometimes the product grade, its origin (and sometimes Canadian status), and occasionally (only a limited number are permitted) claims about the health value and nutrient function of the food. As discussed above, consumers rely extensively on product labels for information, but although there have been recent improvements in several of these areas, others remain suboptimal. As well, little information on the food production process and possible contaminants is provided.

There is also some evidence, as when trans fat labelling was introduced in Canada, that label information causes firms to change their formulations to generate a more positive label, and potentially better sales. Yet, such realities still appear to be insufficient to provoke more profound regulatory improvements to drive manufacturing changes.

Nutrition labels are the most direct information linking consumer health and product purchasing. Until recently, nutrition labelling was not mandatory in Canada, except in cases where a specific nutritional claim was being made. Where labelling was provided, it was often incomplete, and/or difficult for the consumer to read and interpret. Regulations introduced in 2003 made nutrition labels mandatory in Canada for all prepackaged foods (with some exceptions) on 12 December 2007 [37]. For large manufacturers (those with more than $1 million in sales/year) an earlier deadline was imposed: 12 December 2005 [38].

Health Canada estimated at the time that this new nutrition labeling could help save as much as $5 billion in tax dollars over the next 20 years through a reduction in rates of diet-related diseases such as cancer, diabetes and heart disease because of dietary improvements among the population [39].

The new labels contain a Nutrition Facts Table with a standard format and which lists information on calories and 13 ingredients (fat, saturated fat, trans fat, cholesterol, sodium, carbohydrate, fibre, sugars, protein, vitamin A, vitamin C, calcium and iron) in standard units [16]. The table also lists the % Daily Value of these ingredients, which is the proportion of the daily recommended intake of the ingredients (as determined by Health Canada) that exists in the food.

The new regulations also control the nutrient content (e.g., fat free) and disease risk reduction claims (e.g., A diet rich in vegetables and fruit reduces risk of some types of cancer) and therapeutic health claims that can be made. Nutrient content claims can only be made in cases where the amount of the ingredient in question meets a regulated standard. Diet-related health claims are only allowed, with stipulations, in relation to five combinations of foods and their proven health benefits: sodium and potassium and high blood pressure; calcium and vitamin D and osteoporosis; saturated and trans fats and heart disease; fruits and vegetables and cancer; carbohydrates and dental caries [16].

The new labels are a notable improvement on the old ones. As Health Canada stated, "Prior to these regulations, nutrition labeling was optional with a few exceptions. The format of the table (the way it looked) was not consistent, and if nutrition labeling was provided, information was given on only a few nutrients. Not all nutrient content claims were regulated and the criteria for some of the existing claims did not reflect the latest science. Diet-related health claims were not allowed in Canada, before these regulations" [40].

There remain, however, some exceptions that limit the utility of the labels in certain food categories. Nutrition information is not usually provided for fresh meat, poultry, seafood, fruits and vegetables. There are exceptions in the fat and cholesterol labeling regulations which mean that not all foods are required to display their contents in a Nutrition Facts Table. Foods which are not required to display the Table include (but are not limited to) [16]:

- A variety of cow and goat milk products sold in refillable glass containers
- Raw meat and poultry (unless it's ground)
- Raw fish and seafood
- Food sold only in the retail establishment where it is prepared and processed from its ingredients
- Individual servings of foods that are sold for immediate consumption (e.g., sandwiches or ready-made salads)
- Clerk-served foods which are packaged at the time of sale (these are exempt from having any label at all since they are not considered "prepackaged").

The last three exceptions are important ones since they mean that nutrition information is not available for fast foods (donuts, french fries, hamburgers), one of the greatest sources of calories, saturated and trans fats in the Canadian diet. The first three are also of concern since animal products are often high in fat.

Serving size information is confusing. Nutrition information must be based on one serving size, but a serving size is only loosely defined as "an amount of food which would reasonably be consumed at one sitting by an adult" [16]. Reference serving sizes are suggested, but frequently are not followed by manufacturers. Consequently, within a particular food category, serving sizes may vary significantly across brands, making it very difficult for consumers to compare. Furthermore, the rules are sufficiently vague to allow manufacturers to set unrealistically low serving sizes which means problematic ingredients (e.g., salt or fat) may appear to be of lower content than would be in a reasonable amount consumed during one eating event.

Ultimately, in terms of population health objectives, Canadian product regulations make it difficult for consumers to understand how to relate individual nutrient levels to overall diet requirements that optimize health. The link between, for example, nutrition labels and Canada's Healthy Eating Guidelines is not direct except in some specific cases.

4.5.2.1.2   Ingredients Listings—Food Additives, Processing and Storage Products

Product ingredients have received some attention from consumers over the years, in part because of media reports about health problems associated with specific ingredients, particularly preserving agents, food dyes, flavour enhancers, and fat and sugar substitutes.

Current ingredient list rules specify that "ingredients must be listed in descending order of proportion by weight, as determined before they are combined to make the food. The exceptions are spices, seasonings and herbs (except salt), natural and artificial flavours, flavour enhancers, food additives, and vitamin and mineral nutrients and their derivatives or salts, which may be shown at the end of the ingredient list in any order" [16]. In other words, the way in which these components are listed is not consistent with other components. Unfortunately, in contrast to many other jurisdictions, Canada does not require Quantitative Ingredient Declarations (QUID).

Other exceptions of importance include:

- not all prepackaged multi-ingredient foods require an ingredient list, including those packed from bulk at retail "(exception: mixed nuts, and meat products packed by a retailer which contain phosphate salts and/or water)" [16].

- prepackaged individual portions served with meals or snacks by a restaurant or airline or servings prepared by commissaries and sold in canteens or vending machines; and prepackaged meat or poultry products or by-products barbequed, roasted or broiled on the retail premises.
- certain foods and classes of foods may be listed by class names (rather than specific names) including vegetable oil, colour, flavour and artificial flavour, spices, and milk ingredients.
- many foods when used as ingredients of other foods are exempt from a declaration of their components including many fats, sweeteners, jams and flours "unless the components belong to a determined list of allergenic ingredients (such as MSG, aspartame, peanut oil)" [16].
- certain food preparations and mixtures, including flavours and seasonings, are exempt from a declaration of most of their components.

Many of these exemptions appear to be for the convenience of the manufacturer in that they allow more readily for ingredient substitution based on price and availability considerations.

4.5.2.1.3   Point of Purchase

Point of Purchase (POP) materials have been used in-store for years by manufacturers for a number of purposes:

- to bring consumer attention to a new product;
- to encourage brand switching by offering a cents off coupon for purchase;
- to encourage trial of a partner product (may include a coupon);
- to promote a family of brands through a consumer contest;
- to encourage increased product purchase through coupon ad pads (often requiring multiple purchases or for next-time purchase);
- to encourage increased usage through alternative product uses (e.g., baking soda for cleaning carpets);
- to improve shelf presence in a brand category where there are many competitors;
- to promote a new product benefit.

Using POP to deliver nutrition and health messages is much more unusual. In the dominant view of POP marketing, consumers are only attentive in a store for a few seconds/item, so the message on any promotional/educational material needs to be short, with a small number of very legible words to communicate the message (e.g., ad pads). Delivering health messages is not the same as product promotion. The store is not the best place for educating, but it can be used to remind people of something, or to direct people somewhere else for more information. To be effective, the material needs to stand out from the products it will be placed beside and "grab" people as they walk by. There have been some successes in changing actual purchasing patterns dating back to the 1980s [41], with shelf labels having the following characteristics:

- bold, easy to see graphics
- writing at a grade 6/7 level
- shelf price labels with nutrient information right on it
- choose more often/choose less often categories

Retailers are unlikely to be willing to pay for such materials, given how POP has traditionally generated additional profits. Manufacturers are unlikely to pay, as health messages will not be product specific, and would likely focus more on less processed foods.

On the positive side, many retailers and manufacturers will see participation in a health program as part of being good corporate citizens. They may recognize that generic promotion of fruits, vegetables, complex carbohydrates and dairy products will be positive for store and product reputations. Well respected health organizations (Canadian Cancer Society, Canadian Heart and Stroke Foundation) have subsequently participated in third party endorsement programmes. The regulations pertaining to third party endorsements of retail food products are outlined in chapter 8.11 of CFIA's 2003 Guide to Food Labeling and Advertising, "Third Party Endorsement, Logos and Seals of Approval." These regulations flow from the "Policy on the Use of Third Party Endorsement, Logos and Seals of Approval, Food Division, Consumer and Corporate Affairs Canada, March 1991" and were created for organizations that provide "health and nutrition information" for a single food item or brand, but not to "groups or classes of food" [16]. Chapter 8.11 of the Guide to Food Labeling and Advertising requires that the food label "clearly explains the reason for the appearance of the third party's name, statement, logo, and so on" [16], applied to organizations making a health or nutrition claim. The policy was not designed for environmental claims, but could presumably be applied as long as it wasn't confusing to consumers [42].

4.5.2.1.4   Advertising

Although advertising can contribute to market efficiency by providing consumers with information, it can also be part of an insidious process of misinforming or only partially informing the public [43]. Advertising promotes the feeling that happiness is associated with the purchase of goods and services. It shifts consumer focus from needs to wants by redefining basics needs as wants. It proposes consumption as a cure for anxiety and fear, and redefines serious social issues as personal problems that can be solved by buying products [44]. The costs of such misinformation have been borne by the public, directly in product prices, and indirectly in lost government tax revenues, because advertising expenses have regularly received preferential tax treatment [45]. The public and taxpayers also bear the costs of ill-health that result from consumption of many of these products. Some studies have suggested that advertising is not often cost effective, and that it contributes to waste, monopoly and higher prices [43].

Although advertising regulations exist, the focus is on preventing fraud and not on the provision of full product information. As well, Canadian regulations have been weakened over the past few decades. It used to be that food TV commercial scripts and preferably story boards had to be reviewed in advance by Industry Canada and it had the authority to request modifications and even reject TV commercials. Advertisers could not make changes without resubmitting. A recommendation for the prevention, treatment or cure of a disease or ailment would not be permitted unless approved by Health Canada [46]. In the late 1990s, responsibility shifted to the advertising industry. Advertising Standards Canada (formerly the Canadian Advertising Foundation) developed an industry Code of Ethics which states that "no commercial message containing a claim or endorsement of a food or non-alcoholic beverage to which the Food and Drugs Act and Regulations apply may be broadcast unless the script for the commercial message or endorsement has been approved by the Food and Beverage Clearance Section of Advertising Standards Canada and carries a current script clearance number." [47] As part of this reformulation of regulatory oversight, no mandatory requirements were imposed for review of print advertising. Labels could voluntarily be submitted to the federal government for advice.

When responding to calls to restrict advertising, the advertising industry has argued that its influence is overestimated. This appears to be a disingenuous argument. If advertising is not effective in influencing people's choices, then why would companies spend money on it? And there has been evidence at least since the 1980s that particular forms of advertising influence dietary choices, particularly among young people [48]. For example, Engelhard et al. suggest that "exposure to food advertisements significantly and directly affects consumption of fattening food by both children and adults" [49].

The conclusion for many is that current advertising generally runs counter to government efforts to promote healthy eating. The failure to restrict advertising, particularly to young people, means that government is ensuring the failure of its own efforts. A coherent health promotion strategy would require that government restrict the ability of the private sector to offer messages that contradict its own.

4.5.2.2 EFFICIENCY STAGE CHANGES REQUIRED

1. Remove nutrition labelling exemptions so that product or shelf labels are required for currently exempt fresh foods and for many fast food items.
2. Construct messages to link nutrition panel information to Canada's Food Guide. One such as "This food is highly nutritious but also high in fat. Suggested serving size for a healthy adult is maximum xx servings per day, the serving size being xx ounces" (amounts depending on product) would provide consumers with guidance, and, in combination with details about fat content, would offer a more comprehensive way of informing consumers about their food.
3. Rewrite certain Food and Drugs Act Regulations so that fat production and distribution is discouraged, and consistent labelling of fat content encouraged:
    a) Change all prepared meat food definitions so that the product can contain no more than 25% fat by weight.
    b) Change dairy product food definitions so that maximum fat contents are specified for each type of cheese.
    c) Change all product labelling systems so that the label contains both the grams of fat and the percentage of calories consumed as fat (consistent with Canada's Healthy Eating Guidelines)
    d) Require labelling of all fatty ingredients.
    e) Require restrictions in trans-fatty acids (TFA).
4. Canada now largely requires TFA labelling, but in Europe, the industry has made great strides in removing TFA from their processes [50], and of course, the regulatory environment regarding TFA is generally stricter than in North America and was implemented at an earlier stage. For example, the Netherlands reduced TFA content in the mid-90s, to less than 1% in most margarine. Dutch scientists projected that,

as a result of this measure, TFA consumption for the average person would decline by 4 grams/day (compared to 1980s levels), and that coronary disease incidence could fall by 5% [51].

5. Remove ingredient listing exemptions and add the functions that non-nutritive ingredients play in the food product (e.g., preservative, emulsifier, etc.). Implement QUID (Quantitative Ingredient Declarations) on all packaged goods.

6. Continue current efforts to make serving sizes consistent and realistic and require that all packaging describes how many serving sizes are contained in the packaging unit in which the food is sold.

7. Use more shelf talkers/ad pads in supermarkets as health promotion vehicles. Ad pads can work well when used to remind consumers of a campaign that they would already be familiar with through another medium (e.g., television, direct mail or outdoor advertising), when the pads are placed next to the product of the campaign and when the message contained on the ad pad (and its "look") is consistent with that of the familiar campaign.

4.5.2.3   SUBSTITUTION STAGE CHANGES

1. The Science Council of Canada proposed in the 1970s that advertising of nutritionally-questionable products be curtailed by government intervention [52]. This could be one component of an integrated strategy to promote an optimal diet and eliminate or restrict any advertising that constitutes a barrier to achieving this goal. One possible requirement might be that food products that are clearly undesirable or peripheral to an optimal diet be labelled as such.

2. Another approach is to implement a comprehensive marketing ban on low/no nutrition foods. At least 50 countries now regulate television advertising aimed at children, and within Canada, Quebec banned all print and broadcast advertising to children in 1980 [49]. Even broader marketing bans may now need to be considered in the digital marketing era.

3. Require that Canada's Food Guide be placed on all packaging labels with sufficient size to accommodate it. Weston's Wonder Bread was one of the first Canadian products with such a label in the 1990s.

4. Create new systems for adding messages to labels that tell consumers how a food product complies with the government's healthy eating guidelines (e.g., "Eating this product several times a week is consistent

with Canada's Guidelines for Healthy Eating" or something to that effect); this might also be achieved with a colour coding system (e.g., different colours for high, medium and low compliance). Such attributable messages have existed on tobacco products for years. Current back of package nutritional labels are confusing and of limited effectiveness. Colourful, graphic labels on the front of packages (FOP) may be more effective in influencing point of purchase decisions. The UK introduced a FOP graphic traffic-light image with key nutritional information to help buyers distinguish between healthy and less healthy products [53]. After it was introduced, Tesco supermarket saw sales of less healthy prepared meals drop by up to 41%, while those with healthier profiles more than doubled [49]. At least one Canadian retailer/producer has voluntarily introduced FOP nutritional information.

4.5.2.4   REDESIGN STAGE CHANGES

Integrated nutrition profiling systems, with application across a variety of information mechanisms, including product labels, POP and advertising are made mandatory. At least 4 are in play in the USA [54], with the Guiding Stars programme [55] being one example that, in some evaluations, has had an immediate positive impact on purchasing behaviour [56]. Such programmes are currently largely implemented by third parties, but should become mandatory dimensions of information dissemination with appropriate revisions to regulation and guidances. The programmes could be offered by governments as multi-sectoral collaborations, drawing on the expertise and resources of a range of actors to be more effective.

4.5.3   PROMOTING SUSTAINABILITY

Local/sustainable food systems are thought to counter many of the negative effects of the industrial, global model, including improved regional economic development, environmental improvements, and a higher quality food supply [57,58,59,60,61]. This heightened interest in such food systems has many analysts exploring the many policy obstacles and opportunities to enhance their development, including examining the role of consumer information systems. Clearly there are many substantial and contentious elements to any discussion of food system sustainability, but looking at the use of terms associated with local/sustainable serves as a proxy for this larger debate and since it is currently top of mind for many consumers, especially those meeting the citizen-consumer categorization, we focus on it here. The information architecture for identifying local/sustainable systems is, however, poorly developed in Canada. We briefly review current rules on 6 terms that have historically been used by producers and manufacturers to infer the sustainability of their products, and then later propose a series of changes to enhance their utility for the citizen-consumer movement.

4.5.3.1 DEFINING LOCALITY

Local foods are not automatically sustainable by nature of their locality [42], but there is evidence that local and sustainable production together create larger sustainability impacts than they do individually [62]. Because of this, the term local has appeal as a purchasing criterion within the citizen-consumer movement [42].

An obvious concern of regulators is how to define a term for use on labels and a review of the literature reveals that local has no specific definition [57] and is contentious, changing with each supply chain and region. One food system actor may view the term "local" as their surrounding community, another as a provincial or regional parameter [57,63,64]. The conceptual terrain has been set out by Kloppenburg et al. [65] in their arguments in favour of a foodshed analytical frame, but no specific distance has been attached to that concept. The most common modern definition is by sub-national political boundary, e.g., state, province, region or geographic feature. Earlier definitions focused more on cultural boundaries or unique eco-regional features, e.g., terroir. With the possible exception of Quebec, Canada does not have a tradition of terroir.

Given this murky terrain, Hinrichs suggests a "diversity receptive localization" as a template to support and protect local industries as this type of localization encourages "the richness of a place while keeping in mind the rights of a multi-faceted world. It cherishes a particular place, yet at the same time knows about the relativity of all places" [66]. Her approach takes into account that local economies would not remain fully autonomous from global trade. Hinrichs uses the example of an artisanal cheese maker who would not likely sell enough of his or her product within the limits of his or her locality to financially survive [66]. Also, many food items now considered staples cannot be grown in a Canadian climate, for example citrus, coffee and chocolate.

A small number of surveys of producers involved in local food distribution have been undertaken, lending credence to the idea that it is primarily the nature of the current supply chain that informs the definition of local among current food system actors. One study examining 3 counties in Washington State [64] found that 80 to 240 km was generally considered the local food distribution chain with several country boundaries crossed. But in a county with extensive direct market channels and premium niches, the county was more likely to be considered local. Those with limited direct marketing outlets were likely to have a larger view of local and were also more likely to include other supply chain actors in their definition. Only a small percentage of producers defined local by the quality of the social relations or the freshness attributes of the product. Consumers, on the other hand, are more likely to employ a shorter distance to define local than producers [57], which may in part be a result of more limited knowledge of supply chain dynamics.

Given that most of the current literature highlights the importance of spatial definitions of local, some converging lines of analysis suggest that a local definition in the 160–200 km range is both conceptually legitimate and operationally viable at this stage in the relocalization process. Kloppenburg et al. [65], in their conceptual argument, emphasized that a functioning foodshed provides a diversity of products from a diversity of suppliers, within the context of the area being examined. Given the nature of Canadian settlement and agricultural development, few regions of the country would be able to provide that diversity within a 50 km distance. A wider range provides greater possibilities.

Although globalization is a centuries-old phenomenon in the food system, it's most recent expression dates to events of the 1970's [31]. Fresh food now travels 50% further than it did in the late 70s. It was much more common for supermarkets in major urban centres to be supplied primarily by farms within a few hours (within 200 km) [67].

Another way to evaluate the term local is to contrast existing local and global food supply chains. The average North American food molecule currently may travel 2500–4000 km [61]. Full empirical studies of domestic food miles in Canada are lacking, but an Iowa study of local vs. global meal comparisons found that local supply chains are about 5% of global [68]. Five percent may underestimate the situation in many regions of Canada. For example, a supermarket vs. farmers' market comparative study in Toronto found a 50-fold difference, rather than the 20 fold difference reported in the Iowa study [69]. Based on this, assuming a 4000 km distance, 5% is 200 km.

There is currently an extensive, and contested, set of rules regarding Made in Canada labelling [70]. However, given the discussion above, this represents too wide a geography to qualify as a definition of local. There are rules around the use of local and related terms. Local or locally produced [71] is defined by the FDA as:

> "domestic goods being advertised originated within 50 km of the place where they are sold, measured directly, point to point, or meets the requirements of section B.01.012 Food and Drug Act (CFIA information bulletin), which states, "'local food' means a food that is manufactured, processed, produced or packaged in a local government unit and sold only in… the local government unit [or] one or more local government units that are immediately adjacent to the one… in which it is manufactured, processed, produced or package" (FDA 5.1).

According to officials, the CFIA policy pertaining to "local" was instituted to prevent "violations of section 5(1) of the FDA" in the "use of the term "local" on fruits and vegetable" (Food and Drug Regulations C.R.C., c. 870; B.01.012). This definition reflects a supply chain distance approach, consistent with definitions used in some European and U.S. jurisdictions where farmers' markets are central to the conception of local [57]. The distance specified in the CFIA rules is limited because it assumes a direct marketing context for local, and consequently focuses on fruits and vegetables, the most common direct-marketed goods. This represents, however, only one stream of local market development, and for only a limited segment of the food supply.

4.5.3.2   PRODUCTION SYSTEM AND PROCESSING FROM WHICH THE FOOD DERIVES [42]

The current system focuses primarily on product, not the process by which food is produced. In this view, how a food is produced is not relevant unless it changes accepted food safety and nutritional parameters. There are a few significant exceptions to this general rule. Rules for using the label organic, kosher or halal are all process-based. They describe how the food is grown, raised and processed. Canada has regulations about the use of these terms [16], but given consumer concerns and interests, and how these relate to national sustainability objectives, there is a need to widen the use of process terms. Barham ([72], p. 354) has stated:

The focus on process challenges the inability of market economics to take long term impacts into account, and simultaneously calls for conscious social decision making about the direction of the economy, rather than passively waiting for the market to tell us, after the fact, what is optimum. It means that environmental and social impacts will have to be considered ahead of time...The social construction of quality, as it is represented in values-based labeling, can call on aspects of human rationality that move beyond the economistic. To the extent that it incorporates recognition of the responsibility of choice, it breaks free from defining the human being as primarily a self-interest maximizer.

The regulation of 5 key terms describing different approaches to sustainable food production provides information on the current state of affairs and how it might be altered to better facilitate knowledge transfer to citizen-consumers. Though not always emblematic of the most sustainable approaches, these terms are used on many current third party labels or are under consideration, so we use them as indicators of the state of regulatory development:

- Integrated Pest Management (IPM)—in programmes that demonstrate reduced chemical use with alternative pest control methods.
- Ecological—where the certification results in environmental stewardship and enhanced biodiversity.
- Environmentally friendly—where a programme results in reduced packaging and promotes recycling and energy conservation.
- Natural—to describe a minimum standard of animal welfare.
- Sustainable agriculture or sustainable food production—when food producers and processors follow a sustainable food production regime that isn't specifically organic.

In Canada at present, there are no specific regulations that refer to 4 of the 5 terms [42]. However, there are some product-by-product pre-market assessment decisions that would appear to have an impact on some of them. Given the current absence of regulations on their use, producers and processors would be subject to a case-by-case assessment, based on general fraud provisions. Case-by-case interpretations, under conditions of ambiguity, have not proven particularly helpful to sustainable producers in Canada, and the federal government has recently announced that it will eliminate such assessments as a cost saving measure, when not linked to food safety. The 25-year journey to organic regulation serves as one example, as organic producers operated during this period under threat of unfavourable decisions, in the absence of clear rules governing the term. A second unfortunate case involved the suspension of a pesticide reduction labeling programme on the Canadian Prairies, Pesticide-free Production

Canada, due to an unfavourable decision on their pesticide-free label by CFIA officials. The CFIA argued, with some justification, that pesticide-free could not be achieved and was misleading within the programme since producers were allowed to apply pesticides outside the growing season (i.e., for spring and fall cleanup, pre-plant and post-harvest). In a third case, highlighting the benefits of specific term definition over general provisions, CFIA officials offered an interpretation to a national IPM project that the term ecological could not be used because it had already been reserved by regulation in the province of Quebec for use in organic production. The interpretation had a dampening effect on discussions of marketing strategies for the project. It was a rational decision on the part of CFIA, but highlights how sustainable producers can unintentionally undermine each other's efforts when a coherent and comprehensive regulatory framework is lacking [42].

The term natural is more clearly problematic, given current regulations. It has traditionally been reserved for processed food items. However, the CFIA has been working on a "naturally raised" policy in regards to livestock since a consultation paper on the topic was released in 2005. Guidelines were released in late 2011 and are in effect as the CFIA considers public comments from its latest online consultation [73].

"Foods or ingredients of foods submitted to processes that have significantly altered their chemical, physical, or biological state should not be described as natural. 'Natural' can generally be understood as meaning:

- a natural food or ingredient of a food is not expected to contain, or to ever have contained, an added vitamin, added mineral nutrient, artificial flavouring agent, or food additive;
- a natural food or ingredient of a food does not have any constituent or fraction thereof removed or significantly changed, except the removal of water; and/or
- a natural food or ingredient of a food that has been produced through the ordinary course of nature without the interference or influence of humans." [16]

Given this recent decision to reserve the term "natural" in this way, it appears livestock programmes, which are traditionally frequent users of the term, will not be able to use it. It represents a significant departure from their earlier discussion paper [74] on natural and humane livestock production, when it appeared then that the term natural livestock would be linked to existing federally supported Codes of Conduct for the Care and Handling of Farm Animals, developed from the late 80s under the guidance of the Canadian Agri-food Research Council (CARC). Many Canadian animal welfare and natural livestock standards, including those of Local Food Plus (LFP) and the BC Society for the Prevention of Cruelty to Animals (BCSPCA), already reference the CARC Codes of Practice, but the new rules make this less relevant to create legitimacy for the claim.

Although not currently specifically regulated by CFIA, the use of the term "sustainable agriculture" could be problematic in the future, given the CFIA practice of consulting with other departments, in this case AAFC, on unspecified labeling terms. As defined by Agriculture and Agri-food Canada [75], sustainable agriculture is a system that:

- protects the natural resource base; prevents the degradation of soil, water, and air quality; and conserves biodiversity
- contributes to the economic and social well-being of all Canadians
- ensures a safe and high-quality supply of agricultural products
- safeguards the livelihood and well-being of agricultural and agri-food businesses, workers and their families.

Other organizations define the term more consistently with an agroecological interpretation of sustainability [35] and demonstrate a greater commitment to reducing reliance on environmentally problematic processes and inputs.

4.5.3.3 EFFICIENCY STRATEGIES

Requiring immediate attention are existing federal labeling regulations related to the 3 terms that appear to be out of step with other policy-related developments at the federal level and/or conceptual developments in the field—local, natural and sustainable agriculture. Making the labeling rules more supportive of the local and sustainable food movement will require better integration with those concepts and processes.

4.5.3.3.1 Local

If the CFIA wishes to regulate the term local, then it would seem more feasible to set a distance of 200 km rather than the current 50 km and apply it to a wider range of products and marketing channels. An interim measure would be to retain a provincial focus in the short term, and then shift to 200 km within a specified transition period. This again reflects a more realistic interpretation of existing supply chains.

4.5.3.3.2   Natural Livestock Production

Natural livestock production is, admittedly, notoriously difficult to define. It also has an imprecise, but usually significant, relationship with humane livestock production. The term is more closely related, currently, to food preparation and processing than to livestock production, which partly explains the direction that CFIA is pursuing in its regulation of the term.

Linking natural with wild, however, is out of step with other parts of the labeling regulatory apparatus. In the organic standards [76], part of the federal government organic regulatory apparatus, the term employed for wild products is 'wild crafted.' For consistency, it would make sense for CFIA to not now use natural in other parts of its labeling regime for wild crafted products.

It would also be coherent to link the definition of natural and humane livestock production to the existing Canadian Agri-food Research Council (CARC) Recommended Codes of Practice for the Care and Handling of Farm Animals [77]. Developed from the 1980s with numerous scientists, practitioners, and some government and NGO representatives, and currently undergoing a round of revisions under the auspices of the National Farm Animal Care Council (NFACC), these voluntary codes have received fairly widespread acceptance from the agricultural industry, although they are viewed as minimal guidance by many NGOs and humane treatment activists. Nevertheless, they have also been used as foundational material for codes developed by humane treatment NGOs. LFP has also used them as the foundation of their livestock protocol. As has been done in other labeling rules, there is an opportunity here to provide a brief definition of the term and reference the CARC (soon NFACC) codes as the minimal basis for use of the term.

4.5.3.3.3   Sustainable Production

Sustainable agriculture definitions are numerous, but in Canada, they have yet to be directly translated to specific product labels, and there is no current guidance on the term in the Guide to Food Labeling and Advertising [16]. Given the history, discussed above, of ambiguously regulated sustainable production terms, a preliminary assessment of government use of the term, relative to leading sustainability specialists, is important for assessing potential future challenges.

Governments typically employ definitions that account for multiple interests and policy pressures, but are not necessarily rooted in the historical context from which the concept evolved. The early scientific pioneers of sustainable agriculture (e.g., FH King, Sir Albert Howard), had a more profoundly ecological interpretation of sustainability [78,79,80] than is reflected in Canadian government definitions (see discussion above). Numerous sustainability schools of thought have emerged over the years, producing production systems consistent with the early pioneers and more grounded in an understanding of ecology [81]. Much of the scientific literature of today employs an agroecological paradigm [35,36,82,83] that is at odds with the underlying paradigms of the Canadian government approach. The federal government's conception of sustainable agriculture is also out of step with its main trading partners and international agencies, such as the FAO [84]. Labeling regulations consistent with the conceptual foundations of the term and other bodies would bring coherence to its use, both for locally and internationally traded goods. We foresee, ultimately, the need for regulation of a full suite of sustainability-related terms and given that Canada already participates in numerous international reciprocity and harmonization activities (including organic products and pesticide approvals), these terms will need to be applied consistently with trading partners.

4.5.3.4   SUBSTITUTION

For 3 of the 6 terms, IPM, ecological and environmentally-friendly, ambiguity will likely remain until they are more precisely defined. CFIA usually consults the food industry on such matters and would likely need to engage in a process similar to that used with the term natural to develop such precision. Such consultations and rule making usually unfold over several years. However, the difficulty of such consultations is that few participants, including CFIA officials themselves, have a well-rounded understanding of agroecology and how to apply it identifying production systems for consumers [42]. If an agroecological framework is not applied, such consultations are unlikely to produce a useful outcome.

4.5.3.5   REDESIGN

Consistent with a joined up approach to food policy, the redesign stage would focus on integrated and profound changes to existing rules in all the domains addressed in this paper. Using product labels as an example, the state would implement comprehensive product labelling that includes information on environmental and social justice impacts of production, processing and distribution. An example of such a label is provided in Table 4.1. Such labels could be numeric

TABLE 4.1: Hypothetical integrated label for an instant breakfast cereal (adapted from MacRae *et al.* [81])[a].

		Rating [b]
Contents	Whole wheat	
Production	Certified organic, advanced production with biodiversity enhancements	8
Processing	Regular milling: excess heat	5
	No supplements	10
	Moderately strong environmental management system in place	7
Product sourcing and distribution	Local, within 200 km	8 [c]
Social justice	Complies with all safe work regulations	5
	Wages above industry average	7
	No processor involvement in local community development or charitable work	0
Overall score		50/80

[a] Such a label might also contain more specific information, as required by current regulations; [b] Based on a 10-point scale; [c] Based on the following scale: 10—direct sale within 50 km; 8—within 200 km; 6—within 500 km; 4—within province; 2—within Canada; 0—procured internationally.

(as shown here) or colour coded and would clearly require significant analysis to support the rating system. A more visually interesting approach has been proposed by Sustain UK [85], the use of a range of pictograms that quickly convey the degree to which a number of sustainability parameters have been met by the product.

Similar integrated efforts would be required for POP and TV advertising. The environmental features could also be integrated with redesign nutrition elements presented in section 5.2.

4.6   CONCLUSIONS

Health and sustainability are public policy objectives, but Canada's food information rules and practices may not be optimal to support their achievement because the information provided is frequently determined by the marketers of product in a semi-regulated environment. No institution or agency has responsibility for determining the overall coherence of consumer food messages relative to these broader social goals of health and sustainability. Individual consumers do not have the resources to determine easily the completeness of any firm's messages, particularly when faced with the size of food industry advertising budgets. The healthy eating messages of health departments are often competing with contradictory or incomplete messages permitted by the regulatory framework of other arms of government. Investments in programs that successfully promote environmental stewardship in agriculture are undercut in the market because consumers often have too little information to support those efforts with their dollars. This problem exists despite the emergence of "citizen-consumers" who have a broader approach to food purchasing than solely individual maximization. Only recently have some health professionals and sustainable agriculture proponents turned their attention to all these factors and begun to design interventions that take them into account. Significant short, medium and long term regulatory changes are required to facilitate the citizen-consumer phenomenon and better support consumers in their efforts to promote health and sustainability in the Canadian food system.

REFERENCES

1. Toronto Food Policy Council. Making Consumers Sovereign; Discussion Paper #9; Toronto Food Policy Council: Toronto, Canada, 1998.
2. MacRae, R.J. A joined up food policy for Canada. J. Hunger Environ Nutr 2011, 6, 424–457.
3. Tencer, D. Canada Budget 2012: CFIA cuts mean food labelling lies will have to be policed by consumers. Huffington Post. 30 March 2012.
4. U.S. Surgeon General. The Surgeon General's Report on Nutrition and Health; DHHS (PHS) Publication No. 88-50210; U.S. Public Health Service, Office of the Surgeon General: Washington, DC, USA, 1988.
5. Eilers, W., MacKay, R., Graham, L., Lefebvre, A., Eds.; Environmental Sustainability of Canadian Agriculture: Agri-Environmental Indicator Report Series; Report No. 3; Agriculture and Agri-Food Canada: Ottawa, ON, Canada, 2010.
6. Lynch, D.H.; MacRae, R.J.; Martin, R.C. The Carbon and Global Warming Potential impacts of organic farming: Does it have a significant role in an energy constrained world? Sustainability 2011, 3, 322–362.
7. The Organisation for Economic Co-operation and Development (OECD). Agrienvironmental Performance in OECD Countries since 2008: Canada Country Report; OECD: Paris, France, 2008.
8. Tegtmeier, E.M.; Duffy, M.D. External costs of agricultural production in the United States. Int. J. Agric. Sustain. 2004, 2, 1–20.
9. Norwegian Department of Agriculture. On Norwegian Nutrition and Food Policy; Report #32 to the Storting; Norwegian Department of Agriculture: Oslo, Norway, 1975; p. 72.
10. Van Amstel, M.; de Brauw, C.; Driessen, P.; Glasbergen, P. The reliability of product-specific eco-labels as an agrobiodiversity management instrument. Biodivers. Conserv. 2007, 16, 4109–4129.

11. Koc, M.; MacRae, R.J.; Desjardins, E.; Roberts, W. Getting civil about food: The interactions between civil society and the state to advance sustainable food systems in Canada. J. Hunger Environ. Nutr. 2008, 3, 122–144.
12. Victor, P. Managing without Growth: Slower by Design, not Disaster; Edward Elgar: Northampton, MA, USA, 2008.
13. Busch, L. Performing the economy, performing science: From neoclassical to supply chain models in the agrifood sector. Econ. Soc. 2007, 36, 437–466.
14. Howard, P.H.; Allen, P. Beyond organic: Consumer interest in new labeling schemes in the central coast of California. Int. J. Consum. Stud. 2006, 30.
15. Gunningham, N. Reconfiguring Environmental Regulation. In Designing Government: From Instruments to Governance; Eliadis, P., Hill, M., Howlett, M., Eds.; McGill Queen's University Press: Montreal, QC, Canada, 2005; pp. 333–352.
16. Canadian Food Inspection Agency (CFIA). Guide to Food Labelling and Advertising; CFIA: Ottawa, ON, Canada, 2007.
17. Ostry, A. Nutrition Policy in Canada, 1870–1939; UBC Press: Vancouver, BC, Canada, 2006.
18. Blay-Palmer, A. Food Fears: From Industrial to Sustainable Food Systems; Ashgate: Burlington, VT, USA, 2008.
19. Welsh, J.; MacRae, R.J. Food citizenship and community food security: Lessons from Toronto, Canada. Can. J. Dev. Stud. 1998, 19, 237–255.
20. Gabriel, Y.; Lang, T. A Brief History of Consumer Activism. In The Ethical Consumer; Harrison, R., Newholm, T., Shaw, D., Eds.; Sage: London, UK, 2005.
21. The Ethical Consumer; Harrison, R., Newholm, T., Shaw, D., Eds.; SAGE: London, UK, 2005.
22. Micheletti, M. Political Virtue and Shopping: Individuals, Consumerism, and Collective Action, 1st ed.; Palgrave Macmillan: New York, NY, USA, 2003.
23. Johnston, J. The citizen-consumer hybrid: Ideological tensions and the case of Whole Foods Market. Theory Soc. 2008, 37, 229–270.
24. Lockie, S. Responsibility and agency within alternative food networks: Assembling the "citizen consumer." Agric. Hum. Values 2009, 26.
25. Dickinson, R.; Carsky, M.L. The Consumer as Economic Voter. In The Ethical Consumer; Harrison, R., Newholm, T., Shaw, D., Eds.; Sage: London, UK, 2005; pp. 25–38.
26. Micheletti, M. Why More Women? Issues of Gender and Political Consumerism. In Politics, Products, and Markets: Exploring Political Consumerism Past and Present; Micheletti, M., Follesdal, A., Stolle, D., Eds.; Transaction: Piscataway, NJ, USA, 2004; pp. 245–264.
27. Frank, R. In the real world of work and wages, trickle-down theories don't hold up. The New York Times, 12 April 2007. Available online: http://www.nytimes.com/2007/04/12/business/12scene.html?_r=1 (accessed on 14 June 2012).
28. Stolle, D.; Hooghe, M. Consumers as Political Participants? Shifts in Political Action Repertoires in Western Societies. In Politics, Products, and Markets: Exploring Political Consumerism Past and Present; Micheletti, M., Follesdal, A., Stolle, D., Eds.; Transaction: Piscataway, NJ, 2004; pp. 265–288.
29. Jacobsen, E.; Dulsrud, A. Will consumers save the world? The framing of political consumerism. J. Agric. Environ. Ethics 2007, 20, 469–482.
30. Soper, K. Rethinking the 'Good Life': The citizenship dimension of consumer disaffection with consumerism. J. Consum. Cult. 2007, 7, 205–229.

31. Friedman, H.; McMichael, P. Agriculture and the state system: The rise and fall of national agricultures, 1870 to the present. Sociol. Rural. 1989, 19, 93–117.

32. MacRae, R.J. This Thing Called Food: Policy Failure in the Canadian Food and Agriculture System. In For Hunger-Proof Cities: Sustainable Urban Food Systems; Koc, M., MacRae, R.J., Meugeot, L., Welsh, J., Eds.; International Development Research Centre and the Ryerson Centre for Studies in Food Security: Ottawa, ON, Canada, 1999; pp. 182–194.

33. Hill, S.B.; MacRae, R.J. Conceptual framework for the transition from conventional to sustainable agriculture. J. Sustain. Agric. 1995, 7, 81–87.

34. Blas, E.; Gilson, L.; Kelly, M.P.; Labonté, R.; Lapitan, J.; Muntaner, C.; Ostlin, P.; Popay, J.; Sadana, R.; Sen, G.; et al. Addressing social determinants of health inequities: What can the state and civil society do? Lancet 2008, 372, 1684–1689.

35. Altieri, M.A. Agroecology: The Science of Sustainable Agriculture; Westview Press: Boulder, CO, USA, 1995.

36. Agroecology: Researching the Ecological Basis for Sustainable Agriculture; Gliessman, S.R., Ed.; Springer-Verlag: New York, NY, USA, 1990; Ecological Studies Series no. 78.

37. Howlett, M. What is a Policy Instrument? Policy Tools, Policy Mixes and Policy Styles. In Designing Government: From Instruments to Governance; Eliadis, P., Hill, M., Howlett, M., Eds.; McGill Queen's University Press: Montreal, QC, Canada, 2005; pp. 31–50.

38. Health Canada website. Food and Nutrition: Nutrition Labeling. Available online: http://www.hc-sc.gc.ca/fn-an/label-etiquet/nutrition/index-eng.php (accessed on 6 September 2012).

39. CBC news online. 9 May 2007. Food Labels: The Facts about what's in Your Food. Available online: http://www.cbc.ca/news/background/food-supply/index.html (accessed on 6 September 2012).

40. Health Canada. Food and nutrition: FAQs about nutrition labeling. Available online: http://www.hc-sc.gc.ca/fn-an/label-etiquet/nutrition/educat/te_quest-eng.php#1 (accessed on 6 September 2012).

41. Light, L.; Tenney, J.; Portnoy, B.; Kessler, L.; Rodgers, A.B.; Patterson, B.; Mathews, O.; Katz, E.; Blair, J.E.; Evans, S.K.; et al. Eat for health: A nutrition and cancer control supermarket intervention. Public Health Rep. 1989, 104, 443–450.

42. Louden, F.N.; MacRae, R.J. Federal regulation of local and sustainable food claims in Canada: A case study of Local Food Plus. Agric. Hum. Values 2010, 27, 177–188.

43. Singer, B.D. Advertising and Society; Addison-Wesley: Don Mills, ON, Canada, 1986.

44. Wallack, L.; Montgomery, K. Advertising for all by the year 2000: Public health implications for less developed countries. J. Public Health Policy 1992, 13, 204–223.

45. McQuaig, L. Behind Closed Doors: How the Rich Won Control of Canada's Tax System—And Ended Up Richer; Viking Press: Markham, ON, Canada, 1987.

46. Consumer and Corporate Affairs Canada. Guide for Food Manufacturers and Advertisers, Revised Edition; Consumer Products Branch, Consumer and Corporate Affairs Canada: Ottawa, ON, Canada, 1988.

47. Canadian Food Inspection Agency (CFIA). Guide to Food Labelling and Advertising; CFIA: Ottawa, ON, Canada, 1998; pp. 1–4.

48. Taras, H.L.; Sallis, J.F.; Patterson, T.L.; Nader, P.R.; Nelson, J.A. Television's influence on children's diet and physical activity. Dev. Behav. Paediatr. 1989, 10, 176–180.

49. Engelhard, G.; Garson, A.; Dorn, S. Reducing Obesity: Policy Strategies from the Tobacco Wars; University of Virginia and Urban Institute: Washington, USA, 2009; p. 41.

Available online: http://www.urban.org/uploadedpdf/411926_reducing_obesity.pdf (accessed on 19 June 2012).

50. Zorc, A. Nutrition experts divided over trans fats. Community Nutr. Inst. 1996, 26, 4–5.

51. Anonymous. The Netherlands reduces trans fatty acids in foods. Nutr. Week 1995, 25, 3.

52. Science Council of Canada. Canadian Food and Agriculture: Sustainability and Self-Reliance; Science Council of Canada: Ottawa, ON, Canada, 1979.

53. NHS Choices. Available online: http://www.nhs.uk/Livewell/Goodfood/Pages/food-labelling.aspx#Nut (accessed on 6 September 2012).

54. Gerrior, S.A. Nutrient profiling systems: Are science and the consumer connected? Am. J. Clin. Nutr. 2010, 91, 1116–1117.

55. Guiding Stars. Available online: http://guidingstars.ca/about/how-it-works/ (accessed on 6 September 2012).

56. Sutherland, L.A.; Kaely, L.A.; Fischer, L. Guiding Stars: the effect of a nutrition navigation program on consumer purchases at the supermarket. Am. J. Clin. Nutr. 2010, 91, 1090–1094.

57. Canadian Organic Growers. Local and Regional Food Economies in Canada: Status Report; Report to Agriculture and Agrifood Canada; Agriculture and Agrifood Canada: Ottawa, ON, Canada, 2007.

58. Bendavid-Val, A. Regional and Local Economic Analysis for Practitioners, 4th ed.; Praeger: New York, NY, USA, 1991.

59. Planting the Future: Developing Agriculture that Sustains Land and Community; Bird, E., Bultena, G.L., Gardner, J.C., Eds.; Iowa State University Press: Ames, IA, USA, 1995.

60. Lasley, P.; Hoiberg, E.; Bultena, G. Is sustainable agriculture an elixir for rural communities? Am. J. Altern. Agric. 1993, 8, 133–139.

61. Haliwell, B. Home Grown: The Case for Local Food in a Global Market; WorldWatch Institute Paper 163; WorldWatch Institute: Washington, DC, USA, 2002.

62. MacRae, R.J.; Lynch, D.; Martin, R.C. Improving the energy efficiency and GHG mitigation potentials of organic farming and food systems in Canada. J. Sustain. Agric. 2010, 34, 549–580.

63. Allen, P.; FitzSimmons, M.; Goodman, M.; Warner, K. Shifting plates in the agrifood landscape: The tectonics of alternative agrifood initiatives in California. J. Rural Stud. 2003, 19, 61–75.

64. Selfa, T.; Qazi, J. Place, taste, or face-to-face? Understanding producer-consumer networks in "local" food systems in Washington State. Agric. Hum. Values 2005, 22, 451–464.

65. Kloppenburg, J.; Stevenson, G.W.; Hendrickson, J. Coming into the foodshed. Agric. Hum. Values 1996, 13, 33–42.

66. Hinrichs, C.C. The practice and politics of food system localization. J. Rural Stud. 2003, 19, 33–45.

67. Hendrickson, M.K.; Heffernan, W.D. Opening spaces through relocalization: Locating potential resistance in the weakness of the global food system. Sociol. Rural. 2002, 42, 347–369.

68. Pirog, R.; van Pelt, T.; Enshayan, K.; Cook, E. Food, Fuel, and Freeways: An Iowa Perspective on How Far Food Travels, Fuel Usage, and Greenhouse Gas Emissions; Iowa State University, Leopold Center for Sustainable Agriculture: Ames, IA, USA, 2001.

69. Bentley, S. Fighting Global Warming at the Farmers' Market; Foodshare Toronto: Toronto, ON, Canada, 2004.

70. CBC News. 'Made in Canada' rules under review. CBC News Online, 22 May 2010. Available online: http://www.cbc.ca/news/story/2010/05/21/made-in-canada.html (accessed on 18 June 2012).

71. This labeling policy was introduced after the creation of the 2003 Guide to Food Labeling and Advertising; therefore, it is not within the guide, but can be found within the CFIA "Information Bulletins." Available online: http://www.inspection.gc.ca/english/fssa/labeti/retdet/bulletins/bulletinse.shtml (accessed on 20 June 2012).

72. Barham, E. Towards a theory of value-based labeling. Agric. Hum. Values 2002, 19, 349–360.

73. Canadian Food Inspection Agency (CFIA). Guidelines on Natural, Naturally Raised, Feed, Antibiotic and Hormone Claims. Available online: http://inspection.gc.ca/english/fssa/labeti/natall/natalle.shtml (accessed on 20 June 2012).

74. Canadian Food Inspection Agency (CFIA). Consultation Paper on Humane and Natural Livestock Production; CFIA: Ottawa, ON, Canada, 2005.

75. Agriculture and Agri-food Canada (AAFC). Agriculture in Harmony with Nature: AAFC's Sustainable Development Strategy 2001–2004; Minister of Public Works and Government Services: Ottawa, ON, Canada, 2001.

76. Canadian General Standards Board (CGSB). Organic Production Systems: General Principles and Management Standards; CAN/CGSB-32.310-2006; CGSB: Gatineau, QC, Canada, 2006.

77. LivestockWelfare.com. Available online: http://www.livestockwelfare.com/codes.htm (accessed on 1 August 2012).

78. King, F.H. Farmers of Forty Centuries: Permanent Agriculture in China, Korea and Japan; Rodale Press: Emmaus, PA, USA, 1990.

79. Howard, A. An Agricultural Testament; Oxford University Press: London, UK, 1943.

80. Howard, A. The Soil and Health: A Study of Organic Agriculture; Devin-Adair: New York, NY, USA, 1947.

81. MacRae, R.J.; Hill, S.B.; Henning, J.; Bentley, A.J. Policies, programs and regulations to support the transition to sustainable agriculture in Canada. Am. J. Altern. Agric. 1990, 5, 76–92.

82. Wojtkowski, P. Introduction to Agroecology: Principles and Practice; Hawthorn Press: Binghamton, NY, USA, 2006.

83. Warner, K.D. Agroecology in Action: Extending Alternative Agriculture through Social Networks; MIT Press: Cambridge, MA, USA, 2007.

84. SARD and FAO. Available online: http://www.fao.org/SARD/en/sard/2001/index.html (accessed on 6 September 2012).

85. Sustain UK. Pictorial Representations for Sustainability Scoring; Discussion paper; Sustain, the alliance for better food and farming: London, UK, June 2007. Available online: http://www.sustainweb.org/pdf/sustainability_labelling_flowers.pdf (accessed on 6 September 2012).

CHAPTER 5

Food Relocalization for Environmental Sustainability in Cumbria

Les Levidow and Katerina Psarikidou

5.1 INTRODUCTION

Since the 1990s 'sustainable agriculture' has become a mainstream policy concept. Agro-industrial systems have been put onto the defensive for various environmental harms such as soil degradation, vulnerability to pests, greater dependence on agrochemicals, pollution, genetic erosion and uniformity, etc. In response to these criticisms, diverse remedies have been called 'sustainable agriculture' [1]. Originally it meant producers developing alternatives to crop monocultures, e.g., via less-intensive and agro-ecological methods, as a basis for independence from the agricultural supply industry. However, soon the term 'sustainable agriculture' was recast to mean a future high-yield productivist agriculture based on capital-intensive inputs [2,3].

Those two agendas promote divergent ways to link environmental with economic sustainability [4,5]. As food insecurity has been aggravated by climate change and competing demands for land use, this problem likewise has divergent diagnoses and remedies [6]. From a capital-intensive perspective, the Life Sciences are meant to provide more efficient inputs which increase productivity, while also minimizing resource usage and pollution. By contrast, alternative agendas support less-intensive agri-production methods which replace external inputs with local resources and farmers' skills, as means of adding value to food production.

In the last decade, such alternatives have been increasingly promoted as a basis for support from policy measures and consumers. Through closer relations with consumers, farmers can gain more of the value that they have added to food production, as well as greater incentives for more sustainable methods and/or quality products. These closer relations have been variously called short food-supply chains, local food systems or agro-food relocalization.

This paper explores the following questions about agro-food relocalization initiatives:

- How do they link environmental with economic sustainability?
- How do they expand markets beyond the capacity of individual producers, especially through larger intermediaries?
- How do they initially gain state support—and gain greater independence from conventional food chains?

Those questions will be addressed through a case study, the UK's rural county of Cumbria. Until recently, this region was known for industrial-scale production of standard food and drink, importing raw materials from elsewhere, rather than for territorial characteristics favoring local specialty products. Thus Cumbria may have general relevance to other unfavorable contexts.

The next section reviews literature on agro-food relocalization—first in Europe generally, and then in the UK, including relevant policies of the New Labour government (1997–2010). Subsequent sections focus on the Cumbria case—regional policy support for food relocalization, cooperative support networks, and larger-scale marketing of local or territorial brands. Finally, the Conclusion locates the case in its wider context and suggests implications for prospects elsewhere.

5.2   FOOD RELOCALIZATION: IMPETUS AND MEANS

Since at least the 1990s, European food retail and processing have become increasingly concentrated in supermarket chains or at least controlled by them. They have subordinated agricultural producers to standard requirements, narrowly defined 'quality' criteria, marginalized local independent retailers and separated producers from consumers [7, 8]. This has been called food delocalization, reinforcing a more globalized and industrialized agri-food system whose production methods became more dependent on capital-intensive inputs [9]. In the past decade, however, various difficulties have led some farmers to develop less-intensive methods, while also circumventing supermarket chains.

5.2.1 SUSTAINABILITY IMPROVEMENTS VIA RELOCALIZATION

UK agro-food relocalization has had many stimuli. Food scares, especially the 1996 mad cow scandal, highlighted hazards of agro-industrial systems and generated endemic distrust among consumers. Another stimulus was the cost-prize squeeze, whereby farmers' production costs increased faster than their incomes from product sales. Together these pressures have stimulated various strategies, some towards higher-value products and more diverse economic activities:

- As the cost-price squeeze tightens, farm families all over Europe are innovating and finding new opportunities; becoming rural entrepreneurs, adding extra value to their produce, switching from intensive to more sustainable farming, and becoming pluriactive ([10], p. 11).
- Along those lines, more sustainable methods can mean: substituting farmers' skills and local resources for external inputs, substituting grass for animal feed, recycling organic waste, re-integrating animal and crop production, etc. These various responses have been theorized as endogenous development strategies:
- [These are based] on locally available resources, making full use of the ecology, labor force and knowledge of an area as well as those patterns that have developed locally to link production and consumption. Endogenous development strategies are therefore more concerned with integrating the farm and farm household into its own local area and environment ([10], p. 9).
- Environmental sustainability has become a core theme in endogenous rural development—in policy terms, called integrated rural development. This concept was appropriated from the global South for the European context, where closer relationships with consumers have become an important means for producers to gain trust and to gain more of the value that they have personally invested. Bypassing supermarket chains as intermediaries, these more proximate relations construct diverse values and meanings of 'quality' or 'local,' especially by connecting producers through co-operative relations and a sense of shared identity ([11], p. 399). This greater

proximity-has become popularly known as short food-supply chains, local food systems, or simply local food. In academic literature, this greater proximity has been theorized as re-localization, re-spatialization and re-connection [11]; likewise as alternative food networks, or AFNs [12].

Given the dominant marketing chains, which are generally linked to the productivist model of global agri-industrial production, AFNs encompass various alternatives:

- For those operating at the 'production' end of the food chain, the notion of 'difference' becomes critical to the process of reconnection: creating a difference in 'quality' between specific products and mass-produced products; creating a difference between geographical anonymity in food provenance and territorial specificity; and creating a difference in the way certain foods are produced ([13], p. 118).
- For such differences in quality, the 'local' becomes a central concept linking spatial and social meanings. Local food systems facilitate a 'resistance to agro-food distanciation.' They allow actors to rework 'power and knowledge' relationships, especially through physical and social proximity between producers and consumers ([14], pp. 24-25). They are also seen as 'seeds of social change,' which 'seek to construct and portray alternatives to the construction and reproduction of hegemonies of food and agriculture in the conventional food system' ([12], p. 62). They 'present both critique-opposition to the existing food system and an alternative vision of socio-ecological relations embedded in food' ([12], p. 61).

The concept agro-food 'relocalization' expresses such processes building greater proximity than in conventional food chains. Producers go beyond distinctive products, towards distinctive co-operative, proximate relationships— e.g., farmers' markets, box schemes and community-supported agriculture. In many cases relocalization enhances social cohesion, community development and consumers' knowledge about food production methods, especially for reducing external inputs [15].

Through collective farmers' marketing, cooperation has taken many forms. According to an EU-wide study, these include:

- pooling volume in order to increase bargaining power further along the value chain, while also ensuring compliance with regulations;
- promoting high-quality distinctive food products, e.g., using organic or animal-friendly production methods; and/or

- promoting regional food products by linking each product with the territory—e.g., via distinctive landscape values, ecological resources, farmers' skills, product taste, etc.—as a means to valorize territorial resources and strengthen local economic development [16].

Such initiatives depend upon various enabling factors. These include: policies in support of market differentiation and labeling; knowledge and research centers on rural development; proximity to urban centers; location of farms in tourist areas; new opportunities in public procurement; quality orientation and involvement among consumers; and new kinds of food networks ([16], p. 264). Collective marketing has been generally grounded in territorial networks, featuring strong links with rural development initiatives.

Territorial networks have been promoted especially by the LEADER program (*Liaison entre actions de développement de L'économie rurale*). This combines three novel features: a territorial basis (rather than a sectoral one); use of local resources; and local contextualization through active public participation. Through Local Action Groups, participants have built collective capacities and established structures to sustain the local development impetus beyond the initial support [17].

Through the Leader program, local mobilization and capacity building have been supported through vertical relations of governance. Leader-type measures have been incorporated into the Rural Development Program (RDP), i.e., Pillar 2 of the Common Agricultural Policy (CAP) from 2007–2013; there Leader has been meant more as a delivery mechanism for the main measures than as a measure in its own right. In this context,

> the LEADER experiment can be seen as so significant, both in its achievements, despite its modest budget, and also in terms of its very substantial unrealized potential to address and explore solutions to issues at the heart of both neo-endogenous rural development and place-shaping concepts ([18], p. 7).

In these ways, the Leader program has provided crucial stimulus for efforts to relocalize food.

5.2.2 FOOD SECURITY AS AN EXTRA IMPETUS

For sustainable development strategies, a more recent impetus has come from global food insecurity, resulting partly from climate change and partly from greater global competition for land use. Since the 1970s Europe's productivist system has increased its dependence on animal feed imports, especially as soya production became industrialized in Latin America, in turn supporting European meat production. Likewise fertilizer use has been dependent on imports of increasingly scarce phosphates. Food insecurity has raised the stakes for Europe's significant imports, especially of intensive inputs, including energy in various forms. Like unsustainable agriculture in general, however, food insecurity has divergent diagnoses which have been characterized as productivist versus sufficiency perspectives [6].

From a productivist perspective, food insecurity arises from low or variable production levels, which must be made more efficient via technological innovation which can use and recycle renewable resources. For example, 'To guarantee future food security, output needs to be boosted significantly.' Moreover, 'biotechnology is a fundamental requirement for long-term regional (and perhaps global) food security' ([19], pp. 38-40). When managing natural resources, 'the whole supply chain must also be sustainable, to ensure food security, supply sufficient quantities of renewable raw materials and energy, reduce environmental footprints and promote a healthy and viable rural economy' ([20], p. 9). This perspective conflates more efficient inputs and productivity with food security.

From a contrary diagnosis, agro-industrial systems aggravate Europe's dependence on imported inputs and vulnerability to stress. Therefore, according to a coalition of NGOs and farmers, Europe needs efforts towards extensification and food self-sufficiency, also known as food sovereignty [21]. According to a related declaration of agricultural experts, the EU faces food insecurity from dependence on imports, especially soya as animal feed, thereby undermining environmental sustainability abroad as well as in Europe ([22], p. 9). Similar proposals have come from a broad coalition of civil society organizations:

We call for a progressive shift from industrialised agriculture towards a sustainable form of farming, which sustains productive farming everywhere, builds on the regional and local diversity of farming and economies, makes far lighter use of non-renewable resources, respects animal welfare, puts good agronomic sense and agro-ecological innovation at the heart of farming decisions, and achieves a wide range of positive environmental, social and economic outcomes, linked to the vitality of rural areas ([23], p. 7).

Some proposals incorporate elements of both those perspectives—or even attempt to reconcile them. In particular, 'sustainable intensification' has been a flexible concept for incorporating diverse perspectives. According to a UK Royal Society report,

... we must aim for sustainable intensification—the production of more food on a sustainable basis with minimal use of additional land. Here, we define intensive agriculture as being knowledge-, technology-, natural capital- and land-intensive. The intensity of use of non-renewable inputs must in the long term decrease ([24], p. 46).

Necessary means include 'agroecological processes such as nutrient cycling, biological nitrogen fixation, allelopathy, predation and parasitism,' as a means to minimize external inputs ([24], p. 17). Although the Royal Society report embraces biotechnology as an essential tool, it also proposes closer linkages between research and farmers' knowledge:

Linking biological science with local practices requires a clear understanding of farmers' own knowledge and innovations. There are past examples where science has seemingly offered 'solutions' to a problem but without success, because of a poor fit with local circumstances and a lack of local engagement with end-users at an early stage in the innovation process ([24], p. 18).

Government policy has given mixed messages on the role of imports in UK food security [25]. On the one hand, it has proposed to reduce such dependence:

Around two-thirds of energy consumption in UK agriculture is in the form of indirect inputs such as fertilizer, tractors and animal feed. Reduced fertilizer use is a key driver of the overall decline ([26], p. 95).

On the other hand, it has regarded large-scale imports as essential for the long term: 'Food security is enhanced by diversity of supply of both domestic and overseas production' ([26], p. 84).

A more recent Foresight report reiterates sustainable intensification, again encompassing diverse elements. Although it proposes greater investment in all technologies, the report also proposes 'a redirection of research to address a more complex set of goals than just increasing yield.' Moreover, 'Demand for the most resource-intensive types of food must be contained' [27]; the latter potentially means a shift to less-intensive methods. The UK's Coalition government has endorsed sustainable intensification, with an emphasis on 'the developing world' [28], so this has unclear relevance to European agriculture.

Amidst those diverse problem-diagnoses and solutions for food security, neo-endogenous rural development can find extra grounds for its shift towards agroecological methods and renewable resources. The next section examines efforts towards food relocalization in the UK—its advances, limitations and tensions, especially from contradictory policies. This provides a wider context for the Cumbria case study that will follow.

5.3   UK FOOD RELOCALIZATION: PROSPECTS AND LIMITATIONS

In UK agro-food practices, relocalization has made significant advances but has encountered many obstacles. Environmentally more sustainable production methods have been made more economically viable, partly through direct sales which gain higher prices, especially for 'quality' products. But significant expansion needs producer-led, larger-scale intermediaries to maintain proximity with consumers in wider markets. The New Labour government espoused many societal benefits of food relocalization, as well as a territorial approach to rural development, e.g., as a means to sustainable agriculture. But these pathways have remained marginal in funding priorities and related policies in England (which has its own Rural Development Program). Such tensions will be explored first in relevant literature and afterwards through the Cumbria case study.

5.3.1   NEED FOR PRODUCER-LED INTERMEDIARIES

Several pressures and opportunities have generated alternatives to the dominant agro-industrial food system. In addition to the cost-price squeeze, stimulus has come from food scares, especially the 1996 mad cow scandal, the late 1990s controversy over GM food and the 2001 foot-and-mouth epidemic. That crisis led the government to offer grants for food producers to start anew, e.g., via organic conversion [29]. As the Common Agricultural Policy (CAP) decoupled single-farm payments from production levels, instead coupling payments with on the area cultivated, this basis too provided an incentive for higher-quality production. For example, farmers' knowledge was enhanced through agro-ecological methods replacing purchased external inputs, alongside direct sales to help finance less-productive methods.

Such shifts have stimulated efforts to relocalize food chains. Through higher-value territorially branded products, i.e., locality foods, producers have gained more of the value that they have personally invested, and a special territorial identity has been spatially extended to more consumers [30]. Early pioneers were Southwest England and Wales [31-34]. But specialty products remain vulnerable to incorporation within conventional food chains ([9], p. 30); supermarkets have perpetuated producers' dependence and kept them distant from consumers.

Beyond greater financial gain for producers, local food networks have had broader aims—developing a sense of community integration, keeping alive traditional knowledge, and re-establishing trust between producers and consumers. Beyond locality foods at premium prices, some producers build closer links with consumers for simply local food, 'lying within the interstices of the mainstream' ([33], p. 564). Local food networks have been promoted as a pathway to a more sustainable agriculture. In addition to shorter transport distances, these networks also help to finance less harmful production methods, e.g., organic and integrated systems. They can enhance 'economic behavior mediated by a complex web of social relations,' linking consumers with products of environmentally beneficial production methods ([34], p. 334).

A special case has been organic food. Supermarkets chains have been procuring and selling over 70% of organic products [35]. Many organic producers have faced a dilemma: the organic vision (as well as their income) was being lost through conventional chains, but they had difficulty in selling much produce by other means. Moreover, supermarket chains were throwing them into competition with cheaper imports.

As an initial solution, some organic growers created more direct relations with consumers, e.g., via box schemes. Along these lines, organic food gained a special role in linking local products with sustainable development through consumer preferences and loyalty [36]. As a novel form of direct sales, UK farmers' markets have also expanded significantly since they began around 1997. A decade later, farmers' markets were being held at 550 locations, creating 9,500 market days and 230,000 opportunities for stallholders per year. By 2006 total annual turnover was estimated at £220 m (€250 m). These opportunities have been jointly created by local authorities, community groups, stallholders, producer cooperatives and companies managing the markets [37].

However, organic farmers still encountered limits of direct sales and so identified a need for new intermediaries. These would be 'something in between traditional markets and the multiples,' i.e., the supermarket chains, as a necessary means to reach more consumers ([38], p. 452). This gap has become a general difficulty for alternative food networks: 'AFNs potentially face difficulties when trying to distance themselves from the conventional food chain, given the current shortfall of intermediaries able to cope with alternative forms of production, i.e., local abattoirs, transporters, wholesalers' ([39], p. 253). For this reason, larger intermediaries have been sought for reconnecting producers with more consumers and socially embedding food products.

5.3.2 UK POLICY: CONTRADICTORY ASPECTS

Given the efforts to link environmental with economic sustainability via food relocalization in the UK, these modest initiatives encounter various limitations in direct sales. They also face contradictory policies, which espouse benefits of local food but relegate all responsibility to market actors.

Food relocalization has gained prominence in policy circles. A government advisory body, the Curry Commission, recommended economic regeneration by reconnecting people with food production: 'Reconnect our farming and food industry; to reconnect farming with its market and the rest of the food chain; to reconnect the food chain with the countryside; and to reconnect consumers with what they eat and how it is produced.' It distinguished between local food, defined simply by geographical proximity, and 'locality food,' e.g., specialty or territorial brands which can gain advantage also in longer-distance supply chains ([40], p. 6). However, as a main conduit for relocalization, the Curry report emphasized supermarkets and food processors. Some have stocked lines of territorially branded food or simply 'local food'; this arrangement perpetuates farmers' dependence, long food chains and the cost-price squeeze.

The New Labour government took up the Curry Commission proposals on food relocalization in relation to sustainable agriculture and reconnection with consumers. It espoused many benefits of local food, but without significantly changing its policies to facilitate such development. These contradictory messages span several policy areas—special support measures, a territorial approach to rural development, and public procurement—which will be surveyed in turn.

In the context of the CAP reform, which has gradually delinked subsidy from production levels, the Department for Environment Food and Rural Affairs (DEFRA) has celebrated new market opportunities for linking environmental and economic sustainability:

> For the farming and food sector this [CAP reform] presents real opportunities both to meet the demand for high quality, seasonal or locally sourced produce delivered through strong local food chains and, importantly, to help deliver our future energy needs.... There are consumer trends that could work to the advantage of sustainable food and farming, including the premium paid for fresh food in out-of-home purchases, celebrity chefs and the re-emergence of a food culture within the UK, the rapid rise in interest in seasonal food, in animal welfare and the continued growth of demand for organic products ([41], pp. 2, 28).

However, the Department has given little material support for developing such alternatives. At least on paper, a priority was to support 'the quality regional food sector' in three key areas—trade development, consumer awareness and business competitiveness ([41], p. 18). Three years after announcing this strategy, however, it offered little more than advice to food producers and consumers [42]. By default, the policy has depended on consumers taking responsibility for food choices that enhance sustainability within current food markets.

That scant support from UK programs relates to the EU policy framework, which has tensions between productivist versus alternative agendas. In the former, agriculture poses an opportunity and imperative to enhance economic competitiveness via greater efficiency. The EU's Rural Development Program promotes 'the sustainable development of rural areas,' closely linked with efficiency and modernization [43]. In EU policy on multifunctional agriculture, there is also scope for less-intensive methods and higher-quality products in economically less favored regions, e.g., ones which cannot compete on productivity. The latter remain a marginal niche in the dominant policy, rather than a wider strategy for relinking environmental and economic sustainability.

In implementing the CAP reforms, the Rural Development Program England has moved towards regionalization, which opens up prospects for a territorial perspective that reintegrates farming into rural development. This can increase farmers' income by several means, e.g., regional branding of quality products, adding value to on-farm products, etc. [44]. State funding can combine several sources, e.g., Structural Funds and Leader+, whose bottom-up Local Action Groups propose development agendas.

However, according to an academic critique, UK policy remains oriented to agricultural productivism along with non-agricultural activities: the New Labour government has not seriously considered how to reconstitute an agricultural component of rural development. This failure has several sources. DEFRA has been reluctant to decentralize many powers to the regions, instead setting rules for greater market competition among producers ([32], p. 429). More fundamental has been the New Labour government's commitment to neoliberal policies, seeing market liberalization as a general remedy, while also avoiding responsibility:

> The real ideological politics of agri-food in the UK is to let 'the markets' (increasingly dominated by the corporate retailers, of course) become more 'liberalized' and to avoid any deviations from the principles of European competition policy that might strengthen local and regional protectionism ([32], p. 429).

Consequently, rural development grants have largely favored industrial-scale food activities seeking global competitiveness through greater efficiency. Under the RDP England 2000–2006, for example, the Processing and Marketing Grant (PMG) scheme had a minimum £50,000 per grant, which could only cover 40% of the total project costs, so the applicant must provide at least £75,000. Given these requirements, grants have been available only for medium-size or large businesses seeking to expand. Consequently, substantial RDP funds have gone to businesses which may obtain their ingredients from the cheapest source, regardless of distance, thus favoring long food chains.

A commitment to market liberalization has likewise constrained public procurement of local food, despite its promotion by the Curry Commission. EU rules require state authorities to accept 'the economically most advantageous tender' ([45], Article 26). Such a judgment can include quality and environmental criteria [46], e.g., 'fresh' food and greenhouse gas emissions, but cannot specify 'local.'

Throughout Europe, however, many local authorities have been reluctant to justify locally sourced food, especially if it is more expensive; in practice their contracts favor larger suppliers and long food chains. Despite DEFRA statements supporting local food, UK local authorities face an extra constraint: government policy favors 'aggregated purchasing' from large suppliers in order to reduce costs [47].

For many years public procurement managers in the UK have convinced themselves that they cannot procure food from local producers because this is prohibited by EU regulations, which uphold the free-trade principles of transparency and non-discrimination. In reality, these regulatory barriers are more apparent than real… ([48], p. 23).

Wherever those apparent barriers are taken as real ones, contracts go to multinational companies, e.g., 3663 First for Foodservice or Sodexho in the UK. They can offer lower prices and have expert teams to write tenders for contracts.

Having identified tensions around market expansion and state policies for food relocalization, the next section introduces the Cumbria case study; then subsequent sections explore the tensions already surveyed.

5.4   CUMBRIA CASE STUDY: CONTEXT AND METHODS

5.4.1   AGRICULTURAL FEATURES

With a half-million people, the County of Cumbria is rural, mountainous and partly protected from development by the Lake District National Park. More income is derived from the tourism industry than from agriculture. The wider

North West region has a wetter climate than some other parts of the UK, but sunshine hours are also lower, so conditions are adverse for horticulture or arable production. Agriculture is mainly based on livestock, especially upland hill farming. These geographical characteristics limit agricultural diversity and food self-sufficiency. So does agro-industrial concentration, which has increased since the 1990s. At issue is whether that process can be slowed down, or even reversed, through alternatives which relocalize food production and consumption patterns.

Cumbria may seem an unfavorable context for relocalizing food. Like most of northern Europe, Cumbria has few territorial characteristics favorable to specialty food products. As a centre of industrial food and drink production, moreover, Cumbria imports many raw materials for 'local' producers which are effectively global food factories. These price-competitive, long-supply chains potentially marginalize local small-scale local producers. So efforts towards relocalization there can indicate prospects and difficulties relevant to other unfavorable contexts.

Starting from the questions posed in the Introduction, the above literature survey informs a hypothesis for Cumbria: that environmental sustainability-via-relocalization there depends on new forms of cooperation among producers and with consumers, together developing a territorial identity—based on social commitments to sustainability, more than on specialty foods. The case study explores how agro-food relocalization initiatives link environmental with economic sustainability.

5.4.2   CASE-STUDY METHODS

For this case study, initial data sources were internet-based documents of Cumbria policies and practices: regional development agencies (Cumbria Rural Enterprise Agency, Northwest England Development Authority, their funding programs (Distinctly Cumbrian, Cumbria Fells and Dales, Leader+, etc.) farmers' organizations (Cumbria Farmers' Network, Cumbria Organics), farmer-led cooperative marketing initiatives (e.g., Hadrian Organics, Plumgarths) and small-scale businesses (box schemes and farm shops, e.g., Howbarrow Farm). Those documents provided a basis for 12 semi-structured interviews with practitioners; interviewees mainly represented the above bodies, though a few ran small-scale businesses. We selected initiatives which develop short-supply chains and favor cultivation methods with low external inputs. Most interviews were conducted in the working environments of the practitioners (e.g., farms, farm shops or offices), thus giving us a greater familiarity with their everyday practices. Snowballing contacts was an extra method for identifying relevant practitioners and their networks; interviewees mentioned links or analogies with other food initiatives, e.g., via self-description and anecdotes. Given their work commitments, only some practitioners were willing to give us their time for interviews, thus limiting our choices. Interviews were carried out during 2008–2009, with some follow-up in 2010–2011.

Interview questions were adapted from a larger research project about alternative agro-food networks (see Acknowledgements section; also note 15). Drawing on concepts in the literature on AAFNs, we investigated how practitioners discursively position themselves and their activities in relation to conventional food chains. Interview questions asked how Cumbria initiatives differ from conventional food chains—as regards their aims, knowledge, production methods, networks, producer-consumer relations, etc. All interviewees described such differences, with various levels of enthusiasm and emphases.

Whenever an interview question asked about 'alternative' aspects, however, interviewees generally dissociated themselves from the term. Some even gave the term 'alternative' pejorative meanings—e.g., outsiders, marginal, weird, oppositional—as noted in previous studies ([39], p. 253). Some interviewees expressed an aim to be seen as mainstream or to become mainstream—i.e., to be normal, larger and successful—though without being incorporated by conventional food chains.

All interviews were recorded, transcribed and then analyzed along two lines: (i) practitioners' understandings of differences between Cumbria food networks and conventional food chains; and (ii) practitioners' meanings of terms such as local food and Cumbria food; the latter was done initially by searching for those terms. Although 'local' could simply denote a short geographical distance, many interviewees gave the term broader social, cultural or political meanings. In this way, we could better identify interviewees' various aspirations and strategies.

Our draft analyses were circulated for comment from interviewees and other practitioners. Preliminary results were pre-circulated for a stakeholder workshop on 'Cumbria food networks' held in April 2009. This attracted eight local participants; one had been previously interviewed and two were later interviewed, especially about how regional agencies support farmers' cooperation. Workshop discussions provided data for further analysis and insights for further research. In March 2010 a summary was circulated for comment to fifteen key individuals, including all interviewees and workshop participants; some provided comments clarifying points.

Through the above methods, this paper analyzes the language used by interviewees and other practitioners, as an extra insight into their practices. An analytical focus is their meanings of local food and/or Cumbria food, especially how they link environmental and economic sustainability. Empirical results are juxtaposed with academic research and relevant policy documents—at national, regional and local level.

5.5 CUMBRIA'S FOOD RELOCALIZATION: MUTUAL SUPPORT AND FAVORABLE POLICIES

Regional policy changes have facilitated more environmentally sustainable production methods and producer cooperation towards a Cumbrian food culture. Together these practices build agro-food relocalization, as described in this section. Such practices have been expanded beyond direct sales, towards larger intermediaries, as described in the next section.

5.5.1 IMPETUS FOR FOOD RELOCALIZATION

Aggravating the cost-price squeeze, economic disruptions have stimulated changes in agro-food systems, especially in Cumbria. After the 2001 foot-and-mouth crisis, many Cumbrian farmers were compensated for loss of livestock. Some food producers used this opportunity to diversify agro-food business models, e.g., towards organic or other higher-quality products. The New Labour government provided substantial funds for economic regeneration, especially for organic conversion, thereby avoiding agrochemical inputs in such farms. Around the same time, the CAP single farm payments were being decoupled from production levels. With a weaker incentive for high productivity, farmers faced a challenge and opportunity: to reduce external inputs, while also remaining financial viable.

As an extra way forward, direct sales were initiated by Cumbrian farmers attempting to gain premium prices and closer relations to consumers. They established farm shops, farmers' markets or vegetable box schemes, with visitors who often made return visits because they enjoyed the food as fresh, different, local, etc., according to a survey of consumer attitudes in Northwest England [49]. On-farm shops and farmers' markets complemented each other; both outlets attracted similar consumers who recognize quality food from familiar, trusted suppliers [50].

For many producers, 'local' food means building a Cumbrian food culture around both local and locality or specialty products. The latter build and benefit from a territorial identity in local markets, as promoted by the regional authority as well as by cooperative marketing. Some 'Cumbrian' food has been highlighted as distinctive along several lines, e.g., organic and biodynamic cultivation methods, or rare-breed animals with a special taste. With the rise of these specialty labels, the region remains the target market, and there seems little danger of imitators. Cumbria has requested a Protected Geographical Indication (PGI) for only two products—Cumberland Sausage and Herdwick Lamb [51].

Food relocalization has given various meanings to 'local' or 'Cumbria' food, according to our analysis of documents and interviews. For many food producers, those terms mean: local resources rather than external inputs, revitalization of local knowledge, regional branding, mutual support and interdependencies, producer cooperation, greater social proximity to consumers, their support for local agriculture and an economy supporting local businesses. These meanings have been elaborated in producers' support networks, especially Cumbria Organics and Cumbria Farmers Network.

Cumbria Organics includes farmers, growers, processors, retailers and consumers. With over 70 members, it carries out education, promotion and organic production. It was set up in 1999 'to provide self-help support to the increasing number of local farmers converting to organic production' [52]. It helps to develop supply chains and provides technical information for producers interested in farming to organic standards. It builds a wider food culture through public outreach, e.g., via stands at local festivals, a model farm, books, leaflets, games, quizzes and farm walks. Cumbria Organics also help members to obtain contracts for public procurement, e.g., by explaining ways to fulfill requirements of tenders.

Despite their common commitment to organic methods, Cumbria Organics' members have disagreed about whether to supply supermarkets, which often supplement more direct sales. Early on, membership meetings became difficult:

> Instead of really addressing the marketing issues, we end up getting involved in arguments between these two camps…. So we have [separate] meetings for the direct selling businesses, who were selling direct to the consumer and want to develop those alternative networks, and then for the people who sell to the wholesale markets, to help them find contacts and find out about prices [53].

This internal division indicates a missing option: large intermediaries which can maintain producers' control over pricing and labeling across longer supply chains.

Encompassing all agricultural methods, the Cumbria Farmers Network was established in 2005 to promote producer cooperation. With changes in the CAP, moving payments away from a production basis, farmers could 'still produce food in an environmentally sustainable way—not necessarily a financially sustainable way, of course' [54]. Dependent on CAP single farm payments, farmers had become socially isolated. So the Leader program facilitated farmers' cooperation, especially in production and marketing skills; such skills have helped to regain control over food chains: 'Our members have become more aware of the benefits of working cooperatively and really have taken control of their own futures, rather than allow the government or market forces to dictate' [54].

Moreover, Cumbria Farmers Network has helped to build a regional 'food culture' by raising consumers' awareness of food quality, sources and production methods.

> We are trying to make people more aware of how food is produced and when it is produced and what the product is. We have farm open days, when the consumer can come along and look at how the food that they are going to buy and eat is being produced and how it links to the environment [54].

'Promoting local farms to local people' has become a key expression, giving social meanings to the word 'local' [54]. In these ways, conventional farmers too have benefited from and extended the efforts by organic farmers towards relocalization. Shorter food supply chains reduce transport costs and energy inputs, relative to conventional chains, as well as helping to finance more sustainable practices that may be less productive. How do these initiatives gain state support—and gain greater independence from conventional food chains?

5.5.2   POLICY SUPPORT FOR LOCAL SUSTAINABLE FOOD AND PRODUCER COOPERATION

Those efforts towards food relocalization have depended upon state support of various kinds, especially at an early stage. Such support includes: promotion of a Cumbrian food culture, producer networks for mutual support, collective marketing, infrastructure for farmers' markets, processing equipment for individual producers, etc. Together these measures help to enhance the economic viability of environmentally more sustainable production methods and so facilitate their continuation and/or adoption by more farmers.

As the major distributor of relevant funds, e.g., from Structural Funds and the Rural Development Program (RDP), the Northwest Development Authority applies a broader understanding of 'local' food than in food relocalization: 'By provenance, we mean the product must demonstrate a link in some way to the place of its production, the production method or the people who produce it' ([55], pp. 13-14). Thus 'local' can mean simply a local site of production, even if the ingredients have a distant source, as in the global food factories prominent in Cumbria. At the same time, the agency has promoted territorial branding, e.g., through regional competitions for 'fine foods' and a consumers' guide [56].

Along the latter lines, the Northwest England RDP helps farmers to 'reconnect with their consumers,' especially through high-quality regional foods and shorter food chains. This strategy links environmental, economic and social sustainability:

> The goal will be to demonstrate a true community-led approach to rural regeneration across all Axes.... The Forestry Commission and Natural England will engage with the new Leader Groups as the approach offers sustainable, holistic and integrated public benefit delivery.... Key here are the bottom-up approach and the autonomy that the partnerships need to be given ([57], p. 16).

Opportunities also exist to develop locality food brands, offering farmers the chance to add value to commodity production. Potential may also exist to link economic and environmental assets through food branding; especially in areas of high landscape character ([57], p. 8).

Along those lines, Cumbria's RDP has offered small grants for equipment to small businesses, e.g., to expand from domestic kitchens to larger-scale production, or for on-farm processing. For example, the scheme funded The Pie Mill to acquire equipment for pastry rolling and chilling, especially as a means to produce the small-size pies required by schools and old people's homes. Along environmental sustainability lines, RDP grants also have helped farmers to develop renewable energy—e.g., via small wind turbines, solar power, water capture, anaerobic digestion, biogas boilers, etc.—thus minimizing environmental burdens from energy usage. These grants have been crucial because producers need to wait several years before recouping the investment through lower energy costs. For farmers seeking or receiving grants, however, there have been heavy administrative requirements—e.g., detailed reporting requirements and lengthy inspection visits—especially in relation to the small amount of money made available. These burdens deter other small-size businesses from applying for small grants.

5.5.3   SUPPORT FOR PRODUCER COOPERATION

Since its inception, the Leader program has sought to develop a bottom-up approach by involving local stakeholders in Local Action Groups. As in New Labour policy for the RDP, the Cumbria Fells and Dales Leader strategy has promoted economic diversification, including non-agricultural activities. At the same time, it promotes agricultural production methods which minimize environmental burdens and offer quality characteristics recognized by consumers:

> There will also be the wider agenda: local branding/marketing (including adding value and local processing); collaborative working (e.g., cost sharing and "share to rear"), improving animal welfare actions, agri-food tourism; role of farmers in public goods delivery and whether rewards can be obtained for High Nature Value Farming systems; consideration will also be given to increasing the take-up of organic farming; and there will be a further exploration of the creation of a hill farming brand ([58], p. 69).

In particular, organic agriculture is promoted as a means to enhance food security and minimize environmental burdens:

> Food security is beginning to be a concern again, and issues to do with the environmental costs of production (especially the price of oil and the related costs of animal feed) may well drive further growth in the local produce economy. There is strong national consumer demand for organic production but a static or slightly increasing domestic supply ([58], p. 25).

The strategy includes a 'scheme to support co-operative ventures to reduce product miles and increase availability and productivity of local produce' ([58], p. 50). Given the importance of tourism, hotels and restaurants have great potential for promoting local food products. According to a Leader manager:

> Overall the Leader program works with the agricultural production sector, focusing on food security and shorter supply chains. The local food economy is supported via production and processing grants. The promotion of local produce into the tourism sector is supported by some revenue investments. Leader also provides some funds for community-supported agriculture (CSA), producers' markets and similar initiatives [59].

However, the Cumbria Leader program has minimal resources for such aims; it receives a small proportion of the overall RDP budget. Moreover, the European Commission found that the Government Office North West (GONW) had applied inadequate controls under the previous RDP and so was asked to return some (according to our interviewees). In response, the NWDA has been closely managing the Local Action Groups within the standard criteria of the RDP England, e.g., economic competitiveness via efficiency measures. This constrains pathways towards relocalization, while reinforcing DEFRA's reluctance to decentralize control over Leader ([32], p. 429). There had anyway been tensions between the RDP's formal criteria and Leader's bottom-up approach, so these may become more difficult in the future.

Beyond grants to producers, regional agencies offer assistance for small businesses to accommodate various legal and quality requirements. Cumbria Rural Enterprise Agency (CREA) has held training seminars for small-scale food and drink producers. It also provides kitchen facilities to help them accommodate strict hygiene regulations, which would otherwise pose a major obstacle.

Support bodies help to develop a territorial identity for quality food. As a strategy for territorial branding, Distinctly Cumbrian has provided business support, aiming 'to strengthen the rural economy of the county by providing advice and grants to rural businesses in the quality food, drink and craft sectors which make added value products able to be marketed and identified as distinctive to Cumbria.' 'Made in Cumbria' has been another economic development initiative of Cumbria County Council, with funding from the NWDA and the European Commission's Objective 2 structural funds. Made in Cumbria has provided support to four of the county's 15 monthly farmers' markets. It also organizes 'Meet the Buyer' events, helping small producers to meet larger buyers—e.g., the National Trust, the Youth Hostels Association, Centre Parcs Oasis, numerous hoteliers and supermarket chains. These events help small-scale producers to gain self-confidence in dealing with buyers.

Regional agencies have also stimulated producer networks for cooperative activities, thus overcoming several obstacles—farmers' social isolation, mutual distrust and historical dependence on CAP funds. Successive Leader programs have facilitated producer cooperation as a crucial means to sustainable development, e.g., via Cumbria Organics, Made in Cumbria and infrastructure for farmers' markets. The Leader program has played an animator role by developing cooperative relations among producers, leadership skills, business skills, advice on equipment, etc. according to Leader managers [59]. These efforts have led to farmers' collective marketing in various forms (see next section).

As the host of the Leader program, the NWDA established Local Action Groups to formulate many measures, while also linking food with a local identity:

> There are several strands where a degree of learning has been identified as a pre-
> requisite. By raising awareness of consumers of the products of the countryside,
> a greater loyalty is engendered and purchasing habits do change. 'Sense of place'

and 'know your place' training packages have helped other areas to create local pride, develop community spirit and also contribute significantly to the local tourism offer ([58], p. 35).

For such developments, many producers acknowledge dependence upon training programs and producer cooperation, at least at an early stage of business development. This basis provides a more grassroots approach to rural development, by targeting community links and local needs, according to Leader managers [60].

As another support measure, public procurement has also favored local food producers through policies of Cumbria County Council (CCC). The Strategic and Commercial Procurement Team 'is committed to responsible procurement, including the use of social, economic and environmental evaluation.' Its commitment to sustainability means 'protecting the environment and taking responsibility for minimizing the wider environmental effects of its purchasing decisions' [61].

Along those lines, the Council's procurement policy has emphasized quality and sustainability criteria. These include carbon footprints, animal welfare, distance travelled to an abattoir, etc. [62]. According to the Principal Buying Manager, 'Winning a tender is not just about price but also product quality, sustainable behavior and a clear indication that a business has processes in place to deliver what they say they will' [63]. Cumbria Rural Enterprise Agency (CREA) has helped local small food businesses to do so [59]. Also the Council has issued tenders for relatively small contracts, which are split up along several lines: product type, locality, production and distribution roles [64]. This structure helps local small-scale producers and distributors to gain contracts, while deterring multinational companies such as Sodexho.

The Council was named winner in the Sustainable Procurement category for the public sector at the Northwest Business Environment Awards in June 2009. The award emphasized how the Council 'thinks about the wider implications of the products and services it buys—particularly the food and drink products it buys for school meals and caring for older people.' Green purchasing need not cost more: the Council has made savings of £3.5 million and 150,000 food miles through 'smarter procurement' [65].

In sum, regional policy and grants have facilitated direct sales promoting quality food of various kinds. Cooperation among producers has built territorial branding—of simply local food as well as special locality foods—together comprising a Cumbrian food culture. These shorter chains also facilitate links between environmental and economic sustainability. The next section analyzes efforts to expand such practices beyond direct sales through larger intermediaries.

5.6 PROXIMATE INTERMEDIARIES FOR RELOCALIZATION

Despite support from Cumbria's regional authorities, smaller-scale producers face several obstacles. These include: statutory paperwork, a standard fee for organic certification regardless of size, and few local abattoirs, thus requiring longer-distance transport, according to interviewees. Individual farmers have limited capacity for direct sales; such activity diverts their attention from farming and requires social skills of communicating product quality to consumers. For medium-size producers to localize sales, the larger-scale market remains a major challenge. For all these reasons, market expansion poses a dilemma, especially for larger-scale organic producers, who have become divided over supplying supermarkets.

As a way forward, producers have developed cooperative or collective marketing along with specialty brands. These activities complement wider efforts to build a Cumbrian food culture, so that consumers learn more about environmentally sustainable production methods and aesthetic qualities of local food, as a basis to favor such products. This section describes producer-led intermediaries which expand markets beyond the capacity of individual producers for direct sales, while also building or maintaining closer relations with consumers.

5.6.1 PRODUCER-LED INTERMEDIARIES

Hadrian Organics is a farmers' co-operative which does direct sales through collective marketing, whereby farmers take turns selling members' products. With initial support from Leader funds, Hadrian Organics facilitated the initial cooperative links. A network of farmers' markets solved a major problem—how to increase sales while maintaining a close connection with consumers, rather than sell to supermarkets: 'We provide local food for local people,' meaning that producers build consumer support for local, cooperatively sold organic food [66].

Producers' co-operation is vital for empowerment of the co-operative. Members emphasized the direct benefits—mutual support on the production side (e.g., in haymaking), personal relationships that help farmers to overcome isolation and territorial branding. The latter has been emphasized by a practitioner: 'Hadrian Organics has a very good name. Our brand is strong and sales are going up' [66]. These rising sales indicate links between social, economic and environmental sustainability.

Howbarrow Organic Farm is a vegetable box-scheme selling produce of theirs and other organic farmers, totaling 1,600 product lines. According to its representative: 'Now we are dealing with the whole shopping experience, we start to compete with the supermarkets. People can choose to use us rather than a supermarket group [chain] which go into areas and close down local retailers' [67].

Its box scheme faced such a threat from plans to open a Booths supermarket in nearby Grange. Regardless of that particular outlet, supermarkets often win the competition for supplies of high-quality products:

> We have very little buying power, even though we deal with a wholesaler who purchases more than us. The supermarkets are such large buyers. Veg is more positive because we are dealing directly with the farmers. There are more and more farmers who want to sell direct [67].

This box scheme illustrates how relocalization efforts encounter more powerful intermediaries and difficulties in shortening supply chains for quality food; supermarkets compete on a similar basis, selling organic and even 'local' branded food.

Another small intermediary, Low Sizergh Barn Farm Shop and Tea Room, illustrates the earlier historical shift. After many years of conventional farming, the tenants converted to organic production in 2001–2002. A decline in farm-gate prices led them 'down the organic route' as a way to add more value and to gain more of that value by shortening the food chain; this change was also encouraged by the consumers' willingness to pay more, as crucial for environmental sustainability. As they said: 'Price is important but people expect they are buying into other benefits such as environmental standards' [68]. Visitors to the Farm Shop become familiar with the food production process, via a dairy below the café, an educational program and nature walks around the farm.

An additional benefit of the Low Sizergh Barn shop stems from its co-operative relations with other local businesses in the county. Although it does not source all its food locally, Cumbria Organics members supply many products, such as fresh organic vegetables from the Growing Well project and flour from the Water Mill, which in turn uses grain from biodynamic cultivation methods. These initiatives see themselves as part of a support network favoring local, seasonal and organic sources [69]. By 2011 feed prices and market forces had led Low Sizergh farm to leave registered organic production, though high standards remain a core priority.

As an unusual example, the Growing Well project develops more environmentally sustainable methods and educates people about them, while also providing social inclusion for people recovering from mental illness. This combination provides a basis for social justice, resource conservation through agriculture and popular education about these issues.

We aspire to be more sustainable, so to use less water and less power... we try to use environmentally friendly products where possible, those sorts of things.... Our social aim is to help people who are recovering from mental illness, build their confidence and their skills, so we do that by involving them in the running of an organic growing business... We also train people in horticulture, so we use the organic business as a way of teaching people about vegetables and we provide school visits to children... [70].

Following their sustainability principles, local shops sell its products as a way to reduce food miles, support the local economy and build stronger, socially just and healthy local communities.

5.6.2 MAINSTREAMING RELOCALIZATION

For medium-size producers to localize sales, the larger-scale market remains a major challenge for mainstreaming relocalization. Although collective marketing creates larger markets, these are limited mainly to individual consumers making special efforts to reach a sales point and paying premium prices. Medium-scale producers remain dependent on conventional wholesalers and supermarket chains, which increasingly sell 'organic' or even 'local' food but often obscure the supplier's identity.

To reach larger markets, the Curry Commission report encouraged food producers to negotiate better contracts with supermarket chains:

> Well-facilitated collaboration can give small farmers access to professional marketing and technical advice. It can also put them in a better negotiating position when dealing with large customers or suppliers. Smaller-scale regional supermarket chains collaborate in exactly this way when negotiating with their suppliers ([40], p. 34).

Along those lines, some Cumbrian producers have cooperated to create large intermediaries which extend social proximity to consumers through larger buyers. Bypassing conventional wholesalers, these intermediaries more directly supply buyers, including supermarkets. Such intermediaries depend on producers sharing skills of production, marketing, distribution, in order to take advantage of funding and sales opportunities.

New larger intermediaries have provided means to mainstream small-scale producers into local food markets. Starting from a food stall, some expand to supply farm shops, retail groups and supermarkets. Expansion becomes dependent upon a local hub, i.e., a site for collecting goods from various producers and distributing them to buyers, by analogy to hubs of each supermarket chain.

As the most prominent example in Cumbria, Plumgarths Farm Shop established a local hub for distributing products from numerous suppliers to large-scale buyers. By 2006 it was supplying 12 Asda supermarket stores with 80 products from 15 local suppliers. This initiative expanded into Plumgarths Food Park, with food processing units and a food-service supply business. Its website promotes meat from rare breeds with special aesthetic qualities, along with grassland grazing, thus contributing to product quality as well as environmental benefits by avoiding animal feed.

This larger market has offered new opportunities for local producers. Previously, many small-scale producers were selling to local suppliers or farmers' markets, while also doing another full-time job. Plumgarths helped them to expand sales and produce food as their main job, some even employing staff. By selling through Plumgarths, producers also gain commercial experience and confidence to speak directly with other buyers [71].

Such intermediaries have supplied more large-scale retailers, thus reducing transport distances relative to multinational suppliers: 'Although retailers traditionally rely on centralized distribution through large depots, most appear to be willing to work with local food hubs or other specialized intermediaries to facilitate sourcing of local food,' according to a report, Supplying Local Food to Mainstream Customers ([72], p. 20). Hubs can distribute single pallets to local pubs and hotels, thus removing need for large-scale distributors. Such hubs initially supplied local high-quality products to Cumbria hotels. Later they made a similar request to Pioneer, a medium-size distributor which combines many local producers to supply large buyers, including public authorities [64].

Several other hubs have been created or supported by small-scale producers whose local brands are recognized by consumers. For example:

- Herdwick Lamb is supplied directly to Booths' supermarkets from a coordinated group of 10–20 farmers using a local slaughter facility.
- Lakes Free Range Eggs Company brings together many small producers on a traceable basis, so that customers can 'Trace your egg.' It supplies large outlets such as McDonalds, which has attempted to localize its supply chains towards traceable suppliers.

The Plumgarths founders advocate more local hubs—partly to enhance economic sustainability, and partly to address climate change:

More widely, local food is considered to provide benefits to all of the aspects of sustainability. If food travels shorter distances, and particularly if it is distributed to major customers through local food hubs or other intermediaries as recommended in this report, it can help to reduce the carbon emissions which are the principal driver of climate change… Local food production, distribution and sale help to promote sustainability ([72], p. 4).

They also see links between food security and localization, meaning local diversity and practices that reduce dependence on imports: 'We conclude that food security means not putting all your eggs in one basket. Local food contributes to the diversity of the food chain, and its further development will strengthen security' ([72], p. 18).

In several ways, then, producer-led intermediaries mainstream relocalization, while also linking environmental and economic sustainability. They provide ways for higher-quality products and/or environmentally sustainable production methods to gain better remuneration, thus making them economically more sustainable. By shortening the food chain to large buyers, new intermediaries offer producers extra local markets, more control over branding, fewer middlemen and/or less dependence on supermarkets. The shorter chain maintains producers' territorial and specific identity, along with greater proximity to consumers than conventional intermediaries. Marketing emphasizes organic methods or grassland-fed animals, partly as a basis for consumers to pay premium prices. Other environmental benefits (e.g., from on-farm energy production, nutrient recycling) remain less visible; they may apply to only some producers.

All these practices strengthen farmers' incentives to use local renewable resources, rather than commercial inputs such as animal feed or chemical fertilizers, in the face of great commercial pressures from the conventional agro-food system. But farmers' collective marketing remains vulnerable to many uncertainties and difficulties. As three examples: New or larger markets depend on extra investment, whose financial return depends on successful strategies to gain from added value. Farmer-led intermediaries must make special efforts to obtain and maintain contracts with larger buyers, who could easily find cheaper suppliers. As meat prices rise, some farmers may sell more produce on the open market and withdraw from a farmer-led intermediary, thus potentially undermining its role. Amidst such difficulties, the Leader program staff provide somewhat intangible but crucial support—by targeting grants at farmer-led intermediaries, advising them, and strengthening individual farmers' commitments to them [73].

5.6.3 FUTURE OF SUPPORT MEASURES

After the 2010 UK general election, the New Labour government was replaced by a coalition of Conservative and Liberal Democrat parties. Since then, most public expenditure programs have faced great reductions and uncertainties. Even if the government maintains previous budget levels for the RDP and Leader, it may reduce others (e.g., Structural Funds), thus weakening measures that had combined those sources.

Regional Development Agencies will be replaced by Local Enterprise Partnerships (LEPs) by 2012. As their main mandate, LEPs should provide 'strategic leadership in their areas to set out local economic priorities,' including 'partnerships with the private sector.' LEPs will be more numerous and smaller in geographical range than the RDPs, e.g., Cumbria alone, rather than the entire Northwest England. In autumn 2010 the government approved Cumbria's proposal for an LEP. Its agenda for agriculture incorporated many features of the previous RDP, especially support for upland farmers to provide public goods and to gain from the value that they have added. In particular:

- Ensuring our upland farmers are able to contribute to delivery of public benefits such as carbon storage, flood alleviation, biodiversity and access.
- Ensuring farmers secure the added value from the sale of produce ([74], p. 12).

Upland farming was also the focus of a Parliamentary report on rural development. Among its recommendations were the following:

The Government must enable hill farmers to make a financial return from the provision of public goods such as carbon storage and water management. Hill farmers will require access to improved knowledge transfer and extension services to make the most of those opportunities, as well as improving agricultural productivity and sustainability ([75], p. 3).

From those indications, the government will continue support for integrated rural development, to be implemented by a new Cumbrian agency. This role will likewise continue previous tensions over aims, e.g., between higher productivity for competitiveness versus farmers' knowledge for higher quality, as well as between the RDP's formal criteria versus Leader's less tangible facilitation role.

5.7 CONCLUSIONS

In Cumbria's processes of agro-food relocalization, shorter supply chains help to finance environmentally more sustainable practices. These include grass-feeding or locally-sourced feed for livestock; organic or biodynamic cultivation methods, and on-farm production of renewable energy. Many Cumbrian farmers have adopted such methods, often for 'quality' products of various kinds, e.g., organic, territorially branded, rare breeds, etc. These practices reduce transport, energy and agrochemical inputs, relative to conventional chains.

As a pervasive tension, environmentally better methods may lack economic viability. By minimizing external inputs, farmers reduce their costs but may also reduce productivity, relative to agro-industrial methods. As a way forward, producers have sought a higher price by developing greater proximity with consumers and gaining their support via quality brands. This agro-food relocalization helps to link environmental with economic sustainability: producers gain more of the value that they have added to production, while businesses recycle income within the local economy.

These short supply chains reduce transport costs as well as dependence on food imports. Self-sufficient production methods avoid or reduce dependence on external inputs, e.g., animal feed, chemical fertilizer and energy. The region also breeds animal stock with special qualities (e.g., taste and scrapie resistance) for cross-breeding with stock elsewhere in the country and Europe. These activities contribute to food security, though significant advances are inherently limited by Cumbria's geographical terrain, given its scant arable land.

The new linkages between sustainability and security have several motives, as expressed by the term 'local' or 'Cumbria' food. Their various meanings include: local resources rather than external inputs, revitalization of local knowledge, regional branding, mutual support and interdependences, producer cooperation, greater social proximity to consumers, their support for local agriculture and an economy supporting local businesses. These meanings are elaborated through cooperative relations of many kinds, most formally via farmers' membership organizations.

Within those agro-food relocalization processes, this paper has focused on two generic issues—efforts to create larger yet proximate intermediaries to expand local markets, and state support measures during the New Labour government (1997–2010). Regional authorities have devised measures favorable to relocalization, though within a contradictory policy framework (see Section 3.1).

Cumbria's efforts have pursued distinctive aims by operating within the tensions of national policy. England's RDP emphasizes greater competitiveness via productive efficiency; likewise Cumbria's RDP favors large-scale 'local' processors, even if their raw materials come from distance sources. At the same time, a regionalization policy has opened up opportunities for other approaches: Cumbria regional agencies also fund agro-food relocalization as a means to link environmental and economic sustainability, as well as food security (though the latter has only modest prospects in this territory, given its scant arable land).

Regional development agencies have provided support of various kinds for a Cumbrian food culture more closely linking producers with each other and with consumers. After the 2001 crisis over foot-and-mouth disease, such agencies offered grants for organic conversion and small-scale processing equipment. Funds also facilitated initial networking and cooperation among producers, a common infrastructure for farmers' markets and territorial branding for a range of quality products. These modest funds were crucial for overcoming the social isolation which had resulted partly from farmers' historical dependence on CAP funds. Training programs and kitchen facilities help small-scale producers to overcome bureaucratic obstacles, e.g., hygiene regulations. Such grants have come from a combination of sources, e.g., Structural Funds and the Leader+ program; the latter has been crucial for collective capacity-building and producer cooperation which can continue after the initial funding [16]. Moreover, the Council's tenders for public procurement effectively favor local food suppliers, thus circumventing UK government policy on 'aggregated purchasing.'

With such help from regional authorities, especially at an early stage of development, producers have shortened food chains by expanding direct sales and establishing new intermediaries for larger markets. Producers gained a higher price via farmers' markets, especially for organic or territorially branded products, although also simply for 'local' food. However, some producers had limited capacity or other difficulties to expand such activities; for example, they lacked skills to deal directly with consumers or to comply with various regulations. Dissatisfied with supermarket chains, some organic producers established collective marketing, especially via box schemes. Also small-scale producers established 'farm shops' selling other produce. These arrangements give a higher price to producers, who maintain their own product identity with consumers. New intermediaries deal with trading and hygiene regulations, apply for grants, provide public information about food and explain food production to consumers.

Those intermediaries had a limited niche market, attracting mainly affluent consumers. Medium-size farms remained dependent upon wholesalers and/or supermarket chains; the latter often incorporating supplies into its own-brand label, thus distancing consumers from the producer. A way forward has been a much larger producer-led intermediary distributing quality food to large buyers. Those initiatives combine several roles—pooling volume in order to increase bargaining power, promoting high-quality distinctive food products, and promoting regional food products—roles which may characterize separate initiatives elsewhere in Europe [15].

In developing more environmentally sustainable production methods and short food chains in Cumbria, modest success has been due to several related factors: producer cooperation, consumer support, and regional agencies orienting rural development to farmers' livelihoods. Agency-farmer partnerships and farmers' cooperation together have enhanced many factors that enable agro-food relocalization [15], thus promoting neo-endogenous rural development through place-shaping concepts [18]. Many initiatives have gained economic independence from their start-up grants, though some still depend on collective infrastructural support, e.g., for farmers' markets. Regional agencies cite UK policy and the RDP England to justify these measures. Yet they undergo tensions with the national policy orientation towards agricultural productivism plus non-agricultural activities, as well as tensions with DEFRA's centralized control over the RDP [32].

In summary, this case study has verified our earlier hypothesis: Sustainability-via-relocalization has depended on more proximate, cooperative relations among producers and with consumers. Together they developed a territorial identity around sustainability, as expressed by various meanings of the terms 'local' or 'Cumbria' food, more than around specialty foods per se. As an extra condition for success, state agencies have provided training in marketing skills and modest funds for equipment or common infrastructure. As the most important support, though somewhat invisible, the Leader program has facilitated cooperative relations among farmers so they can develop mutual trust, learn from each other's skills, develop collective marketing and confidently deal with any difficulties that may arise. These success factors imply a general recommendation for agro-food relocalization: such efforts should promote long-term farmers' cooperation through bottom-up programs such as Leader, with staff experienced in these activities.

Although Cumbria's agro-food relocalization initiatives remain marginal, they resist the trend towards delocalization. In attempting to link social, economic and environmental sustainability, these initiatives also indicate modest means towards food security. Cumbria is readily typified as an industrial-scale food and drink producer, importing many raw materials from elsewhere, rather than a region with territorial characteristics favoring local specialty products. So its advances towards agro-food relocalization indicate potential for expansion elsewhere, including contexts which may seem unfavorable.

REFERENCES

1. Pretty, J. Participatory learning for sustainable agriculture. World Dev. 1995, 23, 1247–1263.
2. Levidow, L. Cleaning up on the farm. Sci. Cult. 1991, 2, 538–568.

3. Peterson, T.R. Sharing the Earth: The Rhetoric of Sustainable Development; University of South Carolina Press: Columbia, SC, USA, 1997.
4. Lang, T.; Heasman, M. Food Wars: Public Health and the Battle for Mouths Minds and Markets; Earthscan: London, UK, 2004.
5. Vanloqueren, G.; Baret, P.V. How agricultural research systems shape a technological regime that develops genetic engineering but locks out agroecological innovations. Res. Pol. 2009, 38, 971–83.
6. Sustainable Food Consumption and Production in a Resource-Constrained World. Available online: http://ec.europa.eu/research/agriculture/scar/foresight_en.htm (accessed on 31 March 2011).
7. Poole, R.; Clarke, G.; Clarke, D. Growth, concentration and regulation in European food retailing. Eur. Urban Reg. Stud. 2002, 9, 167–177.
8. Stichele, M.V.; Young, B. The Abuse of Supermarket Buyer Power in the EU Food Retail Sector; Agribusiness Accountability Initiative (AAI): Amsterdam, The Netherlands; March; 2009.
9. Watts, D.C.H.; Ilbery, B.; Maye, D. Making reconnections in agro-food geography: Alternative systems of food provision. Prog. Hum. Geog. 2005, 29, 22–40.
10. O'Connor, D.; Renting, H.; Gorman, M.; Kinsella, J. The Evolution of Rural Development in Europe and the Role of EU Policy. In Driving Rural Development: Policy and Practice in Seven EU Countries; O'Connor, D., Renting, H., Eds.; Royal Van Gorcum: Assen, The Netherlands, 2006; pp. 145–166.
11. Renting, H.; Marsden, T.; Banks, J. Understanding alternative food networks: Exploring the role of short food supply chains in rural development. Environ. Plann. A 2003, 35, 393–411.
12. Allen, P.; FitzSimmons, M.; Goodman, M.; Warner, K. Shifting plates in the agrifood landscape: The tectonics of alternative agrifood initiatives in California. J. Rural Stud. 2003, 19, 61–75.
13. Ilbery, B.; Morris, C.; Buller, H.; Maye, D.; Kneafsey, M. Product, process and place: An examination of food marketing and labelling schemes in Europe and North America. Eur. Urban Reg. Stud. 2005, 12, 116–132.
14. Feagan, R. The place of food: Mapping out the 'local' in local food systems. Prog. Hum. Geog. 2007, 31, 23–42.
15. Karner, S., Ed.; Alternative Agro-Food Networks (FAAN): Stakeholders' Perspectives on Research Needs. Local Food Systems in Europe: Case Studies from Five Countries and What They Imply for Policy and Practice, 2010. Available online: http://www.faanweb.eu (accessed on 31 January 2011).
16. Knickel, K.; Zerger, C.; Jahn, G.; Renting, H. Limiting and enabling factors of collective farmers' marketing initiatives: Results of a comparative analysis of the situation and trends in 10 European countries. J. Hunger Environ. Nutr. 2008, 3, 247–269.
17. Ray, C. Introduction to special issue. Sociol. Ruralis 2000, 40, 163–171.
18. Shucksmith, M. Disintegrated rural development? Neo-endogenous rural development, planning and place-shaping in diffused power contexts. Sociol. Ruralis 2010, 50, 1–14.
19. European Technology Platform Plants for the Future: Strategic Research Agenda 2025. Part II; EPSO: Brussels, Belgium, 2007.
20. Becoteps. The European Bioeconomy in 2030: Delivering Sustainable Growth by Addressing the Grand Societal Challenges, 2011. Available online: http://www.europabio.org/EU%20 Projects/bioeconomy-2030-becoteps-final-white-paper.pdf (accessed on 1 March 2011).

21. European food declaration: Towards a healthy, sustainable, fair and mutually supportive Common Agriculture and Food policy, European Platform for Food Sovereignty. 2009. Available online: http://www.epfs.eu/uploads/campaign/09%2012%2005%20European%20 food%20declaration.pdf (accessed on 31 January 2011).

22. Boussard, J.-M.; Trouvé, A. Proposal for a new European agriculture and food policy that meets the challenges of this century. 2010. Available online: http://www.europeanfood-declaration.org (accessed on 31 January 2011).

23. Agricultural and Rural Convention. A Communication from Civil Society to the European Union Institutions on the future Agricultural and Rural Policy. 2010. Available online: http://www.arc2020.eu (accessed on 1 March 2011).

24. Royal Society. Reaping the Benefits: Science and the Sustainable Intensification of Global Agriculture 2009.

25. UK Food Security Assessment: Our Approach; Department for Environment, Food and Rural Affairs (DEFRA): London, UK; August; 2009.

26. UK Food Security Assessment: Detailed Analysis; Department for Environment, Food and Rural Affairs (DEFRA): London, UK, August 2009.

27. Foresight: The Future of Food and Farming: Executive Summary; The Government Office for Science: London, UK, 2011.

28. Department for Environment Food Rural Affairs (DEFRA). Agricultural revolution needed to fight food shortages. Available online: http://www.defra.gov.uk/news/2011/01/24/food-shortages/ (accessed on 24 January 2011).

29. Winter, M. The policy impact of the foot and mouth epidemic. Polit. Quart. 2003, 74, 47–56.

30. Ilbery, B.; Kneafsey, M. Producer constructions of quality in regional speciality food production: a case study from South-west England. J. Rural Stud. 2000, 16, 217–230.

31. Marsden, T.K.; Smith, E. Ecological entrepreneurship: sustainable development in local communities through quality food production and local branding. Geoforum 2005, 36, 440–451.

32. Marsden, T.K.; Sonnino, R. Rural development and the regional state: Denying multifunctional agriculture in the UK. J. Rural Stud. 2008, 24, 422–431.

33. Morris, C.; Buller, H. The local food sector: A preliminary assessment of its form and impact in Gloucestershire. British Food J. 2003, 105, 559–566.

34. Ilbery, B.; Maye, D. Food supply chains and sustainability: Evidence from specialist food producers in the Scottish/English borders. Land Use Policy 2005, 22, 331–344.

35. Smith, E.; Marsden, T.K. Exploring the limits to growth in UK organics: Beyond the statistical image. J. Rural Stud. 2004, 20, 345–357.

36. Seyfang, G. Ecological citizenship and sustainable consumption: Examining local organic food networks. J. Rural Stud. 2006, 22, 383–395.

37. Farmers' Markets in the UK: Nine Years and Counting; FARMA: Southampton, UK, 2006.

38. Smith, A. Green niches in sustainable development: The case of organic food in the United Kingdom. Environ. Plann. C: Government and Policy 2006, 24, 439–458.

39. Venn, L.; Kneafsey, M.; Holloway, L.; Cox, R.; Dowler, E.; Tuomainen, H. Researching European 'alternative' food networks: Some methodological considerations. Area 2006, 38, 248–258.

40. Farming and Food: A Sustainable Future; Policy Commission on Farming and Food, Cabinet Office: London, UK, 2002.

41. Sustainable Farming and Food Strategy: Forward Look; Department for Environment, Food and Rural Affairs (DEFRA): London, UK, 2006.

42. Food Matters 1 Year on: Progress over the Last 12 Months; Department for Environment, Food and Rural Affairs (DEFRA): London, UK, 2009.

43. Council Regulation No 1698/2005 of 20 September 2005 on support for rural development by the European Agricultural Fund for Rural Development (EAFRD). Official Jnl. European Union 2005, 277, 1–40.

44. Marsden, T.K.; Smith, E. Ecological entrepreneurship: Sustainable development in local communities through quality food production and local branding. Geoforum 2005, 36, 440–451.

45. Directive 2004/18/EC of the European Parliament and of the Council of 31 March 2004 on the coordination of procedures for the award of public works contracts, public supply contracts and public service contracts. Official J. L. 2004, 134, 114–156.

46. Buying Green: A Handbook on Environmental Public Procurement; Commission of the European Communities: Brussels, Belgium, 2004. Available online: http://ec.europa.eu/ environment/gpp/pdf/buying_green_handbook_en.pdf (accessed on 31 January 2011).

47. Gershon, P. Releasing Resources to the Front Line: Independent Review of Public Sector Efficiency; Crown Copyright: London, UK, 2004.

48. Morgan, K.; Sonnino, R. Empowering consumers: The creative procurement of school meals in Italy and the UK. Intl. J. Consum. Stud. 2006, 31, 19–25.

49. Archer, G.P.; Sánchez, J.G.; Vignali, G.; Chaillot, A. Latent consumers' attitudes to farmers' markets in North West England. British Food J. 2003, 105, 487–497.

50. Youngs, J. A study of farm outlets in North West England. British Food J. 2003, 105, 531–541.

51. Northwest Regional Development Agency (NWDA). More resources to grow rural economy. 315 Magazine 2008, 15, 20–21.

52. About Cumbria Organics. 2009. Available online: http://www.cumbriaorganics.org (accessed on 31 January 2011).

53. Interview Cumbria Organics, 26 August 2008

54. Interview Cumbria Farmers Network, 15 August 08.

55. Food Northwest. Fulfilling the Potential of Local Food in England's Northwest, 2008, Available online: http://www.foodnw.co.uk/downloads/food_nw_local_food_supply_ chain_action_plan.pdf (accessed on 1 March 2011).

56. Food Northwest. NW Food Lovers' Guide; 2010.

57. Rural Development Programme for England, Implementation Plan for England's Northwest 2007–2013; Northwest Regional Development Agency (NWDA): London, UK, 2008.

58. Cumbria Fells and Dales Local Action Group, Development Strategy, Rural Development Programme for England. 2008. Available online: http://www.cumbria.gov.uk/business/ rdpe/fellsanddales/fellsanddales.asp (accessed on 1 March 2011).

59. Interview Cumbria Fells & Dales Leader, 12 March 2010

60. Interview Cumbria Fells and Dales Leader, 28 April 2009.

61. A Guide to Doing Business with Cumbria County Council; Cumbria County Council (CCC): Carlisle, UK, 2009. Available online: http://www.cumbria.gov.uk/elibrary/ Content/Internet/536/654/1085/1087/3817013821.pdf (accessed on 31 March 2011).

62. Interview Cumbria Fells and Dales Leader 08 May 2009

63. Food Technology Centre Team Support Small Businesses for County Contracts; Cumbria Rural Enterprise Agency (CREA): Kendal, Cumbria, 2008. Available online: http://www.crea.co.uk/news/county_contracts.htm(accessed on 31 March 2011).

64. Register of Procurement Contracts; Cumbria County Council (CCC): Carlisle, UK, 2009. Available online: http://www.cumbria.gov.uk/scprocurement/contractsregister/contractsregister.asp (accessed on 1 March 2010).

65. Cumbria Awarded for Green Procurement Excellence; Cumbria County Council (CCC): Carlisle, UK, 2009. Available online: http://www.cumbria.gov.uk/scprocurement/awards.asp(accessed on 31 January 2011).

66. Interview Hadrian Organics, 14 August 2008.

67. Interview, Howbarrow Farm, 15 August 2008.

68. Interview Low Sizergh Barn 13 August 2008.

69. According to several interviewees.

70. Interview Growing Well, 03 September 2008.

71. Interview Cumbria Fells and Dales Leader, 18 January 2011.

72. Brown, H.; Geldard, J. Supplying Local Food to Mainstream Customers; Westley Consulting Limited: Ashley, UK, 2008.

73. Interview Cumbria Fells and Dales Leader, 28 February 2011.

74. Cumbria Local Enterprise Partnership; Cumbria Partners: Carlisle, UK, 2010.

75. Farming in the Uplands; Third Report of Session 2010–11, HC 556. House of Commons, Environment, Food and Rural Affairs (EFRA) Committee: London, UK, 2011.

CHAPTER 6

Plant Breeding for Local Food Systems: A Contextual Review of End-Use Selection for Small Grains and Dry Beans in Western Washington

Brook O. Brouwer, Kevin M. Murphy, and Stephen S. Jones

6.1 INTRODUCTION

In the USA, the alternative agriculture movement has arisen in response to per-ceived failures of industrial agriculture associated with intensification of human land-use (Foley et al., 2005). The alternative agriculture movement, which has been described as 'new,' 'sustainable,' 'regenerative' and 'agroecological' agriculture, seeks to develop production models and food systems which sup-port local economies, reduce impacts on the environment, deliver high quality food directly to consumers and strengthen food sovereignty (Hamilton, 1996; Feenstra, 1997; Kloppenburg et al., 2000; Pearson, 2007; Gliessman, 2012; Holt-Giménez and Altieri, 2012). The complexity of goals associated with alter-native agriculture was addressed by Kloppenburg et al. (2000) who worked with a group of 'competent, ordinary people' to expand the definition of a sustainable food system, concluding, '[i]t is through honoring and understanding the multiple dimensions of motivation and intent that people bring to the transformation project [of food system sustainability] that it can actually be brought to fruition.' The importance of sustainable production practices that increase productivity and maintain ecosystem services has also been recognized, including the potential contributions of plant-breeding programs (Tilman et al., 2002).

As part of the alternative agriculture movement, Hamilton (1996) called for increased transparency in consumer food markets by accompanying food products with, '… not just traditional information as seen with the nutritional labels, but other information as well, such as who raised it, how was it raised, what is in it, what variety is it and what is its history?' This is in direct contrast to the dominant model of food production in which commodities are fetishized by the obfuscation of the underlying components of production (Marx, 1906). As Allen and Kovach (2000) point out, 'When consumers consider a strawberry in the market, … they tend to regard its value as a function of its material characteristics, not as a function of processes that include backbreaking labor, soil erosion, and public research investment.' This is particularly true in the case of staple commodity crops, such as wheat, which are often distanced from the end-user by additional layers of processing, such as milling and baking (Hills et al., 2013b).

6.2   DEFETISHIZATION OF COMMODITY FOOD

Commodity defetishization is a process of revealing the underlying social connections and production impacts of food as described by Allen and Kovach (2000) in the context of organic agriculture. Another example of this process is punk cuisine in which, 'Mainstream American food, with its labor and natural components cooked beyond recognition, is countered with the raw and rotten foods of punks; foods that are ideally natural, home grown, stolen, discarded, and uncommodified' (Clark, 2004). In punk culture, food can be cleansed of its commodity character by dumpster diving on the one hand or growing it yourself on the other hand. This approach to food consumption fosters a direct connection to food production and the waste stream, which is absent from the commodity food distribution system.

Celebrity chefs may also facilitate a defetishization of food by celebrating the stories of food producers and the exceptional quality of products. This has been demonstrated by the success of farm-to-table restaurants such as Chez Panisse, which pioneered a cuisine, 'based on the finest and freshest seasonal ingredients that are produced sustainably and locally' (Chez Panisse).

In his recent book, *The Third Plate*, chef Dan Barber advocates for developing regional cuisines. Such an approach would bring attention to crops which are important for the functioning of the whole agricultural system, not just the high value cuts of meat and flashy vegetables (Barber, 2014). Through such efforts chefs are able to directly engage consumers with description and sensory experience of food placed in a regional context.

6.3 LOCAL AGRICULTURE, LOCAL-WASH AND TERROIR

Local food has arisen as a central theme of the alternative agricultural movement and is championed as a means to build connection between producers and consumers (Kloppenburg et al., 1996; Feenstra, 1997; Wilkins, 2005; Ostrom, 2006; Adams and Salois, 2010). In the USA, the success of local agriculture is reflected in the expansion of direct marketing to consumers. For example, the number of farmers markets increased 63% between the years 1994 and 2000 (Payne, 2002) and the value of direct-market sales increased 105% between 1997 and 2007 (Martinez et al., 2010). However, as has been discussed in regards to the industrialization of organic production (Buck et al., 1997; Allen and Kovach, 2000), this success has also opened the messages of local agriculture to co-option (DeLind, 2011).

The term 'local-wash' was recently introduced to represent the disconnects between usage of the term local in advertising and objectives of the alternative food movement (Cleveland et al., 2015). Local food is often defined in terms of a specific distance or political boundary. Food system scholars have pointed out the limitations of such strict spatial definitions. Cleveland and his co-authors (Cleveland et al., 2015) cite cases where the assumed goals of local agriculture are contradicted, or not defensibly linked, to the reality of local food production. Consumer, and supply chain intermediary, perception of local food has been found to be flexible, depending on the product in question and location of individuals purchasing the food (Selfa and Qazi, 2005; Hills et al., 2013b; Carroll and Fahy, 2015). Despite these challenges, local agriculture has potential to contribute to a defetishization of food systems through the strengthening of regional food culture.

The concept of terroir links food quality directly to a sense of place. In the European Union, terroir is supported by the legally defined Protected

Designation of Origin which can include products, 'whose quality or characteristics are essentially or exclusively due to a particular geographical environment with its inherent natural and human factors' (European Union, 2012). In an evaluation of landrace bean varieties, researchers in France found end-use quality was influenced by interactions between the effects of environment and bean variety providing direct support for the notion of terroir (Florez et al., 2009). Similar concepts of terroir are developing in the USA. In an exploration of artisan cheese production in Wisconsin, Bowen and Master (2014) noted that strengthening the alternative food movement will come from a deeper connection to values of 'heritage and territory.'

In order to reground the local food movement, DeLind (2011) argues that we need to connect local food with 'the art of place-making,' and 'find ways to keep them both connected and vital, ways to marry poetry and science.' While organic farmers, chefs and punk activists have all taken on roles in the alternative agriculture movement, plant breeders are underutilized in achieving the goals of local agriculture.

6.4 PLANT BREEDING FOR LOCAL FOOD SYSTEMS

Plant breeding is often described as the art and science of crop improvement, and more recently as a business (Bernardo, 2010). Plant breeding is contextualized by the boundaries that define the selection of new crop varieties, which in many ways reflect boundaries of local food systems. These include biological boundaries enforced by precipitation regimes, day length, temperature extremes and pest pressure, as well as cultural boundaries such as available infrastructure and consumer preference. Acknowledging the complexity of interactions between these factors will require a reassessment of the plant-breeding process and the interpretation of genotype × environment interaction (Desclaux et al., 2008). Public plant breeders may be subject to political boundaries as universities are beholden to regional stakeholders and taxpayers, which may influence variety selection priorities (Dawson and Goldberger, 2008).

In the past decade, there has been a decline in regional public plant-breeding programs, as well as consolidation of the private seed industry and the use of increasingly restrictive forms of intellectual property protection (Frey, 1996; Guner and Wehner, 2003; Stuber and Hancock, 2007; Pardey et al., 2013). This commodification of plant breeding and the seed system, in general, is viewed as being at odds with the goals of the alternative agriculture movement (Kloppenburg, 2005). In contrast, there has also been increased interest in the development of varieties specifically for organic production (Lammerts van Bueren and Myers, 2012). This has been the focus of the recently established Student Organic Seed Symposium and may stimulate a revival of interest in public plant breeding at Land Grant Universities (Luby et al., 2013). Selection for organic agriculture is targeted to a production system, while potentially complimentary; it does not inherently address the needs of local food production.

To improve understanding of how plant breeding can be incorporated into efforts to directly support local agriculture and facilitate a defetishization of commodity food production, we review four approaches to crop end-use selection: (1) Centralized; (2) Consumer-based; (3) Farmer-based; and (4) Participatory. These models will be discussed within the context of selecting barley (Hordeum vulgare), wheat (Triticum aestivum) and dry beans (Phaseolus vulgaris) for Western Washington.

6.5   SMALL GRAINS AND DRY BEANS IN WESTERN WASHINGTON

Washington State is divided physically and culturally by the Cascade mountain range, creating two geographically distinct regions. Western Washington has a marine climate and dense urban population centers. The current food culture and production system found in the 19 counties located west of the Cascades have evolved out of an early recognition of the productive potential of the region as well as consistent access to local markets. In 1916, a regional farm paper featured an article calling for the building of more farm communities in the region, proclaiming, 'The time is coming when the Puget Sound basin will be one of the richest most intensely cultivated and densely populated sections of the agricultural world … supply[ing] the daily food necessities for the industrial centers and commercial cities … for the reason that location cannot be changed' (Shomaker, 1916).

Today the agricultural economy in Western Washington is characterized by high-value products, such as dairy, berries, fresh market vegetables, vegetable seed, floriculture, nursery-stock and shellfish aquaculture produced on diverse small- to medium-size farms. According to the USDA 2012 Census of Agriculture, the annual farm-gate value of agricultural products produced on the 16,345 farms in the 19 counties west of the Cascades is US$1.44 billion (WSDA, 2014). There are 113 farmers markets registered with the Washington State Farmers Market Association, 80% of which are located in Western Washington (Washington State Farmers Market Association, 2014), reflecting regional interest in local food.

Despite this interest there remains a gap between production and consumption of key staple crops, including legumes and small grains (American Farmland Trust, 2012). Recent efforts to re-localize staple crop production have focused primarily on the potential of adding value by overcoming barriers to local milling and baking (Hills et al., 2013a, b). Increasing the value of these crops is critical because small grains have a relatively low economic value yet they remain an integral part of regional production systems. Primarily grown in rotation with higher value crops, small grains can help break disease and weed cycles, maintain soil organic matter and diversify income sources.

In addition to adding value to existing rotational crops, there is a need to introduce new crops to local farming systems. In 2010, a regional frozen pea industry was lost when the last processing facility left Western Washington (McMoran, 2009). This effectively eliminated a major legume crop from regional cropping systems. Dry beans have been identified as a potential crop for Western Washington, which may help fill this gap (Wagner et al., 2006). However, barriers to adoption include access to equipment, and availability of regionally adapted varieties (Brouwer et al., 2014). These are barriers which plant breeding can help to overcome.

The maintenance and incorporation of cereal and legume crops as a part of Western Washington cropping systems will help maintain and increase crop system diversity while providing a local supply of staple crops. High land prices and a shrinking agricultural land base due to demand for residential and commercial development have placed increasing pressure on agriculture in Western Washington (Canty et al., 2012). Increasing the value of grains and legumes within the rotation system is one way to maintain economically viable farms capable of withstanding this pressure. Appropriate equipment, soil management and development of infrastructure and markets are critical components of a local food system. Here we focus on the largely overlooked potential of plant breeding to support such a local agricultural system.

6.6   MODEL 1. CENTRALIZED END-USE SELECTION: STRIVING FOR LOCATION-NEUTRALITY

Centralized plant breeding relies on clearly defined and easily measured selection criteria to achieve rapid genetic gains toward a goal. This approach has been compared with an accelerated form of evolution (Cleveland and Soleri, 2007). In selection, breeders must define a reasonable region of adaptation (Ceccarelli, 1989); however, there is an economic incentive for breeding programs to select for broadly adapted varieties, because this will increase the potential acreage and market of these varieties (Atlin et al., 2001).

Broad-adaptation, including both agronomic and end-use quality, is an important criterion for breeding commodity crops, which must meet specific quality parameters, even when grown across large geographic regions. End-use selection in wheat exemplifies how centralized plant breeding works. Wheat is the most widely grown crop (in terms of global land area) and must meet well-established quality standards for the concentrated milling and baking industries (USDA-FAS, 2014). In a review of wheat breeding for end-use quality, Bushuk (1998) wrote, 'The quality of protein in future wheat cultivars will have to be location-neutral.' This approach to breeding separates production, place and culture, creating a placeless commodity product.

Breeding barley for consistent malting quality is another case where specific end-use parameters have shaped breeding strategies and the resulting crop diversity. Within the USA barley is primarily used for animal feed and brewing purposes with a small percentage utilized for food (Baik and Ullrich, 2008). The malting and brewing industry is highly concentrated with the top ten malting companies accounting for 44% of global malt production (Punda and Prikhodko, 2009). Two international brewers control over 210 distinct brands (Kenney, 2013) and the top four firms control over 40% of global sales (Howard, 2013). To ensure that malting barley is of consistent quality for such large-scale processing, new varieties must meet American Malting Barley Association (AMBA) standards prior to entering production. Guidelines for barley breeders, developed by AMBA, include 18 quality parameters such as kernel size, activity of starch degrading enzymes and extractable sugar. These guidelines conclude with the additional note that 'malted barley must provide desired beer flavor' (AMBA, 2014). This raises the question of who is defining beer flavor and how might it be influencing the breeding process.

Historically AMBA guidelines have been divided into two categories, two-row and six-row, referring to the morphology of the barley spike. In 2014, an additional category was included for all-malt brewing, a style of beer production that utilizes malted barley as the sole source of starch. The Craft Brewers Association also published a white paper advocating the development of varieties better suited to the needs of the craft brewing industry, which typically relies on all-malt brewing methods (Brewers Association, 2014). The authors note that, 'as a result of the recent varietal progression ... the sensory profiles of their flagship brands have evolved over time, drifting towards lower overall flavor impression and/or complexity.' This request for a diversification of end-use quality is of particular relevance to Western Washington where there are over 150 licensed breweries (Washington State Liquor Control Board, 2014). Even if there is an increase in selection of barley varieties for all-malt brewing, meeting a nationally defined quality parameter does not promote regional differentiation essential for supporting local food systems.

To meet strictly defined malt quality parameters, barley breeders have relied on conservative approaches to variety development such as advanced cycle breeding, in which elite lines within a breeding program are reused as parents for developing new varieties. Although this approach has resulted in gains in agronomic performance of North American barley while continuing to meet malting quality targets (Rasmusson and Phillips, 1997), there has been a reduction in genetic diversity within both six- and two-row germplasm (Martin et al., 1991; Horsley et al., 1995; Matus and Hayes, 2002; Condón et al., 2008). This reduction in diversity may increase vulnerability to biological and abiotic stress (Condón et al., 2008), and limit options for end-users interested in alternative styles of production. A centralized breeding process generates a self-fulfilling cycle in which selection criteria are established, the genetic diversity of breeding programs is reduced to meet these criteria, which reduces the opportunity for identifying novel end-use qualities, and industry standards are not challenged to expand; thus the selection cycle is repeated.

In major bean-producing regions of the USA, consolidation of dry bean processing has driven a restructuring of bean production and variety selection to meet industry needs (Bingen and Siyengo, 2002). Canning quality is the primary end-use parameter for dry beans and is the focus of centralized breeding efforts (Heil et al., 1992). Bean breeding efforts are primarily focused on major market classes such as pinto, black, navy and kidney, though programs will occasionally release non-commodity colored patterned types. Introducing beans to Western Washington will require not only breeding for local agronomic adaptation, but also selecting varieties with unique patterns desirable in local markets (Wagner et al., 2006). Given the lack of regional processing facilities, and consumer demand for bulk dry beans, canning quality is not an important quality parameter for Western Washington.

Breeding for increased yield of crops grown in Western Washington, while selecting for a commodity end-use, such as large-scale malt production in the case of barley or canning in the case of beans, does not serve to achieve the goals of a robust local food system. Relative to other parts of the state or country where lower land values, established processing and transport infrastructure allow for production on a larger scale, Western Washington growers are placed at an economic disadvantage when they enter the commodity market. This model of centralized breeding for commodity production and end-use uniformity may be very effective in terms of rapid genetic gain; however, it adds relatively little to the local food economy and culture.

In an early critique of public research devoted to meeting specific commodity food processing needs, Hightower (Hightower, 1972) wrote:

"But mechanization means more than machinery for planting, thinning, weeding and harvesting. It also means improving on nature's design—breeding new food varieties that are better adapted to mechanical harvesting. Having built machines, the land grant research teams found it necessary to build a tomato that is hard enough to survive the grip of mechanical 'fingers'..."

Hightower's critique shows how centralized variety selection specifically for industrial production and processing is perceived by the public as being out of touch with the cultural and culinary aspects of agriculture. Heirlooms are a common response to this model of varieties bred for mass production, storage and processing.

6.7   MODEL 2. CONSUMER-BASED END-USE SELECTION: THE HEIRLOOM DILEMMA

Generally defined as a crop grown in a region for over 50 years, heirlooms are often strongly connected to a local culture or even the history of a single family (DeMuth, 1998). The concept of heirloom as presented here is similar to 'conservation variety' which was added to European seed legislation in 2008 (Bocci, 2009). The term heirloom is used in the present paper because of its role in US food markets, while recognizing the complexity of defining varieties (Chable et al., 2008). Concerns regarding loss of agro-biodiversity stimulated the development of citizen movements in Europe to provide a framework for the production and marketing of locally adapted crop varieties (Bocci and Chable, 2009).

In the USA, heirloom tomatoes have emerged as an icon in the local food movement (Jordan, 2007) and recently heirloom apples have been highlighted as an emerging culinary trend (National Restaurant Association, 2014). In the case of apples, local agricultural markets have been credited with maintaining a wider diversity of varieties (Goland and Bauer, 2004). Interest in heirlooms is not limited to high-value fruits and vegetables.

Efforts to bring uniqueness to staple crops have included the reintroduction of heirloom cultivars, which are being promoted for their unique flavor, quality profiles and cultural significance. 'Sonora' wheat has been celebrated in the re-localization of wheat in Arizona (Morris et al., 2013). This variety is featured in the Arc of Taste, an initiative of the Slow Food movement with the goal of 'collecting small-scale quality productions that belong to the cultures, history and traditions of the entire planet' (Slow Food International). Similarly, the 'Red Fife' presidium was established in Canada to, 'relaunch Red Fife by introducing it to artisan bakeries' (Slow Food International). While heirloom varieties offer an avenue to connect food to a place and tradition, they are subject to fetishization by removal from the geographic and historic context that gives them agronomic, cultural and culinary value.

'Scots Bere' is a barley landrace which was historically cultivated and milled into flour in parts of the UK. It is still cultivated and milled on a small scale in parts of the UK and used in the production of special bannock bread (SASA). Scots Bere also features in the history of early settlement in the Pacific Northwest and was likely one of the first barley varieties introduced to the region (Scheuerman, 2013). Trials conducted in 2012 at the Washington State University Northwestern Washington Research and Extension Center in Western Washington showed that the yield of Scots Bere was 33% of the highest yielding variety, and Scots Bere was the lowest yielding variety in the trial (data available online at: http://plantbreeding.wsu.edu/Organic-Spring-Barley-2012.pdf, note that Scots Bere was designated 'Tolmie Bere' for this trial). Similarly, organic trials conducted in Eastern Washington between 2004 and 2006 showed that Red Fife and Sonora yielded 17 and 67%, respectively, as much grain as the highest yielding modern wheat included in the trial (Murphy et al., 2008a, b). If the yield of heirloom varieties is well below half the yield of modern varieties, are consumers prepared to pay twice as much for these crops?

Though infrequent, heirloom varieties optimally suited to specific farming systems or environments that have been ignored by modern breeding programs can be utilized by growers. For example, the wheat 'Canus' (released in 1934) had grain yields higher than or comparable with modern varieties when grown on three organic farms in Western Washington (Murphy et al., 2011). In addition to desirable agronomic qualities, Canus had high concentrations of Ca, Cu, Fe, Mg, Mn, P and Zn in the seed, indicating its potential usefulness in local markets as a value-added and nutritionally dense variety.

The yield potential of heirloom dry beans has also been investigated. Variety trials conducted in Michigan found that heirloom bean varieties were not agronomically competitive, even when grown in low input organic production systems (Heilig and Kelly, 2012). This is contrary to trials conducted in Western Washington, which have identified heirloom beans with agronomic qualities comparable with modern varieties (Brouwer et al., 2014). These findings may reflect the lack of an active dry bean breeding effort in Western Washington. Similarly, the wheat variety Canus, described above, was evaluated in comparison with modern spring wheat varieties specifically selected for production in Eastern Washington, prior to active wheat breeding efforts in Western Washington.

Even when heirloom varieties are identified with desirable end-use qualities and agronomic performance, the question remains, can end-use qualities be improved through selection? Reluctance to improve the culinary value of heirlooms may be due to public perception that plant breeding is detrimental (or even considered by some to be actively dangerous, as in the case of genetic engineering) to food quality (Bredahl, 2001). In a study of European consumers, Guerrero et al. documented the perception that, 'the application of innovations may damage the traditional character of Traditional Food Products' (Guerrero et al., 2009). In the USA, plant breeding has been critiqued in the popular press as a form of innovation which has damaged the character of crops, in particular reducing their nutritional quality (Robinson, 2013).

Given the generally low-yield potential, promoting the commercial production of heirloom varieties has the potential to be actively detrimental to local food systems. An exception to this may be when varieties have been actively selected and maintained by farmers within a region, such as is occurring with the peasant seed movement in Europe (Bocci and Chable, 2009).

6.8 MODEL 3. FARMER-BASED END-USE SELECTION: DEVELOPMENT OF FOLK VARIETIES

Much of the current global agro-biodiversity has developed through farmer-based maintenance of landraces and the active selection of 'folk varieties' (Berg, 2009). Farmer-based selection is a process in which varieties are actively maintained or bred by a farmer or community without outside assistance from formally trained plant breeders (Berg, 2009).

Generally, farmer-based selection is limited in regions, or in crops, where there is a well-established variety selection and dissemination system. In Europe, the development and marketing of farmer-selected varieties is actively restricted by regulations of variety release and propagation (Bocci and Chable, 2009). Despite these challenges there is active farmer-based conservation and selection of crops. A combined ethnobotanical and population genetics study of a diverse wheat population in France found that farmers maintained and selected diverse subpopulations from a single population-variety, consistent with reports of seed diffusion and propagation (Thomas et al., 2012).

While the establishment of conservation varieties (as addressed in Model 2) may offer increased access to diverse, unique and locally adapted germplasm, the peasant seed movement acknowledges the evolutionary process of crop variety development and challenges the strict definitions of breeding to include farmer-based and participatory methods (Chable et al., 2008; Bocci and Chable, 2009).

In Tibet, barley is central to agriculture and cuisine, and specific varieties are prized for the preparation of common dishes. 'Lhazi Ziqingke,' a purple seeded barley is preferred for preparation of Chang, a fermented drink, and the pale colored 'Garsha' is preferred for the preparation of Tsangpa, a roasted barley flour (Tashi et al., 2012). In the highland region of Ethiopia, Shewayrga and Sopade (2011) identified 15 barley landraces and 20 main barley foods and drinks associated with those varieties. These distinct landraces and colored grains are examples of selection for 'perceptual distinctiveness' which enables farmers and consumers to quickly associate specific varieties with agronomic, and end-use qualities (Gibson, 2009). In local food economies perceptual distinctiveness helps farmers rapidly differentiate their products and build connects between specific varieties and consumers.

The incentive for farmers to select varieties may be greater in Western Washington because the distance from major seed distributors and breeding programs located in Eastern Washington makes obtaining well-adapted varieties relatively difficult. While barley seed is commercially available, the desire to maintain and improve locally adapted varieties has resulted in some Western Washington farmers maintaining their own varietal blends, which may eventually develop into distinct farmer selected varieties (Sam McCullough, pers. comm.). In this case, selection is an informal process in which planting a mixture of pure line varieties allows for undirected outcrossing to occur at low rates between distinct genotypes. Over time natural selection and consistent replanting of the highest quality seed may lead to gradual shifts in the genetic composition of this population as has been reported in studies of evolutionary breeding (Murphy et al., 2005).

While such long-term farmer-based selection can lead to the development of varieties with specific agronomic and end-use qualities, increasing the rate of genetic gain requires the introduction of novel genetic material through seed exchange (Ellen and Platten, 2011; Pautasso et al., 2013). In Western Washington a handful of organizations such as Olympia Seed Exchange, (http://www.olympiaseedexchange.org/), Lopez Community Land Trust Seed Library (http://www.lopezclt.org/seed-security-initiative-and-seed-library/) and the Orcas Island Seed Library (http://www.orcaslibrary.org/seeds.html) facilitate the stewardship and exchange of seed.

In a description of informal seed exchanges in the UK, Ellen and Platten (2011) describe the accidental 'leakage' of germplasm from research stations to producers. Such a transfer of germplasm from researchers to farmers is a potentially valuable source of new genetic diversity for farmer-based selection. This leakage could be deliberately facilitated through direct collaboration between plant breeders and farmers. Additionally, the use of directed crosses, particularly in self-pollinated species, has the potential to greatly increase genetic gain; this may be carried out by farmers or through a participatory process as described below.

6.9   MODEL 4. PARTICIPATORY END-USE SELECTION: FARMER–BREEDER–CHEF COLLABORATION

Murphy et al. (2005) describe an evolutionary participatory breeding approach to selection for low-input systems in which on-farm selection is supplemented with segregating populations generated by university researchers. Such direct farmer research collaborations may be particularly effective for improving agronomic qualities; however, additional perspectives are necessary for the selection of complex end-use qualities such as baking and brewing. The farmer–breeder–chef collaborative approach seeks to leverage the agronomic skills and knowledge of the farmer, the culinary perspective of the chef and the logistical skills of the breeder to develop varieties with local adaptation and end-use potential.

Cultural and culinary attributes of crop varieties have been recognized as important aspects of farmer adoption of new varieties in developing countries (Morris and Bellon, 2004). In the context of low-input systems in developed countries, the need to achieve acceptable end-use quality is generally addressed by initiating participatory breeding projects with high-quality parents, due to the difficulty of actively selecting for end-use quality early in a breeding program (Murphy et al., 2005). Integrating end-use selection fully into participatory plant breeding will require a recognition of the complexity of interactions between genotype, bio-physical environment, crop management as well as social components of actors, outlets, regulations and societal dynamics as described by Desclaux et al. (2008).

In parts of the USA, farmer–breeder–chef collaborations have emerged as a potential means of improving end-use qualities. The mission of the Culinary Breeding Network, located in Portland, Oregon is: 'to bridge the gap between breeder and eaters to improve agricultural and culinary quality in the Northwest' (Culinary Breeding Network). This project has contributed to the selection and subsequent market success of several new pepper varieties (Waterbury, 2013), demonstrating how engaging the culinary community in the selection process can facilitate rapid adoption of new varieties.

Rivière et al. (2015) describe a participatory breeding program in France that worked to improve productivity of organic bread wheat. In this project, the original parental variety selections were made by a farmer–baker and hybridizations of these parental varieties were performed on-farm. Researchers facilitated quality and genetic analysis of the breeding populations. Early generation selections were made by participating farmers based on agronomic quality (Rivière et al., 2015). In supplementary materials, the authors note that meeting specific end-use parameters were not critical because craft bakers within the community could work with diverse wheat varieties to create marketable bread. In this case, the need for end-use quality was balanced by increased agronomic performance and the creative skills of bakers embedded in a local food system. Surveys of bakers in Western Washington indicated that consistency was a major concern regarding the sourcing of local flour (Hills et al., 2013a). A more direct engagement with bakers during the variety selection process could help overcome this barrier and facilitate the adoption of new locally grown varieties.

The development of malting barley in the Czech Republic reflects an approach to regional plant breeding that is inspired by local culture. Czech plant breeders have effectively increased the agronomic quality of barley and improved certain aspects of malting quality, while still maintaining the specific malting attributes necessary for production of distinctively Czech style beer (Kosar et al., 2004). This process of selecting for specific malting quality characteristics is similar in ways to the breeding guidelines developed by AMBA (2014). However, the Czech model promotes a regional style of barley, which is currently lacking from malting barley selection in the USA. Following a similar approach, Western Washington brewers could facilitate the selection of unique regional varieties.

Italian researchers undertook a project to improve the agronomic qualities of an Italian bean landrace (Almirall et al., 2010). This work resulted in higher yield potential and upright plant architecture desirable for mechanical harvest. Advanced breeding lines were also selected for decreased cooking time, and reduced vulnerability to splitting while maintaining desirable flavor and appearance characteristics. Western Washington heirloom dry beans are a potential genetic resource which could be improved in a similar manner. In these examples, heirloom varieties may serve as a genetic resource for developing a new crop variety that meets local culinary standards. However, the introduction of parental germplasm from other regions and food traditions may be necessary to facilitate genetic advances in agronomic and even end-use quality.

Plant breeding can also expand culinary traditions within a region. In Ethiopia, collaborative efforts between farmers and breeders, including end-use quality evaluation resulted in the release of barley particularly well adapted to the production of injera, a regionally important staple food (Abraha et al., 2013). Prior to the initiation of this breeding project, barley was generally considered inferior to tef (Eragrostis tef) for the production of injera. Through engaging farmers and cooks in the process of selection the breeders were able to develop a variety that expanded the culinary potential of barley. By combining agronomic improvements with such deliberate selection for end-use quality participatory breeders can add to the productivity and diversity of regional cropping systems and cuisine.

In Washington, Oregon and Idaho, a farmer–breeder–chef collaboration is currently underway to select high-yielding and flavorful hull-less, food grade barley with high β-glucan content. As most barley varieties used for food are high-yielding hulled varieties, nutritional compounds found in the pericarp are removed during the pearling process and this product is no longer considered to be a 'whole-grain.' In this farmer–breeder–chef participatory breeding project, organic farmers select among elite hull-less varieties for farming system appropriate agronomic traits of interest (resistance to lodging and disease, early maturity, plant height, weed suppression ability, etc.), breeders select for high yields and high β-glucan content, and chefs select for taste, texture, cooking and baking quality, and unique flavors. As with the Ethiopian example this approach has the potential to add new culinary options to local food systems, while increasing crop productivity.

Participatory wheat breeding and variety selection projects have emerged out of grain re-localization movements across the USA. On-farm selection and agronomic evaluation of wheat populations is underway in Vermont (Darby et al., 2013) as is active evaluation of end-use quality in collaboration with local bakeries (George, n.d.). Within Western Washington, the goal of the Bread Laboratory, located at the Washington State University, Northwestern Washington Research and Extension Center is to, '... combine science, art, curiosity, and innovation to explore ways of using regionally available grains to move the craft of whole grain bread baking and other grain usage forward.' This is achieved through a collaborative approach involving farmers, bakers, millers, maltsters and chefs in the process of variety development and selection. The success of this approach to plant breeding will be put to the test in the near future as varieties of wheat, barley and oats bred specifically for Western Washington are released by the program.

Numerous arrangements have been made in the development of participatory plant-breeding projects (Li et al., 2014). A fundamental aspect of this process is the recruitment of stakeholders early in the process to define goals and develop selection criteria. Selection for end-use quality may be achieved through nutritional or functional testing, and/or organoleptic sensory analysis as described by Vindras-Fouillet and co-authors in the evaluation of wheat varieties (Vindras-Fouillet et al., 2014). Given the potential of information to alter preference (Kihlberg et al., 2005) selection methodologies which can isolate perceived value are critical for making genetic gains in end-use quality.

The logistical challenges and cost of evaluating organoleptic quality on a large scale will require end-use selection to occur at advanced stages of testing. The initiation of a breeding program with parental types that have desirable end-use qualities, or heirlooms with specific cultural significance, will help facilitate the development of a varieties suited to regional end-use. As was noted above in the discussion of Model 2, the selection for culinary quality must be balanced with necessary agronomic qualities. For efficiency, selection may be carried out by specific participants at different stages, for example the farmer may identify lines with suitable agronomic qualities which can then be evaluated by the chef for culinary acceptability. Broader consumer participation in selection could also be facilitated through public events (Vindras-Fouillet et al., 2014). Through this participatory process the role of the breeder could evolve from identifying suitable parent lines to generating diversity through crossing and facilitating selection at different stages of the process.

6.10   DISCUSSION

Because plant breeding operates at the intersections of production, processing and consumption, it has the potential to support local food systems through the development of regionally and culturally appropriate varieties. Achieving this may require approaching the development of end-use criteria in a participatory manner as described in Model 4. In the case of Western Washington, participatory plant breeding has the potential to contribute to ongoing efforts to add value to small grains and dry beans.

The presence of over 150 breweries in Western Washington (Washington State Liquor Control Board, 2014) and the recent establishment of Skagit Valley Malting, a craft malting facility, suggest the potential for adding value to local barley. In particular, the relevance of varieties with non-commodity quality profiles is demonstrated by Skagit Valley Malting successfully utilizing the recently released variety 'Alba,' which did not meet the standards of malting quality established by AMBA (Graebner et al., 2014). Future regional breeding efforts can place a high priority on agronomic performance and disease resistance, while engaging the creativity of end-users in the development of suitable varieties.

The selection of dry beans in Western Washington reflects the complexity of developing varieties which are profitable for producers as well as accessible to the broader community. Efforts to promote consumption of beans in regional classrooms are underway and could represent one target for community engagement (Atterberry et al., 2014). There is also potential to select for a high-end culinary market, as local restaurants have already recognized the value of regional heirloom varieties (Leson, 2013). To meet the goals of food sovereignty associated with local agriculture, quality evaluation could be broadened to include low-income members of the community. Such an approach could also help bring the benefits of public research to 'multicultural and disadvantaged farmers' (Ostrom et al., 2010).

Through a participatory approach, plant breeders can leverage the technical and artistic skill of the culinary community. In doing so, quality selection can be brought out of the laboratory, and laid out for '... public participation in prototyping ... around a public forum like a dinner table' as proposed by Carruth (2012) in a discussion of art–science collaboration. Increasing public engagement may also serve to reinvigorate support for classical plant-breeding education and research (Gepts and Hancock, 2006). This is critical as the number of public programs training plant breeders has been in decline (Frey, 1996; Guner and Wehner, 2003; Stuber and Hancock, 2007) and plant breeding for local food systems will require a much greater number of regionally based plant breeders and associated seed distribution systems.

In the absence of robust public funding alternative sources of revenue will be necessary to support public plant breeding. Revenue may need to be generated from a variety of sources, including royalties, end-user fee or funding from private enterprises. As outputs of the public breeding program can be considered a public commons, maintaining fair and open access to public varieties is an important component of contributing to a local food system. Royalties or user fees may put a disproportionate burden on lower income members of the public as has been demonstrated and critiqued in the case of paying for access to public lands (More and Stevens, 2000). In some cases, the cost associated with development of new varieties may need to be defrayed by civically engaged private enterprises such as bakeries or breweries which may benefit directly from the development of new varieties. Li and co-authors (Li et al., 2014) discuss models of plant genetic resource conservation and participatory plant breeding which engage various public and private actors in the selection, production and marketing of varieties. The appropriate model for a given crop and region will reflect the needs and limitations of local producers, consumers and regulatory institutions.

Globally, germplasm is increasingly viewed as a product to be controlled and marketed by public universities and private companies (Kloppenburg, 2005). Within the USA there has been a trend toward increasingly restrictive forms of intellectual property in the plant-breeding industry (Pardey et al., 2013). The Open Source Seed Initiative (OSSI) is one potential model for redefining understanding of germplasm in a way that would support a participatory approach to plant breeding. This initiative revolves around a seed packet pledge which, 'is intended to ensure your freedom to use the seed contained herein in any way you choose, and to make sure those freedoms are enjoyed by all subsequent users' (OSSI). In an early manifestation of this concept, Kloppenburg described 'Biological open source ... [as] ... a plausible and fecund modality ... for the actual repossession of a relatively autonomous space within which practices and ideas with transformative potential can be enacted' (Kloppenburg, 2010). OSSI offers a vision of a protected, yet open exchange of germplasm suitable for plant breeding for local food systems.

6.11   CONCLUSION: A CULTURALLY BASED APPROACH TO PLANT BREEDING

The models presented here represent a range of approaches to variety selection and plant breeding. Among these, participatory plant breeding is strongly positioned to facilitate a greater connection with local culture, strengthen the local food movement and help to resist co-option, through a defetishization of the food system. Participatory plant breeding has the potential to achieve gains in agronomic and end-use quality while honoring local knowledge and expanding culinary possibilities. Achieving these goals will require an expansion, and decentralization, of public breeding institutions, strengthening of regional seed systems and loosening of intellectual property restrictions on germplasm, and it will require plant breeders to pause and ask, what are we missing?

REFERENCES

1. Abraha, A., Uhlen, A.K., Abay, F., Sahlstrøm, S., and Bjørnstad, Å. 2013. Genetic variation in barley enables a high quality Injera, the Ethiopian staple flat bread, comparable to tef. Crop Science 53(5):2040–2050.
2. Adams, D.C. and Salois, M.J. 2010. Local versus organic: A turn in consumer preferences and willingness-to-pay. Renewable Agriculture and Food Systems 25(4):331–341. [CJO Abstract]
3. Allen, P. and Kovach, M. 2000. The capitalist composition of organic: The potential of markets in fulfilling the promise of organic agriculture. Agriculture and Human Values 17(3):221–232.
4. Almirall, A., Bosch, L., Castillo, R.R., del Rivera, A., and Casañas, F. 2010. "Croscat" common bean (Phaseolus vulgaris L.), a prototypical cultivar within the "Tavella Brisa" type. HortScience 45(3):432–433.
5. AMBA. 2014. Malting Barley Breeding Guidelines Ideal Commercial Malt Criteria [Internet]. American Malting Barley Association, Available from: http://ambainc.org/media/AMBA_PDFs/News/Guidelines_for_Breeders.pdf
6. American Farmland Trust. 2012. Planting the Seeds: Moving to More Local Food in Western Washington [Internet]. American Farmland Trust, Seattle, WA. Available from: http://www.farmland.org/documents/AFTPlantingTheSeedsF2.pdf
7. Atlin, G.N., Cooper, M., and Bjørnstad, Å. 2001. A comparison of formal and participatory breeding approaches using selection theory. Euphytica 122(3):463–475.
8. Atterberry, K., Miles, C., Riddle, L.A., Betz, D., and Rueda, J. 2014. School garden-based pulse nutrition and biology education to increase consumption of targeted foods in K-12 students. HortScience 49(9):S164 (Abstract).
9. Baik, B.-K. and Ullrich, S.E. 2008. Barley for food: Characteristics, improvement, and renewed interest. Journal of Cereal Science 48(2):233–242.

10. Barber, D. 2014. The Third Plate: Field Notes on the Future of Food. Penguin Press, New York, NY.
11. Berg, T. 2009. Landraces and folk varieties: A conceptual reappraisal of terminology. Euphytica 166(3):423–430.
12. Bernardo, R. 2010. Breeding for Quantitative Traits in Plants. 2nd ed. Stemma Press, Woodbury, MN.
13. Bingen, J. and Siyengo, A. 2002. Standards and corporate reconstruction in the Michigan dry bean industry. Agriculture and Human Values 19(4):311–323.
14. Bocci, R. 2009. Seed legislation and agrobiodiversity: Conservation varieties. Journal of Agriculture and Environment for International Development 103(1/2):31–49.
15. Bocci, R. and Chable, V. 2009. Peasant seeds in Europe: Stakes and prospects. Journal of Agriculture and Environment for International Development 103(1/2):81–93.
16. Bowen, S. and Master, K.D. 2014. Wisconsin's "Happy Cows"? Articulating heritage and territory as new dimensions of locality. Agriculture and Human Values 31(4):549–562.
17. Bredahl, L. 2001. Determinants of consumer attitudes and purchase intentions with regard to genetically modified food – Results of a cross-national survey. Journal of Consumer Policy 24(1):23–61.
18. Brewers Association. 2014. Malting Barley Characteristics for Craft Brewers. Available from: https://www.brewersassociation.org/best-practices/malt/malting-barley-characteristics/.
19. Brouwer, B., Miles, C., Atterberry, K., and Jones, S. 2014. Overcoming dry bean production constraints in western Washington. Bean Improvement Cooperative Annual Report 57:23–24.
20. Buck, D., Getz, C., and Guthman, J. 1997. From farm to table: The organic vegetable commodity chain of Northern California. Sociologia Ruralis 37(1):3–20.
21. Bushuk, W. 1998. Wheat breeding for end-product use. Euphytica 100(1–3):137–145.
22. Canty, D., Martinsons, A., and Kumar, A. 2012. Losing Ground: Farmland Protection in the Puget Sound Region [Internet]. American Farmland Trust, Seattle, WA. Available from: http://www.farmland.org/documents/AFTLosingGroundReportWeb.pdf
23. Carroll, B.E. and Fahy, F. 2015. Locating the locale of local food: The importance of context, space and social relations. Renewable Agriculture and Food Systems FirstView 1–14.
24. Carruth, A. 2012. The Future of X: Allison Carruth on Art-Science Collaboration [Internet]. Available from: http://www.theatlantic.com/video/archive/2012/08/the-future-of-x-allison-carruth-on-art-science-collaboration/260676/
25. Ceccarelli, S. 1989. Wide adaptation: How wide? Euphytica 40(3):197–205.
26. Chable, V., Goldringer, I., Infante, T.V., Levillain, T., and Lammerts van Bueren, E. 2008. Report on the definitions of varieties in Europe, of local adaptation, and of varieties threatened by genetic erosion [Internet]. Available from: http://prodinra.inra.fr/record/38488
27. Chez Panisse. n.d. Available from: http://www.chezpanisse.com/about/alice-waters/
28. Clark, D. 2004. The raw and the rotten: Punk cuisine. Ethnology 43(1):19–31.
29. Cleveland, D.A. and Soleri, D. 2007. Extending Darwin's analogy: Bridging differences in concepts of selection between farmers, biologists, and plant breeders. Economic Botany 61(2):121–136.
30. Cleveland, D.A., Carruth, A., and Mazaroli, D.N. 2015. Operationalizing local food: Goals, actions, and indicators for alternative food systems. Agriculture and Human Values 32(2):281–297.

31. Condón, F., Gustus, C., Rasmusson, D.C., and Smith, K.P. 2008. Effect of advanced cycle breeding on genetic diversity in barley breeding germplasm. Crop Science 48(3):1027.

32. Culinary Breeding Network. n.d. Available from: http://culinarybreedingnetwork.word-press.com/

33. Darby, H., Monahan, S., Burke, C., Cummings, E., and Harwood, H. 2013. 2013 Vermont On-Farm Spring Wheat Breeding Trials [Internet]. University of Vermont Extension. Available from: http://www.uvm.edu/extension/cropsoil/wp-content/uploads/2013-Spring-Wheat-Breeding-Trials-final.pdf

34. Dawson, J.C. and Goldberger, J.R. 2008. Assessing farmer interest in participatory plant breeding: Who wants to work with scientists? Renewable Agriculture and Food Systems 23(3):177–187. [CJO Abstract]

35. DeLind, L.B. 2011. Are local food and the local food movement taking us where we want to go? Or are we hitching our wagons to the wrong stars? Agriculture and Human Values 28(2):273–283.

36. DeMuth, S. 1998. Vegetables and fruits: a guide to heirloom varieties and community-based stewardship. Introduction. USDA- Agricultural Research Service [Internet] Available from: http://pubs.nal.usda.gov/sites/pubs.nal.usda.gov/files/heirloom.htm

37. Desclaux, D., Nolot, J.M., Chiffoleau, Y., Gozé, E., and Leclerc, C. 2008. Changes in the concept of genotype × environment interactions to fit agriculture diversification and decentralized participatory plant breeding: Pluridisciplinary point of view. Euphytica 163(3):533–546.

38. Ellen, R. and Platten, S. 2011. The social life of seeds: The role of networks of relation-ships in the dispersal and cultural selection of plant germplasm. Journal of the Royal Anthropological Institute 17(3):563–584.

39. European Union. 2012. Regulation (EU) No 1151/2012 of the European Parliament and of the Council of 21 November 2012 on quality schemes for agricultural products and food-stuffs. Official Journal of the European Union [Internet] Available from: http://eur-lex.europa.eu/LexUriServ/LexUriServ.do?uri=OJ:L:2012:343:0001:0029:en:PDF

40. Feenstra, G.W. 1997. Local food systems and sustainable communities. American Journal of Alternative Agriculture 12(1):28–36. [CJO Abstract]

41. Florez, A., Pujolà, M., Valero, J., Centelles, E., Almirall, A., and Casañas, F. 2009. Genetic and environmental effects on chemical composition related to sensory traits in common beans (Phaseolus vulgaris L.). Food Chemistry 113(4):950–956.

42. Foley, J.A., DeFries, R., Asner, G.P., Barford, C., Bonan, G., Carpenter, S.R., Chapin, F.S., Coe, M.T., Daily, G.C., Gibbs, H.K., Helkowski, J.H., Holloway, T., Howard, E.A., Kucharik, C.J., Monfreda, C., Patz, J.A., Prentice, I.C., Ramankutty, N., and Snyder, P.K. 2005. Global consequences of land use. Science 309(5734):570–574. [PubMed]

43. Frey, K. 1996. National Plant Breeding Study-I: Human and Financial Resources Devoted to Plant Breeding Research and Development in the United States in 1994 [Internet]. Iowa Agriculture Home Economics Experiment Station, Ames, IA. Available from: http://www.csrees.usda.gov/nea/plants/pdfs/frey_report.pdf

44. George, R. n.d. Bake Testing of Vermont Grown Wheat [Internet]. Northern Grain Growers Association, Available from: http://northerngraingrowers.org/wp-content/uploads/flour_test_article.pdf

45. Gepts, P. and Hancock, J. 2006. The future of plant breeding. Crop Science 46(4):1630–1634.

46. Gibson, R.W. 2009. A review of perceptual distinctiveness in landraces including an analysis of how its roles have been overlooked in plant breeding for low-input farming systems. Economic Botany 63(3):242–255.

47. Gliessman, S. 2012. Agroecology: Growing the roots of resistance. Agroecology and Sustainable Food Systems 37(1):19–31.

48. Goland, C. and Bauer, S. 2004. When the apple falls close to the tree: Local food systems and the preservation of diversity. Renewable Agriculture and Food Systems 19(4):228–236. [CJO Abstract]

49. Graebner, R.C., Cuesta-Marcos, A., Fisk, S., Brouwer, B.O., Jones, S.S., and Hayes, P.M. 2014. Registration of "Alba" barley. Journal of Plant Registrations 9(1):1–5.

50. Guerrero, L., Guàrdia, M.D., Xicola, J., Verbeke, W., Vanhonacker, F., Zakowska-Biemans, S., Sajdakowska, M., Sulmont-Rossé, C., Issanchou, S., Contel, M., Scalvedi, M.L., Granli, B.S., and Hersleth, M. 2009. Consumer-driven definition of traditional food products and innovation in traditional foods. A qualitative cross-cultural study. Appetite 52(2):345–354. [PubMed]

51. Guner, N. and Wehner, T.C. 2003. Survey of US Land-Grant Universities for training of plant breeding students. Crop Science 43(6):1938–1944.

52. Hamilton, N.D. 1996. Tending the seeds: The emergence of a new agriculture in the United States. Drake Journal of Agricultural Law 1:7–29.

53. Heil, J.R., McCarthy, M.J., and Özilgen, M. 1992. Parameters for predicting canning quality of dry kidney beans. Journal of the Science of Food and Agriculture 60(4):519–523.

54. Heilig, J.A. and Kelly, J.D. 2012. Performance of dry bean genotypes grown under organic and conventional production systems in Michigan. Agronomy Journal 104(5):1485–1492.

55. Hightower, J. 1972. Hard tomatoes, hard times: Failure of the land grant college complex. Society 10(1):10–22.

56. Hills, K.M., Goldberger, J.R., and Jones, S.S. 2013a. Commercial bakers and the relocalization of wheat in western Washington State. Agriculture and Human Values 30(3):1–14.

57. Hills, K.M., Goldberger, J.R., and Jones, S.S. 2013b. Commercial bakers' views on the meaning of "local" wheat and flour in western Washington State. Journal of Agriculture, Food Systems, and Community Development 3(4):13–32.

58. Holt-Giménez, E. and Altieri, M.A. 2012. Agroecology, food sovereignty, and the New Green Revolution. Agroecology and Sustainable Food Systems 37(1):90–102.

59. Horsley, R.D., Schwarz, P.B., and Hammond, J.J. 1995. Genetic diversity in malt quality of North American six-rowed spring barley. Crop Science 35(1):113.

60. Howard, P.H. 2013. Too big to ale? Globalization and consolidation in the beer industry. In Patterson, M. and Hoalst-Pullen, N. (eds). The Geography of Beer: Regions, Environment, and Society. Springer, New York. p. 155–165.

61. Jordan, J.A. 2007. The heirloom tomato as cultural object: Investigating taste and space. Sociologia Ruralis 47(1):20–41.

62. Kenney, C. 2013. Beer Map: Two Giant Brewers, 210 Brands [Internet]. Available from: http://www.npr.org/blogs/money/2013/02/19/172323211/beer-map-two-giant-brewers-210-brands

63. Kihlberg, I., Johansson, L., Langsrud, Ø., and Risvik, E. 2005. Effects of information on liking of bread. Food Quality and Preference 16(1):25–35.

64. Kloppenburg, J. 2005. First the Seed: The Political Economy of Plant Biotechnology. 2nd ed. University of Wisconsin Press, Madison, WI.

65. Kloppenburg, J. 2010. Impeding dispossession, enabling repossession: Biological open source and the recovery of seed sovereignty. Journal of Agrarian Change 10(3):367–388.

66. Kloppenburg, J., Hendrickson, J., and Stevenson, G.W. 1996. Coming in to the foodshed. Agriculture and Human Values 13(3):33–42.

67. Kloppenburg, J., Lezberg, S., De Master, K., Stevenson, G., and Hendrickson, J. 2000. Tasting food, tasting sustainability: Defining the attributes of an alternative food system with competent, ordinary people. Human Organization 59(2):177–186.

68. Kosar, K., Psota, V., and Mikyska, A. 2004. Barley varieties suitable for the production of the Czech-type beer. Czech Journal of Genetics and Plant Breeding 40(4):137–139.

69. Lammerts van Bueren, E.T. and Myers, J.R., (eds). 2012. Organic Crop Breeding. 1st ed. Wiley-Blackwell, Chichester.

70. Leson, N. 2013. The Monachine bean is truly a Pellegrini family heirloom [Internet]. The Seattle Times. Available from: http://seattletimes.com/html/pacificnw/2020743243_pacificptaste21.html

71. Li, J., Lammerts van Bueren, E.T., Leeuwis, C., and Jiggins, J. 2014. Expressing the public value of plant genetic resources by organising novel relationships: The contribution of selected participatory plant breeding and market-based arrangements. Journal of Rural Studies 36:182–196.

72. Luby, C.H., Lyon, A.H., and Shelton, A.C. 2013. A new generation of plant breeders discovers fertile ground in organic agriculture. Sustainability 5(6):2722–2726.

73. Martin, J.M., Blake, T.K., and Hockett, E.A. 1991. Diversity among North American spring barley cultivars based on coefficients of parentage. Crop Science 31(5):1131.

74. Martinez, S., Hand, M., Da Pra, M., Pollack, S., Ralston, K., Smith, T., Vogel, S., Clark, S., Loren, T., Lohr, L., Low, S., and Newman, C. 2010. Local Food Systems: Concepts, Impacts and Issues [Internet]. USDA, Economic Research Service. Available from: http://www.ers.usda.gov/publications/err-economic-research-report/err97.aspx

75. Marx, K. 1906. Capital: A Critique of Political Economy. Charles H. Kerr and Company, Chicago, IL.

76. Matus, I.A. and Hayes, P.M. 2002. Genetic diversity in three groups of barley germplasm assessed by simple sequence repeats. Genome 45(6):1095–1106.

77. McMoran, D. 2009. Skagit County Agriculture Statistics. WSU Skagit Extension, Mount Vernon, WA.

78. More, T. and Stevens, T. 2000. Do user fees exclude low-income people from resource-based recreation? Journal of Leisure Research 32(3):341–357.

79. Morris, M.L. and Bellon, M.R. 2004. Participatory plant breeding research: Opportunities and challenges for the international crop improvement system. Euphytica 136(1):21–35.

80. Morris, N.R., Zimmerman, E., Bianco, M., and Schmidt, C. 2013. White Sonora Wheat: Heritage grains into the local foods mix. In Nabhan, G.P. (ed.). Conservation You Can Taste: Best Practices in Heritage Food Recovery and Successes in Restoring Agricultural Biodiversity Over the Last Quarter Century. University of Arizona Southwest Center, Tucson, AZ. p. 20–23.

81. Murphy, K., Lammer, D., Lyon, S., Carter, B., and Jones, S.S. 2005. Breeding for organic and low-input farming systems: An evolutionary–participatory breeding method for inbred cereal grains. Renewable Agriculture and Food Systems 20(1):48–55. [CJO Abstract]

82. Murphy, K.M., Dawson, J.C., and Jones, S.S. 2008a. Relationship among phenotypic growth traits, yield and weed suppression in spring wheat landraces and modern cultivars. Field Crops Research 105(1–2):107–115.

83. Murphy, K.M., Reeves, P.G., and Jones, S.S. 2008b. Relationship between yield and mineral nutrient concentrations in historical and modern spring wheat cultivars. Euphytica 163(3):381–390.

84. Murphy, K.M., Hoagland, L.A., Yan, L., Colley, M., and Jones, S.S. 2011. Genotype x Environment interactions for mineral concentration in grain of organically grown spring wheat. Agronomy Journal 103(6):1734–1741.

85. National Restaurant Association. 2014. What's Hot Culinary Forecast [Internet]. National Restaurant Association [cited 2014 Oct 13]. Available from: http://www.restaurant.org/News-Research/Research/What-s-Hot

86. OSSI. n.d. Open Source Seed Initiative: Pledge. Available from: http://www.opensource-seedinitiative.org/about/ossi-pledge/

87. Ostrom, M. 2006. Everyday meanings of "local food": Views from home and field. Community Development 37(1):65–78.

88. Ostrom, M., Cha, B., and Flores, M. 2010. Creating access to Land Grant resources for multicultural and disadvantaged farmers. Journal of Agriculture, Food Systems, and Community Development 1(1):89–105.

89. Pardey, P., Koo, B., Drew, J., Horwich, J., and Nottenburg, C. 2013. The evolving landscape of plant varietal rights in the United States, 1930–2008. Nature Biotechnology 31(1):25–29. [PubMed]

90. Pautasso, M., Aistara, G., Barnaud, A., Caillon, S., Clouvel, P., Coomes, O.T., Delêtre, M., Demeulenaere, E., Santis, P.D., Döring, T., Eloy, L., Emperaire, L., Garine, E., Goldringer, I., Jarvis, D., Joly, H.I., Leclerc, C., Louafi, S., Martin, P., Massol, F., McGuire, S., McKey, D., Padoch, C., Soler, C., Thomas, M., and Tramontini, S. 2013. Seed exchange networks for agrobiodiversity conservation. A review. Agronomy for Sustainable Development 33(1):151–175.

91. Payne, T. 2002. US Farmers Markets – 2000 A Study of Emerging Trends. USDA Agriculture Marketing Service, Washington, DC.

92. Pearson, C.J. 2007. Regenerative, semiclosed systems: A priority for Twenty-First-Century agriculture. BioScience 57(5):409–418.

93. Punda, I. and Prikhodko, D. 2009. Agribusiness Handbook: Barley Malt Beer [Internet]. FAO, Rome. Available from: http://www.fao.org/fileadmin/user_upload/tci/docs/AH3_BarleyMaltBeer.pdf

94. Rasmusson, D.C. and Phillips, R.L. 1997. Plant breeding progress and genetic diversity from de novo variation and elevated epistasis. Crop Science 37(2):303.

95. Rivière, P., Goldringer, I., Berthellot, J.-F., Galic, N., Pin, S., De Kochko, P., and Dawson, J.C. 2015. Response to farmer mass selection in early generation progeny of bread wheat landrace crosses. Renewable Agriculture and Food Systems 30(2):190–201. [CJO Abstract]

96. Robinson, J. 2013. Breeding the Nutrition Out of Our Food [Internet]. The New York Times. Available from: http://www.nytimes.com/2013/05/26/opinion/sunday/breeding-the-nutrition-out-of-our-food.html

97. SASA. n.d. Scottish Landraces and Traditional Varieties: Bere Barley [Internet]. Science and Advice for Scottish Agriculture, Available from: http://www.scottishlandraces.org.uk/barley.htm

98. Scheuerman, R.D. 2013. Harvest Heritage: Agricultural Origins and Heirloom Crops of the Pacific Northwest. Washington State University Press, Pullman, WA.

99. Selfa, T. and Qazi, J. 2005. Place, taste, or face-to-face? Understanding producer–consumer networks in "local" food systems in Washington State. Agriculture and Human Values 22(4):451–464.

100. Shewayrga, H. and Sopade, P.A. 2011. Ethnobotany, diverse food uses, claimed health benefits and implications on conservation of barley landraces in North Eastern Ethiopia highlands. Journal of Ethnobiology and Ethnomedicine 7:19. [PubMed]

101. Shomaker, J. 1916. Why not build more farm communities? The Northwest Farmer 1(28):3.

102. Slow Food International. n.d. Slow Food Presidia: Red Fife Wheat [Internet]. Available from: http://www.slowfoodfoundation.com/presidia/details/989/red-fife-wheat#. VF_kQskYp8E

103. Slow Food International. n.d. The Ark of Taste [Internet]. Available from: http://www. slowfoodfoundation.com/ark

104. Stuber, C. and Hancock, J. 2007. Sustaining plant breeding–National Workshop. Crop Science 48(1):25–29.

105. Tashi, N., Yawei, T., and Xingquan, Z. 2012. Food preparation from hulless barley in Tibet [Internet]. In Zhang, G., Li, C., Liu, X. (eds). Advance in Barley Sciences: Proceedings of 11th International Barley Genetics Symposium. Springer Science & Business Media, Zhejiang, China. p. 151–158. Available from: http://books.google.com/books?id=nbNGA AAAQBAJ&pg=PA151&lpg=PA151&dq=Food+Preparation+from+Hulless+Barley+in+ Tibet&source=bl&ots=lw-OW5GbHY&sig=XrkLXzrRwJrinVacYII_8iYSM4Y&hl=en&sa =X&ei=JIrWU8jDNqO1igLBooGACg&ved=0CC4Q6AEwAg#v=onepage&q=Food%20 Preparation%20from%20Hulless%20Barley%20in%20Tibet&f=false

106. Thomas, M., Demeulenaere, E., Dawson, J.C., Khan, A.R., Galic, N., Jouanne-Pin, S., Remoue, C., Bonneuil, C., and Goldringer, I. 2012. On-farm dynamic management of genetic diversity: The impact of seed diffusions and seed saving practices on a population-variety of bread wheat. Evolutionary Applications 5(8):779–795.

107. Tilman, D., Cassman, K.G., Matson, P.A., Naylor, R., and Polasky, S. 2002. Agricultural sustainability and intensive production practices. Nature 418(6898):671–677.

108. USDA-FAS. 2014. World Agricultural Production [Internet]. United States Department of Agriculture Foreign Agricultural Service. Available from: http://apps.fas.usda.gov/psdonline/circulars/production.pdf

109. Vindras-Fouillet, C., Ranke, O., Anglade, J.-P., Taupier-Letage, B., Véronique, C., and Goldringer, I. 2014. Sensory analyses and nutritional qualities of hand-made breads with organic grown wheat bread populations. Food and Nutrition Sciences 5(19):1860–1874.

110. Wagner, J., Miles, C., and Miklas, P. 2006. Evaluating heirloom dry bean varieties as a niche market crop in the maritime northwest. Bean Improvement Cooperative Annual Report 49:129–130.

111. Washington State Farmers Market Association. 2014. Member Market Directory [Internet]. [cited 2014 Nov 20]. Available from: http://www.wafarmersmarkets.com/washington-farmersmarketdirectory.php

112. Washington State Liquor Control Board. 2014. Washington domestic and microbreweries [Internet]. Available from: http://www.liq.wa.gov/records/public-records-index#Licensing

113. Waterbury, M. 2013. Breeding seeds worth saving [Internet]. Portland Monthly Magazine. Available from: http://www.portlandmonthlymag.com/home-and-garden/at-home/articles/breeding-seeds-worth-saving-september-2013

114. Wilkins, J.L. 2005. Eating right here: Moving from consumer to food citizen. Agriculture and Human Values 22(3):269–273.

115. WSDA. 2014. Agriculture – A Cornerstone of Washington's Economy [Internet]. Washington State Department of Agriculture. Available from: http://agr.wa.gov/AgInWa/docs/126-CropMap2014-W.pdf

CHAPTER 7

Local Selling Decisions and the Technical Efficiency of Organic Farms

Luanne Lohr and Timothy Park

7.1 INTRODUCTION

An emerging agricultural marketing issue is the increased emphasis on the promotion of local food systems that reduce "food miles" and transportation costs while offering consumers the benefits of locally grown food. The American Farmland Trust [1] developed a program for the city of San Francisco to support growing, processing, and consumption of local food (food grown within 100 miles of the Golden Gate Bridge) and claimed that closing the gap between local production and local consumption of food can increase profitability for producers. Agricultural policy makers have suggested that switching to local distribution channels such as farmers' markets or direct sales may allow producers to achieve higher margins and increase their incomes [2].

Darby et al. [3] suggested that marketing foods as "locally grown" is an opportunity for farmers to expand their share of consumers' food budgets and their empirical work demonstrates how consumers value locally grown products. Most of this research focuses on the consumer benefits of locally grown food products. Positive impacts of local sales for producers are mentioned but are not linked to the sustainability of the farm operation or quantified to any degree. Born and Purcell [4] are a prominent exception in warning local planners and economic development specialists to avoid the local food trap‖ that assumes local foods inherently offer desirable qualities such as ecological sustainability, better nutrition, food security, freshness and quality.

The Know Your Farmer, Know Your Food‖ initiative builds on the 2008 Farm Bill to strengthen USDA programs promoting local foods and includes plans to enhance direct marketing and farmers' promotion programs, to support local farmers and community food groups, to strengthen rural communities and to promote local eating. The Know Your Farmer, Know Your Food‖ website (http://www.usda.gov/knowyourfarmer) lists opportunities for farm loan programs such as direct and guaranteed ownership loans for beginning farmers and socially disadvantaged groups, farm storage facility loans, value-added producer grants, beginning farmer and rancher development programs, and technical assistance and marketing services for farmers engaged in local selling.

The primary purpose of this paper is to examine the factors that influence earned income of organic farmers explicitly incorporating the farmer's decision to engage in local selling. Survey data on the volume of organic products delivered by the organic farmers to product buyers within 100 miles of the farm is used to define local sales. The econometric model is based on a stochastic frontier approach that recognizes that farmers may not be able to achieve maximum output that is feasible with a given set of inputs. The approach identifies the role model producers who are the most technically efficient along with farm and demographic factors that enhance efficiency. A primary innovation of the research is the inclusion of the local marketing option in the stochastic frontier model.

The empirical work recognizes that organic farmers traditionally have utilized a variety of marketing channels to ensure the sustainability of their operations including direct to consumer sales, direct marketing to grocery retailers and restaurants, and sales through packers, brokers and food processors. They have participated in the rejuvenation of farmers markets and innovations such as community supported agricultural (CSA) operations. Extension experts such as Stephenson and Lev [5] have noted the wide variety of direct marketing outlets and the new outlets that have developed to promote local selling initiatives.

The evidence suggests that organic farmers are flexible in adapting their marketing patterns to enhance the profitability of their farms and are willing to develop the capacity and technical skills to sell their commodities in local markets. The 4th Organic Farming Research Foundation (OFRF) survey showed that the share of organic products marketed directly to consumers reached almost 50 percent in 2001 [6]. An analysis of the factors that influence the decision to sell locally could assist organic farmers in formulating better selling strategies.

The article is organized as follows. The foundations of the stochastic frontier approach are outlined and the information obtained from the model is presented. The survey and data used to estimate the model along with empirical results are then discussed. Offermann and Nieburg's [7] analysis of organic farms in Europe stressed that economic performance, defined as monetary returns to all farm activities related to organic farming, is an appropriate measure to assess the sustainability and efficiency of organic operations. The paper concludes with a discussion of the implications for producers, marketing specialists, and extension experts when guiding organic producers who are considering decisions to engage or expand their efforts in local selling.

7.2   MODELING EFFICIENCY IN ORGANIC PRODUCTION

Stochastic production frontier models, summarized in comprehensive detail in Kumbhakar and Lovell [8], allow for both technical inefficiency and random shocks that are uncontrolled by producers. Stochastic frontier analysis assumes a composite error term consisting of two random variables. The first element in the composite error, v_i, is a symmetric noise term reflecting random noise which influences farmer decisions and can take on both positive and negative values. The asymmetric inefficiency error term, u_i, accounts for technical and managerial constraints and assumes only nonnegative values. A typical specification for a stochastic frontier model is:

$$\ln y = \ln f\,(x_i, r_i) + v_i - u_i \qquad (1)$$

where y_i represents the observed output (here, earnings from organic production) measure for the ith farm, $f(x_i, r_i)$ is the deterministic frontier with inputs x with farm, demographic, and regional effects denoted by r_i. The v_i are mean zero i.i.d. random variables with a variance of $\Phi 2v$ and are assumed to be independent of u_i and the explanatory variables. Technical inefficiency is represented by the one-sided error term, u_i, following a half-normal distribution with a variance $\Phi 2u$. Techniques for estimating the stochastic frontier by maximum likelihood are presented in Kumbhakar and Lovell [8].

The one-sided error term u_i is interpreted as the distance to the best-practice stochastic frontier represented by $\ln y_i = \ln f(x_i, r_i) + v_i$. The best practice corresponds to the maximum output the organic farmer is expected to produce given inputs that are applied and characteristics of the farm operation and producer that are included as explanatory variables.

The frontier production function for the ith producer is specified using a translog functional form for inputs along with measures of the marketing and environmental constraints facing organic producers. The input variables in the model include full-time labor inputs (FLABR), part-time labor (PLABR), and organic acreage (ACRE). Farm level organizational and environmental factors (r_{ij}) that directly influence production are incorporated into the model. The second-order terms in the translog production frontier in Equation 1 are represented by k(FLABR, PLABR, ACRE) with estimated coefficients of α_{ij}. The logarithm of total value of organic production (PRODVAL) is the dependent variable in the production frontier. The stochastic frontier model is:

$$\ln PRODVAL = \alpha_0 + \alpha_1 \ln(FLABR_i) + \alpha_2 \ln(PLABR_i) + \alpha_3 \ln(ACRE_i) +$$
$$\alpha_{ij} \, k(FLABR_i, PLABR_i, ACRE_i) + \sum_j \gamma_j r_{ij} + v_i + u_i \qquad (2)$$

Technical efficiency is estimated as $TE_i = \exp(-\hat{u}_i)$, which has a value between 0 and 1, with 1 indicating an efficient organic farm and that the producer is located on the frontier. Only the difference between the random error terms $\varepsilon i = vi - ui$ can be observed, requiring a derivation to extract technical efficiency ui from εi. Kumbhakar and Lovell [8] summarize methods to obtain estimates of technical efficiency for each producer by deriving $\hat{u}_i$ from the conditional distribution $E(ui|\varepsilon i)$ which are implemented in LIMDEP ([9]).

The analysis focuses on the impact of short-run adjustments in variable inputs such as the labor inputs and certified acreage on the value of organic production, a model specification which is consistent with the stochastic frontier model in Kurkalova and Carriquiry [10]. A more extensive model could incorporate data on farming and management practices for controlling crop diseases and nematodes, insect pest techniques, and weed control methods. Information on these variables was considered for inclusion as quasi-fixed inputs that proxy for capital and equipment used in the organic operation. The OFRF survey elicited information on adopted practices, derived from responses to three separate questions related to crop disease and nematode management, insect pest controls, and weed controls. In previous modeling efforts, these variables had no statistically significant impacts on the value of production of the organic farmers and are omitted from the analysis. Information on capital used directly in the organic farming operation was not available in the OFRF survey.

7.3 MODELING EFFICIENCY IN ORGANIC PRODUCTION

Analysis on a scale broad enough to accurately reflect the production conditions must be drawn from a national survey that is representative of all organic farmers. Since 1993, the private not-for-profit OFRF has conducted biennial surveys of organic farmers in the U.S. The OFRF queried the entire U.S. certified organic farm population, as identified by organic certifiers in 1997. Lohr and Park [11] have established the representativeness of the data by comparing survey response records with production statistics collected by the U.S. Department of Agriculture. The information on the types of sales outlets used by the farmer is only available from the OFRF survey.

The data combine information on all crops grown organically and all regions in which organic crops are produced in the U.S. Table 7.1 details the descriptions and summary statistics for variables in the model estimated following Equation 1. Natural logs of the variables reported in Table 7.1 were used in estimation where appropriate. The logarithm of total organic farming gross income (PRODVAL) is the dependent variable, consistent with the specification in Fraser and Horrace [12] and Mathijs and Swinnen [13] and many others. Mean gross organic income for the farmers was $146,048.

7.3.1 PRODUCTION INPUTS

The variable production inputs given in Equation 1 are two labor measures and organic acreage, assumed to be entirely under the control of the producer, and both subject to change annually depending on the planned output for that season. The labor input is represented by year-round workers and seasonal farm employees. The majority of organic farm operations (52 percent) relied on both year-round and seasonal workers with 34 percent of farms hiring only seasonal workers and 14 percent using only year-round workers. Employment of both year-round and seasonal farm workers is more closely correlated with farm income than is farm size and the relationship is even stronger for organic farms with higher incomes (over $99,000). Farmers focusing on local sales employ the fewest number of both year-round and seasonal workers.

TABLE 7.1: Variable descriptions and summary statistics (N = 787 farms).

Variable	Description	Mean	Standard Deviation
PRODVAL	Total value of organic production, U.S. dollars (US$)	146,048	607,488
FLABR	Managers and full-time employees	2.6	4.5
PLABR	Part-time employees	5.1	23.6
ACRE	Acreage farmed organically, 1 to 6,000 acres	193.4	680.4
LOCPCT	Organic production sold within 100 miles of the farm operation, percent	57.1	46.3
TRANMIXD	Producer transitioned to organic farming and farms both organic and conventional acreage, 1 if yes	15.8	36.5
FULLPART	Part-time organic farmer, 1 if yes	31.4	46.4
MALE	Producer is male, 1 if yes	79.1	40.6
SOLEFAML	Farm is a sole proprietorship or family-owned, percent of total	92.6	26.1
COOPCRPR	Farm operation is organized as a non-family, partnership, corporation, or cooperative, percent of total	5.0	21.7
WEST	Farm is in SARE Region 1, percent of sample	28.6	45.2
SOUTH	Farm is in SARE Region 3, percent of sample	5.7	23.2
NOREAST	Farm is in SARE Region 4, percent of sample	25.4	49.1
NORCENT	Farm is in SARE Region 2, percent of sample	25.4	43.6

The mean farm size in the sample was 193 acres, with the largest farm in the sample operating at 13,000 acres. Organic farm size is most strongly related to production of field crops with a correlation coefficient of 0.73, followed by vegetable production at 0.21, and fruit, nut and tree crop production at 0.01. Larger farms tend to include field crop and vegetable production.

7.3.2 FARM, DEMOGRAPHIC, AND REGIONAL VARIABLES

Additional explanatory variables in the specification, constituting r_{ij} in the stochastic frontier, are the farm, demographic, and regional factors that influence the production frontier. These are not production inputs, but may alter the inclination and ability of the farmer to respond optimally to production constraints.

The producers indicated the volume of organic products delivered to product buyers within 100 miles of the farm and this information is used to define the producer's use of local sales outlets (LOCPCT). Three main commodity categories were identified based on the production data gathered in the OFRF survey: (1) vegetable, herb, and floriculture products, (2) fruit, nut, and tree products, and (3) grain and field crops. There is no formal definition of local sales but the 100-mile limit is fully consistent with working definitions that are used by the food industry. Whole Foods, the world's largest natural foods retailer, features its Locally Grown promise with a commitment to buying from local producers, particularly those who farm organically and are dedicated to environmentally sustainable agriculture [14]. Produce that has traveled seven or fewer hours by car or truck from the farm to the retail outlet can be labeled "locally grown."

For organic farms engaging in local selling the correlation between organic farm earnings and the hiring of seasonal workers is strongly positive (at 0.82), rising 40 percent relative to the correlation across all organic farms. The positive correlation of earnings with year-round workers is stable and about the same as for all organic farms (0.34). The pattern is in sharp contrast to that observed for farms that do not sell any products in local markets. For these producers earnings are more strongly correlated with year-round workers (0.43) and the relationship between earnings and seasonal workers is quite small (0.16). Organic farms with a local sales emphasis typically employ fewer total workers and hiring rates of both year round and seasonal workers for the local sales farms are smaller compared with other organic producers.

Under the U.S. regulations, farmers may certify as organic less acreage than they farm, leading to parallel organic and conventional systems being managed by the same operator. Only 18 percent of the OFRF respondents reported conducting this type of mixed farming. Farmers who were originally conventional producers but transitioned to organic production accounted for 52 percent of the OFRF respondents, compared with 48 percent who began farming as organic producers. The subset of farmers who transitioned to organic farming but maintained mixed farming operations was about 16 percent of operations. These producers were expected to have more familiarity with extension advisors to locate markets and deal with marketing problems along with an enhanced capability to maintain income levels as they continue to use conventional production techniques.

Farm structure variables for family operated farms (sole proprietorship or family corporations) and corporate organizations reflect the potential flexibility accorded the farmer in making management decisions. Family operated farms offer the greatest management flexibility to the farmer because they involve the least number of other decision makers. Corporations offer the least flexibility and have relatively more demanding financial and reporting requirements. The family farm category accounts for about 92 percent of the organic farms with corporations accounting for the second largest share at 4 percent. Alternative farm structures representing a third, intermediate level of flexibility, including partnerships and property management firms, were grouped in the omitted category. Overall about 32 percent of the organic producers were engaged in farming on a part-time basis.

The USDA has placed increasing emphasis on research, technical and marketing assistance focusing on the management concerns of female farmers. Female farmers comprise 21 percent of U.S. organic farms. Recognizing that more women are choosing to manage their own farms, the Cooperative State Research, Education and Extension Service (CSREES) has stated that a primary policy goal is to provide women farmers with the tools they need to succeed. Female producers tend to sell greater shares of their output in local markets with 67 percent selling to outlets within 100 miles of the farm while 49 percent of males show this level of commitment to local sales.

To assess institutional support and information availability for organic production and marketing systems, we used the four USDA Sustainable Agriculture Research and Education (SARE) regions (see http://www.sare.org/htdocs/sare/about.html for a listing of states in each region). These regions reflect the U.S. government's demarcation for sustainable agriculture extension-research support. A dichotomous variable was created for each region, equal to one if the respondent's farm was in that region, and zero otherwise. In the sample, 29 percent of farmers were in the SARE 1 region (WEST), 40 percent in the SARE 2 region (NORCENT), 6 percent in the SARE 3 region (SOUTH), and 25 percent in the SARE 4 region (NOREAST).

The West region has historically received the strongest institutional support for organic agriculture, being home to two of the nation's oldest organic farm and certifying organizations, California Certified Organic Farmers, and Oregon Tilth. California enacted the first state law to define organic foods in 1982. California and Washington were among the first extension services to conduct outreach and applied research on organic agricultural systems using teams of extension professionals rather than individuals. Thus, the locality-specific research needed for successful organic farming emerged sooner in the West than in the other regions. Estimation results are expected to show higher efficiency in the West region.

7.4   ESTIMATION RESULTS

Coefficient estimates and asymptotic t-ratios for the stochastic frontier shown in Equations 1 and 2 are presented in Table 7.2. Random shocks that influence

TABLE 7.2: Stochastic Frontier Parameter Estimates for Organic Producers (N = 787 farms).

Parameter	Variable	Estimate	T-ratio[a]
Vector α_i—Inputs			
α_0	Constant	10.054*	32.872
	ln(FLABR)	0.248*	1.840
α_2	ln(PLABR)	0.223*	2.313
α_3	ln(ACRE)	−0.041	−0.581
Vector α_{ij}—Interactions			
α_{11}	ln(FLABRSQ)	0.311*	3.963
α_{22}	ln(PLABRSQ)	0.055	0.939
α_{33}	ln(ACRESQ)	0.073*	3.720
α_{12}	ln(FLABR) × ln(PLABR)	−0.125*	−2.722
α_{13}	ln(FLABR) × ln(ACRE)	−0.005	−0.162
α_{23}	ln(PLABR) × ln(ACRE)	0.036	1.607
Vector γ_j—Farm, Demographic, and Regional Variables			
γ_1	LOCPCT	−0.214*	−2.470
γ_2	TRANMIXD	−0.199*	−1.892
γ_3	FULLPART	−0.836*	−9.402
γ_4	MALE	0.174*	1.888
γ_5	SOLEFAML	0.119	0.498
γ_6	COOPCRPR	0.628*	2.193
γ_7	WEST	0.170*	3.050
γ_8	SOUTH	−0.271*	−1.770
γ_9	NOREAST	0.026	0.351
γ_{10}	NORCENT	−0.115*	−2.621
λ	LAMBDA	0.744*	7.798
σ	SIGMA	1.148*	961.506

[a] Asterisk indicates asymptotic t-values with significance at $\sigma = 0.10$ level;

[b] The definition of $\lambda = \sigma_u / \sigma_v$ and $\sigma = (\sigma_u^2 + \sigma_v^2)^{1/2}$

the earnings of organic farmers are assumed to be related to acts of nature such as weather, pest and weed infestations, and yield variability that are endemic to organic production. Zellner, Kmenta, and Dreze [15] noted that when producers maximize expected profit rather than ex post profit, the error terms for inputs can be assumed to be driven by human error and managerial misjudgments which should be unrelated to the error term in the stochastic production function. The restrictions consistent with a Cobb-Douglas functional form which would omit the quadratic and interaction terms for the land and labor inputs were rejected. The calculated χ^2 value of 40.11 exceeds that critical value of χ^2_6 of 12.59 at any reasonable confidence level.

We conducted a Hausman exogeneity test of the LOCSELL measure. Intuitively the random shocks to organic earnings for a farmer would not show a high degree of correlation with the local selling measure. The decision to sell 100% of a commodity through local markets represents a long-term plan of the producer and can be considered as predetermined in the econometric model. The farmer must exhibit knowledge of the local market, develop a business plan, and develop appropriate marketing aids. These decisions require extensive planning and should not be highly influenced by short-term production shocks. Instruments included the marketing aids used by the producer in selling through local outlets (such as farm events and demonstrations, local promotional efforts, featured product samples, and in-store demonstrations), the development, promotion, and sales of value added products on the farm, the age of the producer, and the farmer's experience in growing and marketing organic crops (in years). These measures should be correlated with success in stimulating local sales but are determined prior to the short-term random shocks that influence output and earnings at the farm level. The exogeneity hypothesis for the local selling measure was not rejected. The test statistic confirms that endogeneity is not a significant factor in the specification of the production function as the calculated χ^2 value of 2.28 was below the critical value χ^2_1 at any conventional significance level.

7.4.1 OUTPUT ELASTICITIES AND SCALE MEASURES

Using the estimated model, the coefficient estimates for the natural logs of the inputs are converted to output elasticities. The elasticities are calculated by taking the derivatives of the estimated coefficients with respect to the logarithms of each input measure. The output elasticities measure the change in the value of organic farm production in response to a specified change in the use of an input. The resulting output elasticities indicate that a one percent increase in full-time labor used increases the value of organic production by 0.47 percent, with a smaller impact of 0.26 percent for more part-time labor. The estimated elasticity for full-time labor is statistically higher than the elasticity for part-time labor.

An expansion of the acreage farmed by one percent increases the value of organic production by 0.22 percent. Higher input levels lead to increased organic production value ensuring that the monotonocity condition for the production function is met for each input (full-time and part-time labor and organic acreage). Concavity of the production function is satisfied if the Hessian matrix based on the parameter estimates is negative semidefinite and this condition is satisfied for the model.

Incorporating the interaction terms permits consideration of an overall scale effect on production. The output elasticities for labor and acreage were each evaluated for farms in four acreage size percentiles (0–25th percentile, 26–50th percentile, 51st–5th percentile, and 75th–100th). The sum of the output elasticities is the scale elasticity, which measures how output changes when the producer increases the use of both inputs. The estimated scale elasticity increases with the size of the farm, moving from decreasing returns to scale for smaller farms to increasing returns to scale for the farms with the most organic acreage. Producers with farms exceeding 120 acres show slightly increasing returns to scale. Serra and Goodwin [16] also report increasing returns for organic farmers and are consistent with results from an earlier OFRF survey reported by Lohr and Park [17].

The findings from Table 7.2 show that after controlling for producer and farm characteristics, organic farmers who are involved in local sales achieve lower earnings, an effect which is statistically significant. The significant negative coefficient on the measure of the producer's involvement in local selling indicates that a 1 percent increase in the volume of output sold in local markets is associated with a decline in gross income from organic production of about 21 percent on average. This implies a decrease in organic earnings of about $31,194 dollars for the average organic farm. Our findings identify a negative relationship between an increased emphasis on local sales and earnings by organic farmers. The impact of local sales on net earnings or profitability is not directly measured.

A set of the explanatory variables (TRANMIXD, FULLPART, MALE, and COOPCPR) all generated significant coefficient estimates. The variables that are dichotomous indicators cannot be differentiated to evaluate marginal effects so we use Kennedy's [18] procedure to measure their impacts on the value of organic farm production. Farmers who transitioned to organic farming and maintain mixed farming operations have earned income that is about 18 percent lower than other types of producers, $\exp(-0.199) - 1 = -0.18$. Producers engaged in part-time farming accrue earnings that are about 57 percent lower than full-time organic farmers. This decline is exacerbated when the part-timers decide to engage in local sales as the combined effects of the two variables results in an income decline of about 78 percent.

The model shows that male farmers achieve earnings that are about 19 percent higher than female operators even after controlling for key farm, demographic and regional effects. As they intensify their marketing efforts in local sales outlets, male farmers experience only a slight earnings decline of about 2.3 percent. The earnings decline associated with involvement in local sales is much higher for female farmers compared to males. Experience with organic farming does not appear to be a factor driving the larger earnings decline for female farmers. Using data from the OFRF survey the number of years involved in organic farming is virtually the same (about 12 years) for both the female and male farmers.

The geographic variable estimates compare regional organic production value with the average for all U.S. organic farms. The null hypothesis that the regional effects are jointly equal to zero is rejected at $\alpha = 0.05$ while the positive coefficient for the West region and the negative coefficient for the North Central region are statistically significant. Estimated regional indicator coefficients presented on Table 7.2 were normalized as deviations from the acreage-weighted mean farm coefficient. We use a technique from Krueger and Summers [19] that is implemented to assess the impact of each regional indicator variable directly while avoiding the dummy variable trap. These measures are independent of any arbitrarily chosen base region and provide a clearer interpretation of how the complete set of regional effects influences organic production values.

Farms located in the West region have gross organic production values 17 percent higher than the average U.S. organic farm, consistent with the early and continuing institutional support for organic agriculture in this region. Those farms in the North Central region have gross organic production values about 12 percent below the average U.S. organic farm. This may be due to emphasis on field crop production in the region, which tends to return lower gross income per acre than vegetable and fruit crops.

7.4.2 TECHNICAL EFFICIENCY ASSOCIATED WITH PARTICIPATION IN LOCAL SALES

A test of the importance of technical inefficiency in the model is confirmed by the statistical significance of the estimated δ of 0.744. The ratio of the standard deviation of the producer-specific technical efficiency to the overall standard deviation of the producers' income is 0.356. The interpretation is 35.6 percent of the variation in earnings from organic production across farms is due to technical efficiency.

Table 7.3 summarizes the overall technical efficiency for the stochastic frontier model and an efficiency decomposition using information on participation in local sales. The estimated mean technical efficiency score is 58.7 percent for the entire sample, indicating that the organic farmers are not implementing best-practice production methods to achieve the maximum output levels given the inputs used. The mean output-oriented measure of technical efficiency shows that output could feasibly be increased by 41.3 percent with current input use under current production technology. Using data from the Third Biennial National Organic Farmers' Survey, Lohr and Park [17] estimate a mean technical efficiency score of 78.8 percent. There is some tentative evidence of a decline in technical efficiency of organic producers. The estimates are not directly comparable since the additional information on marketing outlets is only available in the most recent survey conducted by the OFRF and the surveys do not represent panel data with the same set of producers.

Reinhard, Lovell, and Thijssen [20] regarded their calculated mean technical efficiency of 89 percent for Dutch dairy farms as "impressively high." Serra and Goodwin [16] reported a mean efficiency of 94.0 percent for organic crop farms in the Spanish region of Andalucía. Even at this very high level, the efficiency of the organic farms is below that of the conventional farms in the

TABLE 7.3: Efficiency Decomposition for Local Sales Participation (N = 787).

	Observations	Mean	Standard Deviation	Minimum	Maximum
Overall Efficiency	787	58.7	9.4	22.1	83.5
By Percent of Production Sold in Local Markets					
100 percent	361	58.4	8.9	27.2	79.2
Some involvement	157	60.1	8.7	29.1	78.3
0 percent of production	269	58.3	10.1	22.1	83.4
By Percent of Production Sold in Local Markets					
Males					
100 percent	269	58.1	8.8	27.2	79.2
0 percent of production	219	58.8	9.8	24.1	83.5
Females					
100 percent	92	59.2	9.3	33.7	74.9
0 percent of production	41	56.0	11.7	22.1	74.5

sample with estimated scores of 97.0 percent. Estimates of technical efficiency for Finish dairy farms from Kumbhakar, Tsionas and Sipiläinen [21] are approximately 79.6 percent and the organic farms are about 5 percent less efficient than the conventional dairy farms. Organic farming subsidies are prominent in these countries and these payments may induce entry by inefficient farmers.

The technical efficiency estimates suggest that producer involvement in local sales has little impact on observed efficiency on organic farms. Farmers who sell all of their output in local markets have a mean efficiency score of 58.7 percent. This performance is only marginally lower than the score of 60.1 percent attained by organic farmers who sell some but not all of their output in local outlets. This important result suggests that the decision to sell in local markets does not induce distortions in the technical efficiency of organic farmers. The decision to sell in local markets is not a critical factor distorting the performance of organic farmers.

The stochastic frontier model is also useful in providing a clearer view of how gender influences the performance of organic farmers. We examined the technical efficiency of organic farmers who commit to selling all their output in local markets. Table 7.3 shows that the technical efficiency of female farmers is slightly higher than male farmers for this set of producers. Kotcon and Thilmany [22] commented that "...extension programs must be tailored to the unique needs and learning styles of the organic community within appropriate venues to successfully reach growers." The stochastic frontier model shows extension agents and agricultural marketing consultants that female farmers have achieved a level of technical efficiency that matches males but may need additional information and training directed to marketing of organic products.

7.5 CONCLUSIONS

The concept of locally grown food is gaining increasing prominence as a marketing and sales strategy for small farmers and organic producers. Wal-Mart offers local food vendors marketing assistance to supply its flagship stores and Sam's Clubs and presents training sessions to teach prospective suppliers how to win favor with Wal-Mart purchasers. Coley et al. [23] mentioned a diverse set of criteria that can be used to evaluate the impact of local purchasing decisions including implications for biodiversity and landscape, local employment, and fair trade. We focus on a factor that has been neglected in current discussions on local sales by examining how earnings are influenced by marketing strategy, specifically the decision to sell in local markets.

The results provide information for organic farmers to assess how the sustainability of their farm operations is influenced by involvement in local sales. A similar point is made by Gilg and Battershill [24] who emphasized that the sustainability of farm systems can be enhanced by evaluating the economic effects associated with specific marketing practices. Earnings equations that control for producer and farm characteristics reveal that organic farmers who are involved in local sales achieve lower gross farm earnings.

A second key finding is the increasing returns to scale in organic farming based on translog function for full-time labor, part-time labor, and organic acreage. This result might be troubling to family farm proponents. One of the most divisive issues in organic agriculture today is the small vs. large argument, in which it is claimed that increasing farm sizes are hastening the industrialization of the organic sector and undermining the philosophical tenets that distinguish organic from conventional agriculture. The results presented here suggest that economic gains will be realized as farm size increases, creating pressure on organic farmers to expand operations. Protecting the small organic farmer is likely to become a policy issue in the near future.

Overall technical efficiency estimates confirm that organic farmers are not implementing best-practice production methods to achieve the maximum output levels given the inputs used. A useful finding is that producer involvement in local sales has little impact on observed efficiency on organic farms. The decision to sell in local markets is not a critical factor distorting the performance of organic farmers.

Our approach treats all organic producers with a single production function and the specification could serve as a useful starting point for considering more detailed subgroups of producers and specific combinations of marketing outlets. Identifying the most efficient producers and observing and testing their methods are the first steps in developing the information needed to assist less successful farmers. To the extent it is possible to transfer this knowledge and implement new technologies the productivity of the entire organic farm sector can be improved.

The stochastic frontier model is useful in identifying the best performers conditional on a marketing strategy, providing role models who can disseminate successful methods to other producers. Identifying the most efficient producers and observing and adapting their methods are the first steps in developing the information needed to assist new farmers and less successful current farmers. To the extent it is possible to transfer this knowledge and implement new marketing techniques, the productivity of the entire organic production sector can be improved.

The current format of the OFRF survey distinguishes between family and non-family labor but does not indicate how labor is allocated to specific tasks on the operation. Information on labor allocated to tasks such as production, planting and harvesting could be differentiated from work requirements related to post harvesting, processing and marketing tasks in future OFRF surveys.

Additional research is warranted to identify the channel management techniques used by organic producers to maximize returns from local sales efforts. Organic farmers involved in local sales are faced with an array of contracting arrangements and pricing strategies, such as spot markets, short-term forward contracts, or long-term forward pricing. Local marketing channels may feature a variety of pricing strategies, packaging, or service provision requirements. These elements should be investigated to assist organic farmers in maximizing returns when considering local sales initiatives.

In future work the long-term effects of the local sales commitment along with the technical expertise and marketing experience of the organic farmer can be incorporated directly as factors influencing the technical efficiency of the production system through the error term in the stochastic frontier. Key explanatory variables such as local sales involvement, education, and organic farming experience can appear as explanatory variables influencing the technical inefficiency component of the one-sided error term. This is consistent with the approach used by Morrison-Paul, Johnston, and Frengley [25] examining the impact of measures that influence both the technological structure of production and technical efficiency at the farm level.

More advanced models based on the marketing literature could be formulated to understand how farmers develop marketing and sales strategies and allocate their sales across local and distant markets. In future work we are considering methods to examine a distance function approach distinguishing between multiple outputs, such as local and distant sales. Färe and Primont [26] demonstrated that the output distance function is a natural generalization of the production function for multiple outputs. The issue of whether to specify an output distance function or an input distance function will need to be carefully motivated. The output distance function is the appropriate approach for assessing input contributions and substitution patterns of inputs while the input distance function captures output effects and tradeoffs of outputs. These are more advanced issues that can be addressed after an initial assessment of the impact of local sales on the sustainability of the producer's operation.

REFERENCES

1. American Farmland Trust. Shortening the distance from farm gate to dinner plate; Available online: http://www.farmland.org/programs/localfood/farmtofork.asp (accessed on 24 December 2009).
2. Ilbery, B.; Maye, D. Food supply chains and sustainability. Land Use Policy 2005, 22, 331-344.
3. Darby, K.; Batte, M.T.; Ernst, S.; Roe, B. Decomposing local: a conjoint analysis of locally produced foods. Am. J. Ag. Econ. 2008, 90, 476-486.
4. Born, B.; Purcell, M. Avoiding the local trap. J. Plan. Educ. Res. 2006, 26, 195-207.
5. Stephenson, G.; Lev, L. Direct marketing introduction for organic farms. eXtension, 9 December 2009; Available online: http://www.extension.org/article/18376 (accessed on 24 December 2009).
6. Walz, E. Final Results of the Fourth National Organic Farmers' Survey; Organic Farming Research Foundation: Santa Cruz, CA, USA; Available online: http://ofrf.org/publications/ survey.html (accessed on 24 December 2009)
7. Offermann, F.; Nieberg, H. Economic Performance of Organic Farms in Europe. Organic Farming in Europe: Economics and Policy; University of Hohenheim: Stuttgart, Germany, 2001; Vol. 5.
8. Kumbhakar, S.C.; Lovell, C.A.K. Stochastic Frontier Analysis; Cambridge University Press: New York, NY, USA, 2000.
9. Greene, W.H. LIMDEP Version 9.0; Econometric Software: Plainview, NY, USA, 2007.
10. Kurkalova, L.A.; Carriquiry, A. Input- and output-oriented technical efficiency of Ukrainian collective farms, 1989–1992: Bayesian analysis of a stochastic production frontier model. J. Prod. Anal. 2003, 20, 191-211.
11. Lohr, L.; Park, T.A. Choice of insect management portfolios by organic farmers: lessons and comparative analysis. Ecol. Econ. 2002, 43, 87-99.
12. Fraser, I.M.; Horrace, W.C. Technical efficiency of Australian wool production: point5 and confidence interval estimates. J. Prod. Anal. 2003, 20, 169-190.
13. Mathijs, E.; Swinnen, J.F.M. Production organization and efficiency during transition: an empirical analysis of East German agriculture. Rev. Econ. Stat. 2001, 83, 100-107.
14. Whole Foods Market. Locally grown—the whole foods market promise; Available online: http://www.wholefoodsmarket.com/products/locallygrown/index.html (accessed on 24 December 2009).
15. Zellner, A.; Kmenta, J.; Dreze, J. Specification and estimation of Cobb-Douglas production function models. Econometrica 1966, 34, 784-795.
16. Serra, T.; Goodwin, B.K. The efficiency of Spanish arable crop organic farms: a local maximum likelihood approach. J. Prod. Anal. 2009, 31, 113-124.
17. Lohr, L.; Park, T.A. Technical efficiency of U.S. organic farmers: the complementary roles of soil management techniques and farm experience. Ag. Res. Econ. Rev. 2006, 35, 327-338.
18. Kennedy, P.E. Estimation with correctly interpreted dummy variables in semilogarithmic equations. Amer. Econ. Rev. 1981, 71, 801.
19. Krueger, A.B.; Summers, L.H. Efficiency wages and the inter-industry wage structure. Econometrica 1988, 56, 259-293.

20. Reinhard, S.; Lovell, C.A.K.; Thijssen, G. Econometric estimation of technical and environ-mental efficiency: an application to Dutch dairy farms. Am. J. Ag. Econ. 1999, 81, 44-60.

21. Kumbhakar, S.C.; Tsionas, E.G.; Sipiläinen, T. Joint estimation of technology choice and technical efficiency: an application to organic and conventional dairy farming. J. Prod. Anal. 2009, 31, 151-161.

22. Kotcon, J.; Thilmany, D. CSREES Organic Opportunities White Paper; USDA Cooperative State Research, Education and Extension Service: Washington, DC, USA, 2006.

23. Coley, D.; Howard, M.; Winter, M. Local food, food miles and carbon emissions: a compari-son of farm shop and mass distribution approaches. Food Policy 2009, 34, 150-155.

24. Gilg, A.W.; Battershill, M. To what extent can direct selling of farm produce offer a more environmentally friendly type of farming? Some evidence from France. J. Env. Manag. 2000, 60, 195-214.

25. Morrison Paul, C.J.; Johnston, W.E.; Frengley, G.A.G. Efficiency in New Zealand sheep and beef farming: the impacts of regulatory reform. Rev. Econ. Stat. 2000, 82, 325-337.

26. Färe, R.; Primont D. Multi-Output Production and Duality: Theory and Applications; Kluwer Academic Press: Boston, MA, USA, 1995.

PART 4

Organic Food and the Human Element: Consumers & Farmers

CHAPTER 8

Alternative Labeling Programs and Purchasing Behavior Toward Organic Foods: The Case of the Participatory Guarantee Systems in Brazil

Giovanna Sacchi, Vincenzina Caputo, and Rodolfo M. Nayga Jr.

8.1 INTRODUCTION

In contemporary food system analysis, the prior overwhelming interest in issues linked to the globalization, industrialization, and standardization of production processes is now shifting toward new ethical issues that call into question the way in which the food system operates. Alternative Food Networks (AFNs) represent efforts to re-spatialize and re-socialize standardized food production and distribution systems [1,2] and spread new forms of political association and market governance [3]. In particular, AFNs' food production and consumption processes are closely related in spatial terms (i.e., geographic proximity between producers and consumers), economic terms (i.e., a fair price for farmers and an affordable price for consumers because of the intermediaries' elimination), and social terms (i.e., the development of networks based on trust linked to mutual knowledge and to each other's reputation). As such, producers in AFNs grow food in proximity to the people who buy and eat it [1,4,5]. Hence, direct marketing brings farmers and consumers face to face and develops bonds of trust and cooperation [6,7,8].

The increasing global interest in the organic food market runs parallel to the development of AFNs and reflects consumer concerns about environmental issues [9,10,11,12,13,14,15,16] and the mainstream focus on food quality assurance and control [17]. In addition, because of the growing anonymity of the trade in organic products, organic consumers are increasingly criticizing food products that are produced under unknown social conditions. In this regard, several examples illustrate that organic food consumers are willing to pay a price premium to directly support small farmers' initiatives in disadvantaged areas [18] because they view the organic food system as a way to alleviate income inequality.

In this context, regulatory standards and certification models are essential tools guaranteeing the authenticity of organic food products. For instance, the mainstream approach linked to organic certification (the best known third-party certification, regulated in the European Union by EC Regulation 834/2007—28 June 2007) is useful for consumers by providing guarantees regarding production processes and food quality. However, while organic certification programs have contributed to the global expansion of the organic foods market, they have also made organic foods less accessible to small-scale producers and lower-income consumers worldwide, particularly in developing countries [19,20,21,22,23,24,25,26]. For example, the costs associated with third-party certification can constrain small-scale producers from obtaining organic certification and can increase the price of organic products to the detriment of lower-income consumers. In addition, an increasing number of consumers are discontented with the globalization of organic food provision [27]. Consequently, the adoption of alternative quality assurance systems has become an important issue for both producers and consumers.

In an attempt to cope with the costs related to third-party certification adoption, groups of small producers in several countries have begun to rely upon alternative quality assurance systems to differentiate their organic food products. These quality assurance systems are characterized by alternative distribution strategies based on direct marketing that links producers and consumers' demands without third-party intervention. To date, two main alternative guarantee systems, or labeling programs, have appeared: Internal Control Systems

(ICS) and Participatory Guarantee Systems (PGS). These alternative guarantee systems simplify bureaucratic procedures and reduce costs for small producers who are often overwhelmed by the extensive documentation required by third-party certification. They also reduce costs since they do not involve a foreign certification body.

Recently, a number of studies have documented an increasing interest in the issue of alternative certification strategies for organic foods [22,28,29,30,31,32,33,34]. However, none of these studies has focused on consumption behavior patterns. Our study focuses on factors that affect the likelihood of purchasing organic products guaranteed by the PGS program. We first offer an overview of how the PGS labeling program works by referring to one of the oldest networks that has adopted it: the Brazilian Rede Ecovida de Agroecologia, which represents an exemplary case of an AFN in which an alternative organic labeling program has been implemented. We then examine the effects of sociodemographic factors, knowledge about the meaning of PGS labels, and purchasing habits on consumer buying behavior for organic foods that are guaranteed by the PGS labeling program.

This study contributes to the literature on consumer buying behavior toward organic foods in different ways. First, while several studies have investigated the effects of demographics on consumer buying behavior toward organic food certified by third-party bodies [27,35,36,37,38,39,40,41,42,43,44,45,46,47], to the best of our knowledge this is the first study that focuses this issue on organic food guaranteed by the PGS labeling program. In addition, no prior studies have investigated the effects of knowledge about PGS labels on the purchasing habits related to organic PGS products. Consumer buying behavior toward PGS food products may also vary among population subgroups and may depend on purchasing habits. For example, is there any difference in terms of consumer buying behavior across the different consumers who live in rural areas and those who live in urban areas? Does the level of consumers' knowledge about the PGS labeling program affect their buying behavior? Finally, does buying organic products at farmers' markets increase the probability of buying organic PGS products? Knowledge of the relationship between individual characteristics and buying behavior in relation to organic foods guaranteed by the PGS labeling program is useful for the design and implementation of these emerging labeling programs and can be used to tailor information to specific consumer subgroups [47].

The article is organized as follows. The first section discusses organic third-party certification in terms of consumption trends and emerging issues. The following section describes the Brazilian Rede Ecovida de Agroecologia. The third and fourth sections describe the data and the empirical model used to analyze the effects of sociodemographics on consumer buying behavior toward organic foods guaranteed by the PGS scheme. The last two sections discuss the results and conclusions from the study, respectively.

8.2   ORGANIC THIRD-PARTY CERTIFICATION: TRENDS AND EMERGING ISSUES

Originally, organic product conformity was achieved through interpersonal linkages among the stakeholders and a mutual trust relationship based on reputation and geographical proximity. Generally, the advent of market globalization has required steady guarantees to protect the parties involved in organic product trading through the "industrial proofs" defined by Thévenot [48]. This precipitated the rise of standards and certification as a new form of governance in the organic market [49]. This view is supported by Courville [23] of the ISEAL Alliance, the global membership association for sustainability standards, who stated:

> Paradoxically, the regulatory systems that were developed to protect the integrity of organic agriculture including standards setting and conformity assessment systems are now reshaping the organic landscape in ways that threaten many of the values held by the movement that created it. [23] (p. 201)

The organic agriculture movement was born during the last century out of the desire to develop a sustainable, fair, and ecological alternative to the agro-industrial production paradigm. The main strategy was represented by the creation of alternative models of production, distribution, and consumption that focused on a local and cooperative dimension in which all the stakeholders actively participated. Now that organic food has evolved from a small niche to a market segment, some organic activists have raised the issue of whether the organic movement's original principles have been shifting toward procedural standardization [50,51].

In 1980, the International Federation of Organic Agriculture Movements (IFOAM) established basic standards that were adopted worldwide as reference points. The IFOAM basic standards serve "as a guideline on the basis of which national and private standard setting bodies can develop more specific organic standards" [52]. Nevertheless, many developing countries had no regulations for organic production and often, when local standards are issued, they become instruments for control bodies aimed at assuring exports to the European Union and to other Western countries. In general, third-party certification involves independent and officially accredited bodies that are charged with providing independent confirmation that organizations, companies, and farmers adhere to National Organic Standards and norms. Some commentators argue, however, that these rules are based on the standards of the export destination countries [19]. As contended by Sylvander [53], agricultural practices and conditions vary from country to country and may not lend themselves to a priori coding practices. In this context, it is interesting to draw attention to the movements that are currently developing alternatives to standardized organic certification practices and that mirror the real needs and peculiarities of stakeholders and countries. According to IFOAM, alternative ways of certifying organic products can be viewed as tools that allow producers to access the (domestic) market. The most popular alternative labeling programs to third-party certification are the aforementioned ICS and PGS, which IFOAM recognizes and supports.

The mission of the ICS consists of the creation of farmer associations and enterprise networks that voluntarily adhere to common organic production standards. An independent and external certification body then verifies how well the group is functioning or inspects a limited number of randomly selected members. The inspection results, either positive or negative, are then applied to the whole group. The advantage of adopting such a quality assurance model is that it simplifies the certification procedures for small producers, who are mostly unfamiliar with dealing with the documentation required for third-party certification. It is also less expensive compared to the mainstream certification model.

The PGS movement, on the other hand, coordinates its actions toward the establishment of a collective dimension based on a shared understanding of production and distribution principles and on a common agreement of responsibility. PGS incorporates elements of environmental and social education in relation to quality improvement for both producers and consumers. The basic common elements of PGS projects worldwide are: (i) a participatory approach; (ii) social control; (iii) a shared vision and shared responsibility among stakeholders regarding quality, transparency, trust building, and reinforcing mechanisms; and (iv) a non-hierarchical relationship among stakeholders [29,54]. Currently, 113 cases of PGS adoption are recognized in farmer networks, 67 of which are active projects while the remaining 46 are under development. These cases involve 43,280 producers worldwide [55,56]. PGS are mostly used in developing countries such as Brazil, India, and Costa Rica, although several cases also exist in Western countries. Among the most famous networks that have adopted PGS is the Rede Ecovida de Agroecologia (Brazil), Certified Naturally Grown (USA),

Nature et Progrès (France), Keystone Foundation (India), and Organic Farm NZ (New Zealand).

The PGS program makes organic food more affordable to local consumers because of reliance on direct selling and the effect on social control. For instance, Zanasi et al. [31] stated that:

> The role played by social control is of paramount importance in explaining these theoretical assumptions. The more relevant to the community the issue at stake, the higher the level of trust needed. Trust, in turn, is enhanced by social control as a guarantee against dishonest behavior. [31] (p. 57)

These aspects of PGS adoption help reduce intermediaries and transaction costs and also grant a higher share of added value to farmers. In the absence of an alternative procedure for quality assurance, most small disadvantaged producers would not have access to the local market [32,57,58]. Local consumers are also affected in these circumstances because they are unable to purchase organic products. In other words, the main PGS goal is to facilitate the production of organic products by small farmers and to promote local food systems in accordance with organic agriculture principles and production models.

8.3   THE BRAZILIAN REDE ECOVIDA DE AGROECOLOGIA

The Rede Ecovida de Agroecologia (henceforth referred to as "Rede") represents an exemplary case demonstrating the well-structured path that led to the official recognition of PGS within Brazilian national legislation, resulting in the enactment of Law 10.831 of 23 December 2003, regulated by Decree 6323 of 27 December 2007.

Brazilian legislation recognized two formal and one informal guarantee systems for organic quality assurance [59]:

1. Third-party certification, subject to the Conformity Assessment Bodies (Organismos de Avaliação da Conformidade, or OAC);
2. Participatory Guarantee Systems, subject to the Participatory Bodies for Conformity Assessment (Organismos Participativos de Avaliação da Conformidade, or OPAC); and
3. Organizations for Social Control (Organização de Controle Social).

Producers who fall into the first two categories obtain the organic label of the Brazilian System for Evaluating Organic Conformity (SisOrg), identified by the different guarantee system (see Figure 8.1). In the figure below, the first caption

FIGURE 8.1   The Brazilian organic labeling system for organic conformity assessment.

identifies the Participatory Guarantee Systems (in Portuguese *sistema participativo*) while the second caption (*certificação por auditoria*) stands for third-party certification. However, those producers who fall into the third category can only sell their products according to direct marketing strategies.

As of January 2015, the Brazilian Ministry of Agriculture has recognized and authorized 10,719 producers to sell their products as organic. More than half of these producers (6125) are associated with social control organizations and PGS in order to guarantee their products' quality and authenticity (according to the Brazilian national list of organic producers, as of December 2014, 4593 producers are certified by third-party certification bodies, 3096 producers by social control organizations, and 3029 by PGS [60].

Rede has been in operation since the 1980s and has spread throughout three southern states of Brazil (Paraná, Santa Catarina, and Rio Grande do Sul). The birth of the Rede can be associated with the agricultural and environmental movements that arose in the south of Brazil around the 1980s. In response to modernization and with the aim of recovering agriculture's natural foundations several social movements developed such as the Pastoral Land Commission (*Comissão Pastoral da Terra*), the Landless Workers' Movement (*Movimento dos Trabalhadores Rurais Sem Terra*), and the Rural Women Workers' Movement (*Movimento da Mulher Trabalhadora Rural*). Between 1980 and 1990, in parallel with the development of these movements, a series of NGOs concerned about the harmful effects of modern agricultural production appeared in the south of Brazil [30]. The creation of such movements and NGOs led to the Regional Meetings of Alternative Agriculture (*Encontros Regionais de Agricultura Alternativa*), the Brazilian Meetings of Alternative Agriculture (*Encontros Brasileiros de Agricultura Alternativa*), and, contextually, the formalization of production groups and agroecological markets [61].

In the following years, because of the increase in agro-ecological projects, an urgent need arose to improve the organization and refine the message of the network of movements and NGOs. This led to the birth of the Rede and, contextually, to the requirement for quality assurance of the network's products. Nowadays, the network consists of 26 regional groups (nuclei) involving about 180 municipalities, more than 200 farmers' associations and consumer groups, about 100 ecological markets, and 20 NGOs.

The local groups of the Rede represent the primary organizations. Each nucleus consists of a number of family farmers and social actors. These are organized as individuals or grouped into associations and/or cooperatives (e.g., farmers' associations, consumers' cooperatives, processors, small traders, NGOs, and technicians). They currently represent about 3700 operators. As such, the Rede has an extraterritorial nature and coordinates different subnetworks together with local communities. These nuclei mirror the guiding principles of the Rede, which are related to the conservation, maintenance, and diffusion of cultural diversity.

The network is based on a number of principles that include: (i) the implementation of agro-ecology as a basis for sustainable development; (ii) the preservation of typical local or regional products; (iii) the strengthening of popular economic solidarity; (iv) a direct relationship with consumers; and (v) the supply of local and regional products within the framework of food security and food sovereignty [61]. Figure 8.2 shows the Rede and the distribution of its nuclei.

The certification model realized within the Rede can be characterized as participatory because it is developed within the network through an interactive process according to a model of distributed social control. Each group of the Rede must conform to established measures and instructions to obtain its participatory certification. First, an Ethics Council for each operating nucleus must be set up, composed of representatives of each category of actors involved in the Rede. Then, all the production units of the nucleus must fill out a certification request form containing information about the production process. The forms are then sent to the Ethics Council, which analyzes them and requests additional information if necessary. Subsequently, a number of visits (inspections) are made that equals the number of producers that have applied for certification. The Ethics Council then produces a report approving or rejecting the certification for each production unit. If the producers are eligible, the Ethics Council grants them the use of the participatory certification label. If the council rejects the certification request, the rejection report may contain process suggestions and modifications to ensure compliance [62,63].

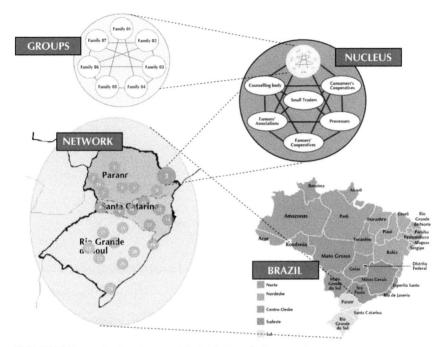

FIGURE 8.2   Nuclei distribution of the Rede Ecovida de Agrecologia.

8.4   DATA

The data used in this study are drawn from responses to a survey instrument administered between January and March 2011 in three nuclei of the Rede: Litoral Solidário, Litoral Catarinense, and Sudoeste do Paraná. Four nuclei of the Rede operate in the areas of analysis (see Figure 8.3).

A sample of 230 adult food shoppers was randomly selected in different cities such as Curitiba, Forianópolis, Francisco Beltrão, Porto Alegre, Torres, and Tubarão (note that the sample distribution over cities is: Curitiba = 12.6%; Forianópolis = 17.4%; Francisco Beltrão = 13.5%; Porto Alegre = 21.7%; Torres = 17.4%; and Tubarão = 17.4%). In order to capture different consumer segments, in each city the interviews were conducted during a two-week period and at different shopping hours. In addition, in line with previous studies [64,65], the interviews were conducted in four different locations, representing different types of food markets such as:

FIGURE 8.3   Cities and nuclei of the Rede Ecovida de Agrecologia considered in the survey.

1. Organic farmers' markets, where both PGS products and non-certified organic foods are sold (Feira Ecológica Lagoa do Violão–Torres; Menino Deus–Porto Alegre; Feira de Produtos Orgânicos do Jardim Botânico and Passeio Público–Curitiba; Feira Agroecológica da Lagoa da Conceição–Florianópolis; Mercado do Produtor–Tubarão; and Feira Agroecológica–Francisco Beltrão;
2. Specialized stores and municipal markets, where both PGS products and third-party certified organic foods are sold (Mercado Municipal de Curitiba–Setor de Orgânicos; Mercado Público Central de Porto Alegre; and Mercado Público de Florianópolis.); and
3. Generic supermarkets, where PGS products are rarely sold.

These market locations were selected so that we could assess the differences in the knowledge and awareness levels of the meaning and function of the PGS labeling program among organic food consumers and examine the degree to which these differences could affect consumer buying behavior. The sample distribution over shopping venue is: supermarket = 11%; municipal market = 11%; specialized stores = 32%; and organic farmers markets = 46%. An overview of the sample technical characteristics is displayed in Table 8.1.

The questionnaire was designed so that we could analyze the effect of sociodemographic factors and purchasing habits on consumer buying behavior toward organic food products guaranteed by PGS. Accordingly, it included two sections.

TABLE 8.1: Sample technical characteristics.

SAMPLE CHARACTERISTICS	Consumers who are responsible for household food purchase (at least 18 years old)
VENUES	Organic farmers' markets; Specialized organic stores; Generic supermarkets.
STATES AND CITIES INVOLVED	RIO GRANDE DO SUL: Torres, Porto Alegre; SANTA CATARINA: Tubarão, Florianópolis; PARANÁ: Francisco Beltrão, Curitiba.
NUCLEI ANALYSED	*Litoral Solidário:*
	20 farmers' groups associated with producers' cooperatives;
	One NGO, the Centra Ecológico Ipê, which offers technical assistance;
	319 families organized in several associations;
	Two consumers' cooperatives that allow daily access to organic food;
	40 suppliers;
	Five organic farmers' markets.
	Litoral Catarinense:
	58 family farmers organized into 11 farmer groups;
	One processor;
	One specialized shop in the city of Florianópolis;
	Three technical consultancy groups;
	One research group;
	Four organic farmers' markets
	Sudoeste do Paraná:
	150 family farmers divided into 15 farmers' associations;
	Two NGOs (Assesoar and Capa);
	One group of researchers and scholars that supports the nucleus action;
	One specialized store;
	Eight organic farmers' markets.
SAMPLE SIZE	230 interviews
SAMPLING	Non-probabilistic sampling
TIMING	January-March 2011

In the first section, we assessed respondents' subjective knowledge about the PGS labeling program by asking if they were aware of its meaning and function. Next, we provided respondents with neutral information about the meaning of the PGS labeling program. Finally, respondents were asked about their shopping habits toward organic foods guaranteed by PGS (e.g., buying behavior, frequency of purchasing, place of purchase, and consumer knowledge about the PGS labeling program). The second section of the survey contained questions on respondents' sociodemographic characteristics (i.e., gender, age, education, residential area, the presence of children within the family, and professional status).

8.5   EMPIRICAL MODEL AND VARIABLE DEFINITION

To investigate the effects of sociodemographic and purchasing habits information on consumer purchase behavior toward organic foods guaranteed by PGS, respondents were asked whether they buy such foods. Given the fact that the dependent variable is represented by a binary response, the model is estimated using a probit model. The probit model is a statistical model based on the cumulative normal probability distribution. As shown by Greene [66], in the binary probit model the probability of choosing an alternative over another one, given a set of factors, can be expressed as:

$$p_i = Prob[Y = 1 \mid X] = \int_{-\infty}^{x_i'\beta} (2\pi)^{-1/2} \exp\left(-\frac{t^2}{2}\right) dt = \phi(x_i'\beta) \qquad (1)$$

where p_i is the probability of consumer i choosing Y; ϕ represents the cumulative distribution of a standard normal random variable; x_i' are the set of independent variables for consumer i; and β is the parameter estimates for the independent variables.

The relationship between a specific variable and the outcome of the probability is interpreted through the marginal effect, which accounts for the partial change in the probability. The marginal effect associated with dummy independent variables on the probability $P(Y_i = 1|X)$, holding the other variables constant, can be derived as:

$$\Delta = \phi(\bar{x}\beta, d = 1) - \phi(\bar{x}\beta, d = 0) \qquad (2)$$

where d is the indicator variable for binary variables in the model. For continuous independent variables, the marginal effect can be derived as:

$$\frac{\partial p_i}{\partial x_{ik}} = \phi(x_i'\beta)\beta_k \qquad (3)$$

In this study, the marginal effects were calculated for each variable while holding the other variables constant at their mean samples. The Maximum Likelihood Estimation (MLE) technique was used to estimate the probit model parameters.

Table 8.2 reports the definition of the variables used in the probit analysis and their mean values.

TABLE 8.2: Description and means of the dependent and independent variables used in the probit analyses.

Variable type	Name	Description	Mean (st. Error)
DEPENDENT-PURCHASING BEHAVIOUR			
Organic- PGS product preference	ORG-PGS	1 if respondent states that he or she has bought organic PGS food products; 0 otherwise	0.643 (0.033)
INDEPENDENT-DEMOGRAPHICS			
Gender	FEM	1 if respondent is female; 0 if male	0.683 (0.030)
Age	AGE	Age (in years)	42.462 (0.858)
Education	EDU	1 if respondent has a degree from high school or lower	0.573 (0.033)
Rural	RUR	1 if respondent lives in a rural area; 0 otherwise	0.141 (0.023)
Suburban	SUB-URB	1= if respondent lives in a sub-urban area; 0 otherwise	0.339 (0.031)
Child	CH	1 if child(ren) is/are present in household; 0 otherwise	0.330 (0.031)
Employed	EMP	1 if the household food buyer is employed; 0 otherwise	0.744 (0.029)
INDEPENDENT-KNOWLEDGE AND PLACE OF PURCHASING			
Knowledgeable	KNOW	1 if respondent states that he or she knows about the meaning of PGS labels; 0 otherwise	0.317 (0.031)
Supermarkets	SUPE	1 if respondent states that he or she usually buys organic food products at a supermarket; 0 otherwise	0.135 (0.023)
Farmers' markets	FARM	1 if respondent states that he or he usually buys organic food products at farmers' markets; 0 otherwise	0.772 (0.028)
Producers	PROD	1 if respondent states that he or she usually buys organic food products directly from producers; 0 otherwise	0.068 (0.017)
Municipal markets	MUNI	1 if respondent states that he or she usually buys organic food products at municipal markets; 0 otherwise	0.099 (0.020)
Specialized stores	SPEC	1 if respondent states that he or she usually buys organic food products directly from producers; 0 otherwise	0.300 (0.031)

The dependent variable is represented by "ORG-PGS," which is equal to 1 if respondents self-reported to have bought organic food products displaying the PGS labeling scheme in the last three months, and 0 otherwise (note that respondents were asked about consumption behavior related to PGS product in general. Ecovida food products are labeled by PGS. Since the data were collected in the south of Brazil where Ecovida operates, survey respondents may have associate PGS products with PGS products offered by Ecovida). The independent variables are represented by a set of the respondents' demographic information such as gender, age, education, the presence of children in the family, residential area (e.g., rural, suburban, and cities), and professional status. Other independent variables are represented by the level of consumer knowledge about the PGS labeling program (subjective knowledge) and the places where the respondents said that they usually buy organic products. As already stated, respondents were given information about the PGS meaning and function to ensure that they could accurately answer the question that is the basis for Equation (1), that is, whether or not they buy organic products certified by the PGS labeling program. With regard to the place of purchase, respondents were asked to indicate where they usually buy organic food products. Different response options were provided to respondents (e.g., supermarkets, municipal markets, farmers' markets, directly from producers, or specialized stores, among others) and they could choose more than one of them.

We have some expectations about what our results would be based on previous literature. With regard to demographic information, we expected that buying behavior toward PGS products is positively related to the area of residency. Specifically, we expected that consumers who live in rural and suburban areas are more likely to buy PGS products than those who live in urban areas. This is because PGS products are developed mostly in rural areas; thus, it is reasonable to assume that respondents not residing in urban areas have more direct contact with producers involved in the PGS network.

With regard to the other demographic information, the findings from past studies on consumer buying behavior toward organic foods show a significant heterogeneity across consumers. For example, while a number of studies found that being younger increases the probability of purchasing organic foods [67,68], others pointed out that older consumers are more likely to purchase organic products because of their greater ability to pay the premium price [37,40,69]. Similarly, a number of studies showed that the presence of children in the family [38,42] and women positively affects the consumption of organic PGS products [36,70]. However, other studies found that these demographic factors have either a negative or no effect on consumer buying behavior toward organic foods [17,71]. As such, no prior expectations were formulated for the effects of demographics on consumer buying behavior toward organic PGS products.

With regard to the effects of purchasing habits on consumer buying behavior toward PGS products, we expected that such effects are positively related to the level of consumer knowledge about the PGS labeling program. Finally, we expected that purchasing organic products in specific places such as farmers' markets would increase the probability of buying organic products guaranteed by the PGS labeling program. This is because organic farmers' markets make local and seasonal organic food available to consumers [72,73]. In this sense, we believe that the easier the access to PGS organic products, the higher the possibility of purchasing them.

8.6   RESULTS AND DISCUSSION

Looking at the demographic information, it can be seen that the majority of interviewees are female (68.3%), with an average age of 42.4. About 74.4% of the main household food buyers are employed. The majority of respondents have a degree at high school level or lower (12.2% have a primary school education, 44.8% have a secondary school education, 27% have a university degree, and 13% have a postgraduate degree). Further, most of the interviewees live in urban areas (52%), while 33.9% of them live in suburban areas in which PGS projects take place and 14.1% live in rural areas. About 34.5% have at least one child in their family.

The results related to participatory certification methods show poor knowledge and awareness of the processes. Most interviewees (68%) claim that they have never heard about the methods, with this figure increasing to 87% for respondents living in urban areas. However, the results change remarkably if we consider the real consumption of PGS products. In spite of a low awareness level of PGS processes, 64% of consumers state that they buy organic products guaranteed by PGS.

With regard to different residential areas, 90% of those living in rural areas usually buy organic PGS products, a trend that is followed by 84% of residents in suburban areas and 43% of respondents living in urban areas. The significant difference between the data on the knowledge of PGS processes and the purchase of PGS products can be attributed to the brief explanation of the meaning and function of the PGS labeling program that follows the question on PGS knowledge.

It is therefore reasonable to assume that respondents realize at this point that they know the system or even realize that they have been purchasing PGS-guaranteed products for years without fully knowing/comprehending the quality control process involved. As mentioned earlier, the purchasing habits were captured by asking respondents where they usually buy organic food products. Alternative response options were provided to them and they could choose more than one response option. As shown in Table 8.2, the majority of the respondents reported that they purchase organic food products mostly at farmers' markets (77.2%), while only a minority purchases them directly from the producers.

The maximum likelihood estimates for the probit analysis are presented in Table 8.3.

TABLE 8.3: Estimates of the probit model and marginal effects.

Variables	Coefficient	Std. Error	|z-statistic|	Marginal effects
Constant	−0.679 ***	0.552	−3.66	-
DEMOGRAPHICS				
FEM	−0.220	0.257	−0.86	−0.056
AGE	0.020 *	0.010	1.92	0.005 *
EDU	0.417	0.269	1.55	0.107
EMP	0.772 **	0.314	2.46	0.198 **
CHILD	0.034	0.245	0.14	0.008
RUR	1.090 ***	0.411	2.65	0.280 ***
SUB-URB	0.769 ***	0.280	2.74	0.198 ***
KNOWLEDGE AND PLACE OF PURCHASING				
KNOW	2.356 ***	0.573	4.11	0.606 ***
SUPE	0.318	0.381	0.83	0.081
FARM	1.154 ***	0.360	3.20	0.297 ***
PROD	−0.557	0.564	−0.99	−0.143
MUNI	0.127	0.396	0.32	0.033
SPEC	−0.143	0.285	−0.50	−0.037
SUMMARY STATISTICS				
LL		−86.984		
Sample Size		230		
McFadden's R^2		0.412		
X^2 (d f= 18)		121.84 (p = 0.0000)		
Correct prediction		79.30		

Note: *, **, and *** indicate statistical significance at the 10%, 5%, and 1% level, respectively.

The null hypotheses that the effects of the independent variables are simultaneously equal to zero are rejected at the 1% significance level. McFadden's pseudo coefficient of determination (R-squared) was calculated at about 0.412 with a correct prediction percentage of 79.30%.

The estimated coefficients and standard errors indicate which factors influence respondents' consumption intentions with regard to organic PGS products. The results indicate that four out of seven demographic variables are statistically significant in their effects on consumer buying behavior toward organic PGS products. These variables are: AGE, EMP, RUR, and SUB-URB. Specifically, and consistent with our expectation, consumers who live in rural and suburban areas are more likely to buy PGS products than those who live in urban areas. According to the marginal effects, those who live in rural areas are 28% more likely and those who live in suburban areas are 19.8% more likely to buy organic PGS products than those who live in urban areas. The reason for this finding could be that consumers who reside in rural and suburban areas are more exposed to PGS publicity from local campaigns and to direct knowledge of producers involved in PGS than those who reside in metropolitan areas. It is also likely that those who reside in non-metropolitan areas grew up around the use of some of these production practices and are therefore more sensitive to ethical issues such as support to local economies and rural areas, issues that are guaranteed by the principles governing PGS.

The findings also show that age is an important demographic indicator for the probability of buying PGS products. As age increases, the tendency to buy organic PGS products rises. Indeed, the findings suggest that the probability of purchasing organic PGS products increases by 0.5% for each year increase in age. This could be due to increasing importance being attached to the characteristics of organic PGS products, as one gets older. In addition, the variable related to the professional status of respondents appears to positively affect PGS product purchasing. According to the marginal effect result, respondents who are employed are 19.8% more likely to buy organic PGS products than others.

Our evidence for gender (FEM), education (EDU), and the presence of children within families (CHILD) indicates that these do not significantly affect the probability of buying organic food that is guaranteed by the PGS labeling program. With regard to gender, our finding is consistent with Van Loo et al. [17] and FMI and AMI [68], who reported no differences in the frequency of buying organic food (chicken) between men and women. Our findings for education and the presence of children within families are consistent with Byrne et al. [72] and Thompson [74], respectively. In the first case, we indeed find that education is inversely correlated with organic purchases, while according to Thompson, the presence of children has no significant effects on purchasing behavior related to organic products.

Turning to the self-reported consumers' knowledge of organic PGS foods (KNOW), the positive and statistically significant coefficient indicates that the higher the organic PGS knowledge that consumers state they have, the higher the probability of buying organic food products guaranteed by the PGS labeling program. Specifically, we find that consumers' degree of awareness of PGS labels is associated with the probability of choosing organic PGS. This evidence is consistent with other studies on consumer choice behavior for organic foods [41,44], which found that product knowledge and awareness about organic food products are associated with consumers' organic buying decisions. Finally, with regard to the place of purchase, only the "farmers' market" (FARM) variable is statistically significant, as we expected. Based on the marginal effects, the results indicate that consumers who usually buy organic products at farmers' markets are about 30% more likely to purchase organic PGS products than those who usually buy organic products elsewhere.

8.7   FINAL REMARKS

Although the organic food market was initially a niche market with products sold through natural food stores and direct-to-consumer markets, organic foods are now traded worldwide and sold in a wide variety of stores such as conventional grocery stores, supermarkets, and hypermarkets. The globalization of the organic food market could also be associated with the role played by the third-party assurance system. This system has represented an increase of trust in organic products worldwide; however, it has also created several problems and barriers for some categories of producers in terms of bureaucratic costs, especially in developing countries. Further, from the consumers' point of view, third-party certification implies "a shift of the credence attribute from the producer to the certifier" [75]. In this regard, several studies have demonstrated consumers' lack of confidence and skepticism about organic food labeling programs guaranteed by a third party [41,76,77].

This trend has also led organic consumers to wonder not only how their food is produced but also where it originates. The growth of organic local movements and alternative organic labeling programs appears to be the organic consumers' reaction to uncertain information about where organic food originates and how it is delivered to the market. The PGS represents an alternative labeling program for organic foods that can contribute to reducing consumer distrust by involving information and knowledge sharing as well as the participation and active involvement of stakeholders.

The present study examines PGS labeling programs by referring to one of the oldest networks that uses it: the Brazilian Rede Ecovida de Agroecologia. In addition, it is the first study in the literature to investigate the effects of a set of factors (gender, age, education, profession, area of residency, presence of children) on Brazilian consumers' buying behavior toward PGS-guaranteed organic food.

Our results indicate that 60.5% of the respondents buy organic PGS products. The findings from our probit analysis suggest that older consumers who live in rural and suburban areas and who are employed are more likely to buy organic PGS products than their counterparts. Further, our results suggest that knowledge of the PGS labeling program significantly increases the probability of purchasing organic PGS products. This is an important finding since it provides evidence that higher self-reported knowledge of the PGS labeling program increases the probability of purchasing organic PGS products. Finally, we also found that consumers who usually shop at farmers' markets for organic products are more likely to purchase organic PGS products than those who usually shop for organic products elsewhere.

In Brazil, PGS runs in parallel with organic legislation and so they are not an alternative in antithesis to third-party guarantee systems. The success of PGS projects in Brazil shows the need to encourage such production processes that optimize results throughout the supply chain, from production to consumption. This study is the first in the literature to analyze consumer buying behavior towards organic products guaranteed by PGS. Thus, it fills a void in the organic foods academic literature by showing the potential consumer interest for PGS-guaranteed organic products. Also, given that producers and policy makers' interest is growing in relation to the promotion of participatory processes in alternative labeling programs [78], our hope is that our findings will encourage more research into PGS certification programs.

The scope of this study was limited in terms of defining the effects of sociodemographic factors, the knowledge of PGS labels, and the purchasing habits of consumer buying behavior for organic foods guaranteed by the PGS labeling program. Future research could perhaps also investigate, using different modeling approaches (e.g., ordered probit, multinomial probit), what food attributes (credence vs. search) affect consumer purchase behavior for organic PGS products by looking at differences across buyer types (e.g., regular vs. occasional buyer). Findings from these studies would help to better characterize consumer preferences for organic PGS products. Moreover, future studies should also test the robustness of our findings across different countries since this would further help us to understand the willingness of organic food consumers to accept and trust alternative quality assurance models across different contexts or cultures.

REFERENCES

1. Renting, H.; Marsden, T.K.; Banks, J. Understanding alternative food networks: Exploring the role of short food supply chains in rural development. Environ. Plan. A 2003, 35, 393–411.
2. Jarosz, L. The city in the country: Growing alternative food networks in Metropolitan areas. J. Rural Stud. 2008, 24, 231–244.
3. Whatmore, S.; Stassart, P.; Renting, H. What's alternative about alternative food networks? Environ. Plan. A 2003, 35, 389–391.
4. Trobe, H.L.L.; Acott, T.G. Localising the global food system. Int. J. Sustain. Dev. World Ecol. 2000, 7, 309–320.
5. O'Hara, S.U.; Stagl, S. Global food markets and their local alternatives: A socio-ecological economic perspective. Popul. Environ. 2001, 22, 533–554.
6. Jarosz, L. Understanding agri-food networks as social relations. Agric. Hum. Values 2000, 17, 279–283.
7. Sage, C. Social embeddedness and relations of regard: Alternative "good food" networks in south-west Ireland. J. Rural Stud. 2003, 19, 47–60.
8. Carolan, M.S. Social change and the adoption and adaptation of knowledge claims: Whose truth do you trust in regard to sustainable agriculture? Agric. Hum. Values 2006, 23, 325–339.
9. Murdoch, J.; Miele, M. "Back to nature": Changing "worlds of production" in the food sector. Sociol. Rural. 1999, 39, 465–483.
10. Murdoch, J.; Marsden, T.; Banks, J. Quality, nature, and embeddedness: Some theoretical considerations in the context of the food sector. Econ. Geogr. 2000, 76, 107–125.
11. Parrott, N.; Wilson, N.; Murdoch, J. Spatializing quality: Regional protection and the alternative geography of food. Eur. Urban Reg. Stud. 2002, 9, 241–261.
12. Aprile, M.C.; Caputo, V.; Nayga, R.M., Jr. Consumers' valuation of food quality labels: The case of the European geographic indication and organic farming labels. Int. J. Consum. Stud. 2012, 36, 158–165.
13. Caputo, V.; Nayga, R.M.; Scarpa, R. Food miles or carbon emissions? Exploring labelling preference for food transport footprint with a stated choice study. Aust. J. Agric. Resour. Econ. 2013, 57, 465–482.
14. Caputo, V.; Vassilopoulos, A.; Nayga, R.M.; Canavari, M. Welfare effects of food miles labels. J. Consum. Aff. 2013, 47, 311–327.
15. Schnell, S.M. Food miles, local eating, and community supported agriculture: Putting local food in its place. Agric. Hum. Values 2013, 30, 615–628.

16. Asioli, D.; Canavari, M.; Pignatti, E.; Obermowe, T.; Sidali, K.L.; Vogt, C.; Spiller, A. Sensory experiences and expectations of italian and german organic consumers. J. Int. Food Agribus. Mark. 2014, 26, 13–27.

17. Van Loo, E.; Caputo, V.; Nayga, J.; Rodolfo, M.; Meullenet, J.-F.; Crandall, P.G.; Ricke, S.C. Effect of organic poultry purchase frequency on consumer attitudes toward organic poultry meat. J. Food Sci. 2010, 75, 384–397. [PubMed]

18. Schmid, O.; Hamm, U.; Richter, T.; Dahlke, A. A Guide to Successful Marketing Initiatives; Research Institute of Organic Agriculture: Frick, Germany, 2004.

19. Harris, P.J.C.; Browne, A.W.; Barrett, H.R.; Cadoret, K. Facilitating the Inclusion of the Resource-Poor in Organic Production and Trade: Opportunities and Constraints Posed By Certification; Rural Livelihoods Department, Department for International Development: London, UK, 2001.

20. Meirelles, L. La Certificación de Productos Orgánicos--Encuentros y desencuentros; Centro Ecológico Ipe: Lapa, Brazil, 2003; (In Spanish).

21. Milestad, R.; Darnhofer, I. Building farm resilience: The prospects and challenges of organic farming. J. Sustain. Agric. 2003, 22, 81–97.

22. Vogl, C.R.; Kilcher, L.; Schmidt, H. Are standards and regulations of organic farming moving away from small farmers' knowledge? J. Sustain. Agric. 2005, 26, 5–26.

23. Courville, S. Organic standards and certification. In Organic Agriculture: A Global Perspective; Kristiansen, P., Taji, A., Renagold, J., Eds.; CSIRO Publishing: New York, NY, USA, 2006; pp. 201–219.

24. Eernstman, N.; Wals, A.E.J. Jhum meets IFOAM: Introducing organic agriculture in a tribal society. Int. J. Agric. Sustain. 2009, 7, 95–106.

25. Sano, D.; Prabhakar, S.V.R.K. Some Policy suggestions for promoting organic agriculture in Asia. J. Sustain. Agric. 2009, 34, 80–98.

26. Constance, D.H.; Choi, J.Y. Overcoming the barriers to organic adoption in the United States: A look at pragmatic conventional producers in Texas. Sustainability 2010, 2, 163–188.

27. Zander, K.; Hamm, U. Consumer preferences for additional ethical attributes of organic food. Food Qual. Prefer. 2010, 21, 495–503.

28. Fonseca, M. Alternative Certification and a Network Conformity Assessment Approach. The Organic Standard; IFOAM: Bonn, Germany, 2004.

29. Fonseca, M.F.; Wilkinson, J.; Egelyng, H.; Mascarenhas, G. The Institutionalization of Participatory Guarantee Systems (PGS) in Brazil: Organic and Fair Trade Initiatives. In Proceedings of the 16th IFOAM Organic World Congress, Modena, Italy, 16–20 June 2008.

30. Radomsky, G.F.W. Práticas de certificação participativa na agricultura ecológica: Rede, selos e processos de inovação. IDeAS 2009, 3, 133–164, (In Portuguese).

31. Zanasi, C.; Venturi, P.; Setti, M.; Rota, C. Participative organic certification, trust and local rural communities development: The case of Rede Ecovida. New Medit. 2009, 8, 56–64.

32. Nelson, E.; Tovar, L.G.; Rindermann, R.S.; Cruz, M.Á.G. Participatory organic certification in Mexico: An alternative approach to maintaining the integrity of the organic label. Agric. Hum. Values 2009, 27, 227–237.

33. Preißel, S.; Reckling, M. Smallholder group certification in Uganda--Analysis of internal control systems in two organic export companies. J. Agric. Rural Dev. Trop. Subtrop. (JARTS) 2010, 111, 13–22.

34. Sacchi, G.; Zanasi, C.; Canavari, M. Modelli alternativi di garanzia della qualità dei prodotti biologici alla luce della Teoria delle Convenzioni. Econ. Agro-Aliment. 2011, 13, 57–80, (In Italian).

35. Wilkins, J.L.; Hillers, V.N. Influences of pesticide residue and environmental concerns on organic food preference among food cooperative members and non-members in Washington State. J. Nutr. Educ. 1994, 26, 26–33.

36. Davies, A.; Titterington, A.J.; Cochrane, C. Who buys organic food? Br. Food J. 1995, 97, 17–23.

37. Schifferstein, H.N.J.; Oude Ophuis, P.A.M. Health-related determinants of organic food consumption in The Netherlands. Food Qual. Prefer. 1998, 9, 119–133.

38. Thompson, G.D.; Kidwell, J. Explaining the choice of organic produce: Cosmetic defects, prices, and consumer preferences. Am. J. Agric. Econ. 1998, 80, 277–287.

39. Chinnici, G.; D'Amico, M.; Pecorino, B. A multivariate statistical analysis on the consumers of organic products. Br. Food J. 2002, 104, 187–199.

40. Cicia, G.; del Giudice, T.; Scarpa, R. Consumers' perception of quality in organic food. Br. Food J. 2002, 104, 200–213.

41. Yiridoe, E.K.; Bonti-Ankomah, S.; Martin, R.C. Comparison of consumer perceptions and preference toward organic versus conventionally produced foods: A review and update of the literature. Renew. Agric. Food Syst. 2005, 20, 193–205.

42. Hughner, R.S.; McDonagh, P.; Prothero, A.; Shultz, C.J.; Stanton, J. Who are organic food consumers? A compilation and review of why people purchase organic food. J. Consum. Behav. 2007, 6, 94–110.

43. Bellows, A.C.; Onyango, B.; Diamond, A.; Hallman, W.K. Understanding consumer interest in organics: Production values vs. purchasing behaviour. J. Agric. Food Ind. Organ. 2008, 6, 1–28.

44. Gracia, A.; de Magistris, T. The demand for organic foods in the South of Italy: A discrete choice model. Food Policy 2008, 33, 386–396.

45. Mintel. Organic Food—US; Mintel: London, UK, 2008.

46. Van Loo, E.J.; Caputo, V.; Nayga, R.M., Jr.; Meullenet, J.-F.; Ricke, S.C. Consumers' willingness to pay for organic chicken breast: Evidence from choice experiment. Food Qual. Prefer. 2011, 22, 603–613.

47. Nayga, R.M., Jr. Sociodemographic Influences on Consumer Concern for Food Safety: The Case of Irradiation, Antibiotics, Hormones, and Pesticides. Rev. Agric. Econ. 1996, 18, 467–475.

48. Thévenot, L. Des marchés aux normes. In La Grande Transformation de L'agriculture: Lectures Conventionnalistes et Régulationnistes; Allaire, G., Boyer, R., Eds.; Economica & INRA: Paris, France, 1995; pp. 33–51, (In French).

49. Jaffee, D. Fair Trade Standards, Corporate Participation, and Social Movement Responses in the United States. J. Bus. Ethics 2010, 92, 267–285.

50. Raynolds, L.T. The globalization of organic agro-food networks. World Dev. 2004, 32, 725–743.

51. Hatanaka, M. McSustainability and McJustice: Certification, alternative food and agriculture, and social change. Sustainability 2014, 6, 8092–8112.

52. Poisot, A.-S. Summary Analysis of Codes, Guidelines, and Standards Related to Good Agricultural Practices; FAO GAP Working Paper Series 2; FAO Agriculture Department: Rome, Italy, 2003.

53. Sylvander, B. Le rôle de la certification dans les changements de régime de coordination: l'agriculture biologique, du réseau à l'industrie. Rev. Écon. Ind. 1997, 80, 47–66, (In French).

54. May, C. PGS Guidelines: How Participatory Guarantee Systems Can Develop and Function; IFOAM: Bonn, Germany, 2008.
55. International Federation of Agriculture Movements (IFOAM). PGS Map 2015. Available online: http://www.ifoam.org/en/pgs-map (accessed on 18 April 2015).
56. International Federation of Agriculture Movements (IFOAM). Global online PGS database 2015. Available online: http://www.ifoam.org/es/global-online-pgs-database (accessed on 18 April 2015).
57. Raynolds, L.T. Re-embedding global agriculture: The international organic and fair trade movements. Agric. Hum. Values 2000, 17, 297–309.
58. International Fund for Agricultural Development (IFAD). The Adoption of Organic Agriculture among Small Farmers in Latin America and the Caribbean; Technical Report. IFAD: Rome, Italy, 2003.
59. Meirelles, L. Regulation of the Participatory Guarantee Systems in Brazil: A Case Study; IFOAM: Bonn, Germany, 2010.
60. Ministério da Agricultura, Cadastro nacional 2015. Available online: http://www.agricultura.gov.br/desenvolvimento-sustentavel/organicos/cadastro-nacional (accessed on 1 March 2015).
61. Rede Ecovida. Uma Identidade que se Constrói em Rede; Caderno de Formação 01. Rede Ecovida de Agroecologia: Lapa, PR, Brazil, 2007.
62. International Federation of Agriculture Movements (IFOAM). Participatory Guarantee Systems: Case Studies from Brazil, India, New Zealand, USA; Technical Report. IFOAM: Bonn, Germany, 2003.
63. Santacoloma, P. Organic Certification Schemes—Managerial Skills and Associated Costs: Synthesis Report from Case Studies in the Rice and Vegetable Sectors; FAO: Rome, Italy, 2007.
64. Grebitus, C.; Colson, G.; Menapace, L.; Bruhn, M. Who cares about food origin? A comparison of hypothetical survey responses and actual shopping behavior. In Proceedings of the Annual Meeting of Agricultural & Applied Economics Association (AAEA), Denver, CO, USA, 25–27 July 2010.
65. Menapace, L.; Colson, G.; Grebitus, C.; Facendola, M. Consumers' preferences for geographical origin labels: Evidence from the Canadian olive oil market. Eur. Rev. Agric. Econ. 2011, 38, 193–212.
66. Greene, W.H. Econometric Analysis, 4th ed.; John Wiley and Sons: Hoboken, NJ, USA, 2000.
67. Govindasamy, R.; Italia, J. Predicting willingness-to-pay a premium for organically grown fresh produce. J. Food Distrib. Res. 1999, 30, 44–53.
68. Zhang, F.; Epperson, J.E.; Huang, C.L.; Houston, J.E. Organic price premiums paid for fresh tomatoes and apples by U.S. households: Evidence from Nielsen Homescan Data. J. Food Distrib. Res. 2009, 40, 105–114.
69. Roddy, G.; Cowan, C.; Hutchinson, G. Organic food—A description of the Irish market. Br. Food J. 1994, 96, 3–10.
70. Food Marketing Institute (FMI). The Organic Shoppers May Not Be Who You Think They Are; Food Marketing Institute Report. FMI: Washington, DC, USA, 2001.
71. Food Marketing Institute (FMI); American Meat Institute (AMI). The Power of Meat—An In-Depth Look at Meat through the Shoppers' Eyes; Joint Report. AMI/FMI: Arlington, WA, USA, 2010.

72. Byrne, P.J.; Toensmeyer, U.C.; German, C.L.; Muller, H.R. Evaluation of consumer atti-
    tudes towards organic produce in Delaware and the Delmarva region. J. Food Distrib. Res.
    1992, 23, 29–44.

73. Dimitri, C.; Dettmann, R.L. Organic food consumers: What do we really know about them?
    Br. Food J. 2012, 114, 1157–1183.

74. Thompson, G.D. Consumer demand for organic foods: What we know and what we need to
    know. Am. J. Agric. Econ. 1998, 80, 1113–1118.

75. Janssen, M.; Hamm, U. The mandatory EU logo for organic food: Consumer perceptions.
    Br. Food J. 2012, 114, 335–352.

76. Aarset, B.; Beckmann, S.; Bigne, E.; Beveridge, M.; Bjorndal, T.; Bunting, J.; McDonagh, P.;
    Mariojouls, C.; Muir, J.; Prothero, A.; et al. The European consumers' understanding and
    perceptions of the "organic" food regime. Br. Food J. 2004, 106, 93–105.

77. Eden, S.; Bear, C.; Walker, G. The sceptical consumer? Exploring views about food assur-
    ance. Food Policy 2008, 33, 624–630.

78. European Commission. A Review of the European Policy on Organic Agriculture--Public
    Consultation, 15/01/2013-10/04/2013; European Commission: Brussels, Belgium, 2013.

Evaluating the Sustainability of a Small-Scale Low-Input Organic Vegetable Supply System in the United Kingdom

Mads V. Markussen, Michal Kulak,
Laurence G. Smith, Thomas Nemecek, and
Hanne Østergård

9.1 INTRODUCTION

Modern food supply systems (production and distribution) are heavily dependent on fossil energy [1] and other non-renewable resources [2]. The global environmental crisis [3,4] and foreseeable constraints on the supply of energy [5] and fertilizer [6,7] clearly show that there is a need to develop food supply systems that conserve biodiversity and natural systems and rely less on non-renewable resources. A similar conclusion is drawn in a report initiated by the Food and Agriculture Organization of the United Nations (FAO) and The World Bank. It emphasizes the need to maintain productivity, while conserving natural resources by improving nutrient, energy, water and land use efficiency, increasing farm diversification, and supporting agro-ecological systems that take advantage of and conserve biodiversity at both field and landscape scale [8].

It has been shown that the food industry in the UK is responsible for 14% of national energy consumption and for 25% of heavy goods vehicle kilometers [9].

The structural development of the food supply system over the past 60 years means that most goods are now distributed through regional distribution centers before being transported to increasingly centralized and concentrated out-of-town supermarkets. This also means that more shopping trips are done by private cars which make up approximately half of the total food vehicle kilometers [10]. In 2002, 9% of UK's total consumption of petroleum products was used for transportation of food [10]. This clearly shows that if the environmental impacts of the food supply system are to be significantly reduced, then it is necessary to view the production and distribution of food together. Direct marketing and local selling of products offers a way for farms to by-pass the energy intensive mass distribution system. Such distribution systems are particularly appropriate for vegetables, which have a relative short lifetime and are most attractive to consumers when they are fresh. On the other hand, depending on the distance travelled and the mode of transport, the local system may be more energy consuming than the mass distribution system [11,12].

The development in food supply systems has also resulted in a push towards producers being more specialized and production being in larger, uniform units [10]. These changes tend to imply reductions in crop diversity at the farm level, which in the long run may cause problems for society. For example, the biodiversity loss associated with these systems has been shown to result in decreased productivity and stability of ecosystems due to loss of ecosystem services [13]. Specifically, biodiversity at the farm level has been shown often to have many ecological benefits (ecosystem services) like supporting pollination, pest and disease control. Therefore, it has been suggested that it is time for a paradigm shift in agriculture by embracing complexity through diversity at all levels, including soil, crops, and consumers [14]. However, high levels of crop diversity may be rather difficult to combine with the supermarket mass distribution system, which at present sell 85% of food in the UK [10]. On the contrary, local based direct marketing has been identified as a driving force for increasing on-farm biodiversity [15].

The sustainability aspects of resource use and environmental impacts of food supply systems can be assessed by Life Cycle Assessment (LCA) [16,17] or emergy assessment [18,19]. Emergy accounting and LCA are largely based on the same type of inventory (i.e., accounting for energy and material flows) but apply different theories of values and system boundaries [20]. In emergy accounting, all flows of energy and materials are added based on the total available energy (exergy) directly and indirectly required to produce the flow. Emergy accounting is particularly suited for assessing agricultural systems since the method accounts for use of freely available natural resources (sun, rain, wind and geothermal heat) as well as purchased resources from the society [18]. LCA draws system boundaries around human dominated processes (resource extraction, refining, transportation, etc.) and includes indirect resources used throughout the supply chain, such as the transport of inputs supplied into the production system. Unlike emergy accounting, LCA disregards energy used by nature and normally also labor. LCA on the other hand considers emissions to the environment in addition to resource use. Due to the differences in system boundaries and scope of analysis, emergy and LCA are complementary methods [21].

We studied the sustainability of a small-scale low-input organic vegetable food supply system by evaluating empirical data on resource use and emissions resulting from production and distribution of vegetables in a box-scheme. This specific case was chosen because the farm is managed with a strong preference to increase crop diversity and to close the production system with regard to external inputs. Combined with the box-scheme distribution system it thus represents a fundamentally different way of producing and distributing food compared to the dominating supermarket based systems. Our hypothesis was that the food supply system of the case study uses fewer resources (especially fewer non-renewable resources) when compared to standard practices. To test this we developed two organic vegetable food supply model systems, low and high yielding. Each system provided the same amount of food as the case study system, and the food produced was distributed via supermarkets rather than through a box-scheme. The case supply system is benchmarked against these model systems based on a combined emergy and LCA evaluation. Therefore, within this study we aimed to evaluate whether it is possible to perform better than the dominating systems with respect to resource use including labor and environmental impacts, and at the same time increase resilience.

9.2 FARM AND FOOD DISTRIBUTION SYSTEM— EMPIRICAL DATA

The case study farm is a small stockless organic unit of 6.36 ha of which 5.58 ha are cropped and a total of 0.78 ha is used for field margins, parking area and buildings. The box-scheme distribution system supplies vegetables to 200–300 customers on a weekly basis.

Data for 2009 and 2010 were collected by two one-day visits at the farm and follow up contacts in the period 2011 to 2013. Data included all purchased goods for crop production and distribution, as well as a complete list of machineries and buildings. The vegetable production was estimated based on sales records of vegetables delivered to consumers for each week during 2009 and 2010 and subsequently averaged to give an average annual production (Table 9.A1). For the years studied, about 20% of the produce was sold to wholesalers. In our analyses, this share was included in the box-scheme sales.

9.2.1 PRODUCTION SYSTEMS

Forty-eight different crops of vegetables are produced (Table 9.A1) and several different varieties are grown for each crop. Crops are grown in three different systems: open field, intensive managed garden and polytunnels, and greenhouses. The open fields are managed with a 7-year crop rotation and make up 5.09 ha of cropping area. The fields are characterized by a low-fertility soil with a shallow topsoil and high stone content. The garden is managed with a 9-year crop rotation and the cropped area is 0.38 ha. In the garden only, walk-behind tractors and hand tools are used for the cultivation. The greenhouse and polytunnels make up 0.10 ha.

The farm is managed according to the Stockfree Organic Standard [22], which means that no animals are included in the production system and the farm uses no animal manure. The farm is in general designed and managed with a strong focus on reducing external inputs (e.g., fuel and fertilizers). An example of this is that the fertility is maintained by the use of green manures. The only fertility building input comes from woodchips composted on the farm and small amounts of lime and vermiculite, which are used to produce potting compost for the on-farm production of seedlings. All seed is purchased except for 30% of the seed potatoes, which are farm saved.

9.2.2 DISTRIBUTION SYSTEM

The distribution is done by weekly round-trips of 70 km, where multiple bags are delivered to neighborhood representatives. Other customers may then come to the representatives' collection points to collect the bags. Customers are encouraged to collect the bag on foot or on bike, and the bags are designed to make this easier (i.e., a wooden box is more difficult to carry). Potential customers are rejected if they live in a location from where they would need to drive by car to pick up their bags, even though they offer to pick up the bags themselves and pay the same price. The neighborhood representatives have some administrative tasks and are paid by getting boxes for free.

9.3   ASSESSMENT METHODS—EMERGY AND LCA

The system boundary in this study is the farm and its distribution system. Cooking, consumption, human excretion and wastewater treatment are excluded from the scope of the analysis. The functional unit, which defines the service that is provided, is baskets of vegetables produced during one year and delivered at consumer's door as average of the years 2009 and 2010. Resource consumption and environmental impacts associated with consumers' transport is included except for transport by foot or bike, which was assumed negligible.

9.3.1   EMERGY ACCOUNTING

Emergy accounting quantifies direct input of energy and materials to the system and multiplies these with suitable conversion factors for the solar equivalent joules required per unit input. These are called unit emergy values (UEV) and given in seJ/unit, e.g., seJ/g or seJ/J. Emergy used by a system is divided into different categories [23] and in the following we describe how they are applied in this study.

Local renewable resources (R). The term "R" includes flows of sun, rain, wind and geothermal heat and is the freely available energy flows that an agricultural system captures and transforms into societal useful products. We include the effect of rainfall as evapotranspiration. To avoid double counting only the largest flow of sun, rain and wind is included.

Local non-renewable recourses (N). This includes all stocks of energy and materials within the system boundaries that are subject to depletion. In agricultural systems, this is typically soil carbon and soil nutrients. In this study we assume that these stocks are maintained.

Feedback from the economy (F) consists *of purchased materials (M)* and *purchased labor and services (L&S)* [23]. M includes all materials and assets such as machinery and buildings. Assets are worn down over a number of years and the emergy use takes into account the actual age and expected lifetime of each asset. The materials come with a service or indirect labor component. This represents the emergy used to support the labor needed in the bigger economy to make the products and services available for the studied system. It is reflected in the price of purchased goods.

Labor and service (L&S). In this study, the L&S component is accounted for based on monetary expenses calculated from the sales price of the vegetables. This approach rests on the assumption that all money going into the system is used to pay labor and services (including the services provided in return for government taxes or insurances). This revenue is multiplied with the emergy money ratio, designated em£-ratio (seJ/£), which is the total emergy used by the UK society divided by the gross domestic product (GDP). Thus the em£-ratio is the average emergy used per £ of economic activity. To avoid counting the service component twice, UEVs assigned to purchased materials (M) are without the L&S component.

Total emergy use (U). The sum of all inputs is designated "U." We use three emergy indicators to reveal the characteristics of the food supply system: (1) Emergy Yield Ratio (U/F), a measure of how much the system takes advantage of local resources (in this study only R) for each investment from the society in emergy terms (F), (2) Renewability (R/U), a measure of the share of the total emergy use that comes from local renewable resources, and (3) Unit Emergy Value, UEV (U/output from system) [23].

9.3.2 LIFE CYCLE ASSESSMENT (LCA)

The LCA approach quantifies the environmental impacts associated with a product, service or activity throughout its life cycle [24]. The method looks at the impact of the whole system on the global environment by tracing all material flows from their point of extraction from nature through the technosphere and up to the moment of their release into the environment as emissions. LCA takes into account all direct and indirect manmade inputs to the system and all outputs from the system and quantifies the associated impacts on the environment.

Impact categories that are relevant and representative for the assessment of agricultural systems [16] were considered: non-renewable resource use as derived from fossil and nuclear resources [25], Global Warming Potential over 100 years according to the IPCC method [26] and a selection of other impacts from CML01 methods [27] and EDIP2003 [28], (i.e., eutrophication potential to aquatic and terrestrial ecosystems, acidification, terrestrial and aquatic eco-toxicity potentials, human toxicity potential). In addition the use of fossil phosphorus was assessed.

The inventories for the LCA were constructed with the use of Swiss Agricultural Life Cycle Assessment (SALCA) models [28], Simapro V 7.3.3 [29] and the Ecoinvent database v2.2 [30]. The following inputs and emissions were based on other studies: life cycle inventory for vegetable seedlings [31]; biomulch [32]; nitrous oxide and methane emissions from open field woodchip composting on the case study farm [33]. The Life Cycle Inventory for irrigation pipeline from ecoinvent was adjusted to reflect the irrigation system of the case farm and the Swiss inventory for irrigation was adjusted to reflect the British electricity mix.

The analysis was carried out from cradle to the consumer's door with respect to the ISO14040 [24] and ISO14044 [34] standards for environmental Life Cycle Assessment. Upstream environmental impacts related to the production of woodchips or manure were not considered. This is following a cut-off approach that makes a clear division between the system that produces a by-product or waste and the system using it. The emissions from livestock farming (associated with the production of manure used in the models of standard practice) are fully assigned to the livestock farmer and the gardener is responsible for the production of woodchips. However, environmental impacts from the transport of both types of inputs to the farm, their storage and composting at the farm and all the emissions to soil, air and water that arise from their application were considered in this study. The results of the impact category non-renewable resource use were investigated in more detail by looking at the relative contribution of particular processes to the overall resource use, because of some similarities with the emergy assessment.

9.4 MODELS FOR STANDARD PRACTICE OF VEGETABLE SUPPLY SYSTEM

The overall aim of developing these models is to assess the resource use and environmental impacts of providing the same service as the case system but in the dominating supermarket based system. The two model systems, M-Low and M-High, express the range of standard practice for organic vegetable production as defined from the Organic Farm Management Handbook [35]. Since the information in this handbook is independent of scale, i.e., all numbers are given per ha or per kg, then the model systems are also independent of scale. Both model systems provide vegetables in the same quantity at the consumer's door (in food energy) and of comparable quality as the case study. The mix of vegetables provided is identical to the case system for the eight crops (two types of potatoes, carrots, parsnips, beetroots, onions, leeks and squash) constituting 75% of the food energy provided (Table 9.1). For the remaining 25% representing 40 crops at the case farm, four crops (white cabbage, cauliflower, zucchini and lettuce) have been chosen based on the assumption that they provide a similar utility for the consumer.

TABLE 9.1: Characteristics of vegetables produced annually in the case system and their counterparts in the model systems.

Case farm crops	Model farm crops	Food energy at consumers (MJ)	Share of total food energy
Storable crops			
Potatoes, main crop	Potatoes, main crop	25,597	34.4%
Potatoes, early	Potatoes, early	8532	11.5%
Carrots (stored and fresh)	Carrots	4635	6.2%
Beetroots (stored and fresh)	Beetroots	4271	5.7%
Onions (stored, fresh and spring)	Onions	3688	5.0%
Parsnips	Parsnips	3555	4.8%
Leeks	Leeks	2902	3.9%
Squash	Squash	2697	3.6%
Cabbages (red-, black-, green-, sprouts, kale, pak choi)	Cabbages, white	5390	7.3%
Cauliflower, broccoli and minor crops (celeriac, fennel, turnips, kohlrabi, rutabaga, daikon, garlic)	Cauliflower	3344	4.5%
Storable crops, total		*64,610*	*86.9%*
Fresh crops			
18 different crops (see Table Al for list of crops)	50% Courgettes	4859	6.5%
	50% Lettuce	4859	6.5%
Fresh crops, total		*9717*	*13.1%*
All crops, total (functional unit)		74,328	100.0%

9.4.1 CROP MANAGEMENT FOR M-LOW AND M-HIGH

M-Low and M-High systems were defined from yields per ha using the range in the Organic Farm Management Handbook [35]. The M-Low farm represents a standard low-yielding farm, the lowest value in the Handbook, and the M-High farm represents a standard high yielding farm, the highest value in the Handbook. The range is shown in Table 9.2 for each crop considered. These yield differences, combined with the food chain losses assumed (see Section 9.4.2), implied that different areas were needed to provide the functional unit, i.e., the average annual amount of vegetables (in food energy) at the consumer's door (Table 9.2).

TABLE 9.2: Yields and corresponding areas needed to provide the amount of vegetables sold in the case system for M-Low and M-High. Areas for the case farm are given for comparison.

	Case (ha)	M-Low		M-Hi	
		Yields[a] (t/ ha)	Areas (ha)	Yields[a] (t/ ha)	Areas(ha)
Potatoes, early		10	0.42	20	0.21
Potatoes, main crop		15	0.83	40	0.31
Carrots		15	0.41	50	0.12
Beetroots		10	0.35	30	0.12
Onions		10	0.35	25	0.14
Parsnips		10	0.21	30	0.07
Leeks		6	0.48	18	0.16
Squash		15	0.17	40	0.06
Cabbage, white		20	0.26	50	0.11
Cauliflower		16	0.23	24	0.15
Zucchini		7	0.88	13	0.47
Lettuce		6	1.73	9.6	1.08
Vegetables	4.02		6.32		3.01
Green manure	1.56		1.58 [b]		0.75 [b]
Field margins and infrastructure	0.78		1.12 [c]		0.53 [c]
Total area	6.36		9.02		4.29

[a] From Organic Farm Management Handbook [35], the lowest and highest yield for each crop; [b] 20% of cultivated area; [c] 14% of cultivated area based on the proportion for the case.

The further definition of the two model systems was based on the assumption that yields are determined by the level of fertilization and irrigation. Therefore, M-Low is defined with a low input of fertilizers and M-High with a higher fertilizer input. The NPK-budgets were calculated based on farm gate inputs and outputs from an average farm with the same crop production and management using a NPK-budget tool from the Organic Research Center [36,37]. Both systems were assumed to have 20% green manure (red clover) in their crop rotation. Input of cattle manure, rock phosphate and rock potash was then modeled such that M-High reached a balance of 90 kgN/ha, 10 kgP/ha and 10 kgK/ha and M-Low a balance of 0 kgP/ha and 0 kgK/ha (Table 9.A2). For the M-Low N balance, the lowest possible value was 54 kgN/ha due to atmospheric deposits and N-fixation.

Further, for M-Low irrigation was only included for the crops for which irrigation is considered essential according to the Handbook [34], whereas for M-High irrigation was also included for crops which "may require irrigation."

Based on the field operations needed for each crop described in Organic Farm Management Handbook [35], the resource use in terms of fuel and machinery was modeled according to resource use per unit process [38] (see Supplementary Material for detailed description of the farming model). The approach for determining yields and resource use was similar to a previous study commissioned by Defra [39].

9.4.2   MODEL DISTRIBUTION SYSTEM

The model distribution system from farm gate to consumer's door was modeled on a crop by crop basis based on published LCA reports for supermarket based food distribution chains [40,41,42] (Table 9.3). The chain is thus assumed to consist of 200 km transport to and storage for 5 days at regional distribution center (RDC), 50 km transport to and storage for 2 days at retailers and 6.4 km transport from the retailer to the customer's home [40] (see Supplementary Material for detailed assumptions.) Transportation from the farm to the RDC is assumed to be in a chilled 32 t truck with an energy consumption of 22.9 mL diesel per euro pallet kilometer [41]. Throughout the system, food waste is taken into account for each crop [42].

The total expenses to labor, service and materials throughout the supply system were estimated based on 12 month average supermarket prices (from March 2012 to March 2013) for each of the vegetables [43]. The prices were adjusted for inflation to reflect average 2009–2010 prices according to the price index for vegetables including potatoes and tubers [44].

TABLE 9.3: Outputs and energy use in model distribution system: from farm to regional distribution center (RDC), to retailer and to consumer's home.

	Farm gate output (t)	Diesel use, transport to RDC (L) [a]	El. use, 5 days storage at RDC (kWh) [b]	RDC gate output (t) [c]	Diesel use, transport to retailer (L) [a]	El. use, storage at retailer (kWh) [d]	NG use, storage at retailer (MJ) [d]	Retail gate output (t) [c]	Gasolin use, transport to home by car (L) [e]	Diesel use, transport to home by bus (L) [e]
Potatoes	16.7	64.6	63.9	11.8	11.5	78.1	346.5	11.5	101.2	2.7
Carrots	6.2	24.0	23.7	4.4	4.3	29.0	128.7	4.3	37.6	1.0
Cabbages	5.3	78.3	77.5	5.3	19.6	34.7	591.7	5.1	45.4	1.2
Cauliflower	3.7	55.4	54.9	3.7	13.9	24.5	419.0	3.6	32.1	0.9
Parsnips	2.1	8.1	8.0	1.5	1.4	9.8	43.5	1.4	12.7	0.3
Beetroots	3.5	13.8	13.6	2.5	2.4	16.6	73.8	2.4	21.6	0.6
Onions	3.5	16.8	16.6	3.2	3.8	21.2	115.6	3.2	28.1	0.8
Leeks	2.9	42.5	42.1	2.9	10.6	18.8	321.7	2.8	24.7	0.7
Squash	2.5	37.7	37.3	2.5	9.4	16.7	285.2	2.5	21.9	0.6
Zucchini	6.1	91.5	90.5	6.1	22.9	40.5	691.4	6.0	53.0	1.4
Lettuce	10.4	219.2	216.9	10.1	53.6	66.8	1620.1	9.9	87.6	2.4
Total	62.9	651.8	645.0	54.0	153.3	356.7	4637.2	52.7	465.9	12.6

[a] The produce is transported 200 km to RDC and 50 km from RDC to retail [40] using 22.9 ml diesel per pallet-km (chilled single drop, 32 t artic) [41].
[b] Electricity consumption in RDC is 0.00059 kWh/l/day [40]. [c] For each crop losses in storage and packaging are taken into account [40]. See Supplementary Material for details. [d] Storage at ambient temperature. Energy use is 0.027 MJ/kg/day (44% electricity for light and 56 % natural gas (NG) for heating) [40].
[e] Based on an average UK shopping trip of 6.4 km with an average shopping basket of 28 kg and where 58% of trips made by private car and 8% made by bus [40]. See Supplementary Material for details.

9.5   RESULTS OF SUSTAINABILITY ASSESSMENT

The service provided by the three systems is a comparable "basket" of vegetables produced during one year and delivered to the consumer's door. This service is measured in food energy and is equal to 74,328 MJ/year as an average of 2009 and 2010 (Table 9.1). This corresponds to the total annual food energy needed for 19–23 people (based on a recommended daily intake of 8.8–11 MJ [45]). The emergy flows are illustrated for the case (Figure 9.1A) and for the model systems (Figure 9.1B). The two diagrams demonstrate clearly the different distribution systems and that in the case the full money flow goes to the farm whereas in the model systems part of money flows to the freight companies, supermarkets, and regional distribution centers (RDC).

9.5.1   EMPIRICAL SYSTEM

The basis of any emergy assessment is the emergy table (Table 9.4) that shows all environmental and societal flows, which support the system. Notably labor and services (L&S) make up 89% of total emergy used by the case (calculated from Table 9.4). As emergy use for L&S is calculated as a function of the emergy use for the national economy, this reflects the national resource consumption rather than the specific business. To avoid distorting the results of the actual farm with the implications of being embedded in an industrialized economy, we consider the emergy indicators both with and without L&S.

The main result of the emergy evaluation for the case system is the transformity of the vegetables, which amounts to 5.20×106 seJ/J with L&S and 5.54×105 seJ/J without (Table 9.5). The Emergy Yield Ratio (EYR) of 1.15 disregarding L&S shows that free local environmental services (R) contribute with only 0.15 seJ per seJ invested from the society. The renewability indicator shows that the system uses 13% local resources when disregarding L&S but only 1% when including L&S. The latter reflects that L&S is considered as non-renewable.

Disregarding L&S, the emergy profiles of the case system are as follows (calculated from Table 9.4). Purchased miscellaneous materials for the cultivation phase contribute 38% of total emergy used. Fuel used for cultivation and electricity used for production of seedlings are the biggest flows with 18% and 11%, respectively. Notably, irrigation contributes 24% of the total flow with the water used constituting the most important element (17%). Likewise, the woodchips, used as soil enhancement and used to produce potting compost,

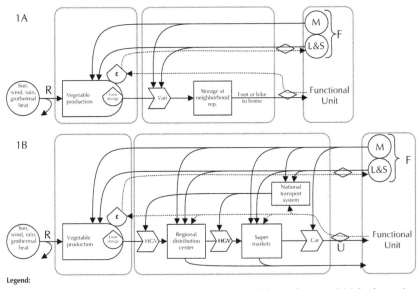

FIGURE 9.1 Material and emergy flow diagrams for the case system (9.1A) and the two model systems M-Low and M-High, which have identical distribution systems (9.1B).

contribute with 10% and farm assets contribute with 7%. The diesel used on the weekly round-trip was estimated to 465 L/year (1.6 × 1010 J, Table 9.4) and it is the major component of the emergy used in the distribution phase (7% of the total emergy used).

Analyzing the case system using the LCA perspective, the processes related to the cultivation phase have a much larger environmental impact than the processes involved in the distribution phase, for all nine categories (Figure 9.2). The LCA impact category non-renewable resource use includes all direct and indirect use of fossil and nuclear fuels converted to MJ. Crude oil in the ground contributes more than 50% of the total raw materials for energy (Figure 9.3). Crude oil is used to produce diesel for operating tractors and pumping the water for irrigation, and to a smaller extent for the manufacture and transport of other inputs.

TABLE 9.4: Use of emergy per functional unit for the three systems: the case, M-Low and M-High. See Tables **A3** and **A4** for notes with details for each item.

	Unit	Case (Unit)	M-Low (Unit)	M-High (Unit)	UEV (seJ/unit)	Case emergy flow (× 10^{14}seJ)	M-Low emergy flow (× 10^{14}seJ)	M-High emergy flow (× 10^{14}seJ)
LOCAL RENEWABLE FLOWS (R)								
1 Sun	J	9.7×10^{13}	1.4×10^{14}	6.6×10^{13}	1.0^a	1.0	1.4	0.7
2 Evapotranspiration	g	3.0×10^{10}	4.3×10^{10}	2.1×10^{10}	1.5×10^{5b}	44.0	62.6	29.8
3 Wind	J	3.9×10^{11}	5.5×10^{11}	2.6×10^{11}	2.5×10^{3c}	9.7	13.8	6.5
4 Geo-thermal heat	J	9.0×10^{10}	1.3×10^{11}	6.1×10^{10}	1.2×10^{4b}	10.8	15.4	7.3
SUM *(excluding sun and wind)*						*54.8*	*78.0*	*37.1*
PURCHASED MATERIALS (M)								
Cultivation Phase								
Miscellaneous materials								
5 Diesel, fields	J	4.0×10^{10}	2.7×10^{10}	1.5×10^{10}	1.8×10^{5d}	73.0	48.9	27.4
6 Lubricant and grease	J	1.9×10^{9}	1.1×10^{9}	5.7×10^{8}	1.8×10^{5d}	3.5	2.1	1.0
7 LPG	J	9.1×10^{8}	6.6×10^{9}	2.3×10^{9}	1.7×10^{5d}	1.6	11.3	3.8
8 Fleece and propagation tray	g	8.8×10^{3}	1.9×10^{5}	9.8×10^{4}	8.9×10^{9e}	0.8	16.7	8.7
9 Electricity	J	1.6×10^{10}	3.8×10^{9}	3.8×10^{9}	2.9×10^{5f}	45.4	10.9	10.9
10 Seedlings	pcs	0.0	3.2×10^{5}	1.8×10^{5}	9.6×10^{9g}	0.0	31.0	17.0
11 Seed	g	2.1×10^{4}	2.9×10^{4}	1.3×10^{4}	1.5×10^{9h}	0.3	0.4	0.2
12 Potato seeds	g	1.1×10^{6}	3.1×10^{6}	1.3×10^{6}	2.9×10^{9i}	30.1	89.4	37.3
SUM						*154.6*	*210.7*	*106.3*

TABLE 9.4: Continued

		Unit	Case (Unit)	M-Low (Unit)	M-High (Unit)	UEV (seJ/ unit)	Case emergy flow (× 10^{14}seJ)	M-Low emergy flow (× 10^{14}seJ)	M-High emergy flow (× 10^{14}seJ)
	Irrigation								
13	Diesel	J	1.5×10^{10}	6.6×10^{8}	4.6×10^{8}	1.8×10^{5}[d]	27.8	1.2	0.8
14	Electricity	J	0.0	1.3×10^{10}	8.9×10^{9}	2.9×10^{5}[f]	0.0	36.9	25.9
15	Ground water	g	3.6×10^{9}	3.0×10^{9}	2.1×10^{9}	1.1×10^{6}[g]	41.5	34.5	24.2
16	Tap water	g	1.4×10^{9}	0.0	0.0	2.3×10^{6}[j]	30.8	0.0	0.0
	SUM						*100.0*	*72.6*	*50.9*
	Soil fertility enhancement								
17	Woodchips	J	3.7×10^{11}	0.0	0.0	1.1×10^{4}[k]	38.8	0.0	0.0
18	Lime	g	2.0×10^{4}	0.0	0.0	1.7×10^{9}[k]	0.3	0.0	0.0
19	Nitrogen (N)	g	0.0	18	2.8×10^{4}	4.1×10^{10}[l]	0.0	0.0	105.7
20	Phosphorus (P2O5)	g	0.0	3.0×10^{4}	7.4×10^{4}	3.7×10^{10}[l]	0.0	27.1	59.3
21	Potash (K$_2$O)	g	6.6×10^{4}	2.5×10^{5}	2.9×10^{5}	2.9×10^{9}[k]	2.3	8.4	9.8
	SUM						*41.5*	*35.6*	*174.8*
	Farm Assets								
22	Tractors	g	1.4×10^{5}	8.2×10^{4}	3.8×10^{4}	8.2×10^{9}[m]	11.7	16.6	5.8
23	Other machinery	g	1.5×10^{5}	2.5×10^{5}	1.3×10^{5}	5.3×10^{9}[m]	8.0	11.4	5.4
24	Irrigation pipe	g	8.9×10^{3}	8.9×10^{3}	8.9×10^{3}	8.9×10^{9}[e]	0.8	0.8	0.8
25	Wood for buildings	J	9.9×10^{9}	9.9×10^{9}	9.9×10^{9}	1.1×10^{4}[k]	1.0	1.0	1.0
26	Glass for buildings	g	7.6×10^{4}	0.0	0.0	3.6×10^{9}[e]	2.8	0.0	0.0
27	Plastic for buildings	g	1.9×10^{4}	0.0	0.0	8.9×10^{9}[e]	5.8	0.0	0.0
28	Steel for buildings	g	2.5×10^{4}	0.0	0.0	3.7×10^{9}[m]	0.9	0.0	0.0
	SUM						*30.9*	*29.8*	*12.7*

TABLE 9.4: Continued

	Unit	Case (Unit)	M-Low (Unit)	M-High (Unit)	UEV (seJ/unit)	Case emergy flow (× 10^14 seJ)	M-Low emergy flow (× 10^14 seJ)	M-High emergy flow (× 10^14 seJ)
Distribution Phase [q]								
29 Diesel	J	1.6	2.8×10^{10}		1.8×10^{5d}	28.6	50.3	
30 Gasoline	J	0.0	1.6×10^{10}		1.9×10^{5d}	0.0	29.6	
31 Electricity	J	0.0	3.6×10^{9}		2.9×10^{5f}	0.0	10.5	
32 Natural Gas	J	0.0	4.6×10^{9}		6.8×10^{4d}	0.0	3.2	
33 Machinery (van)	km	5.7×10^{3}	0.0		2.5×10^{10o}	1.4	0.0	
34 Machinery (truck)	tkm	0.0	1.5×10^{4}		4.1×10^{9p}	0.0	0.6	
SUM						30.0	94.2	
35 **LABOR AND SERVICE (L&S)** [q]	£	8.7×10^{4}	$1.5 > 10^{5}$		4.0×10^{12f}	3451.1	5,854.9	
SUM Purchased materials (M)						357.0	442.9	439.0
SUM Feedback from economy (M + L&S)						3808.2	6297.8	6293.9
TOTAL EMERGY USED (U) with L&S						3863.0	6375.8	6330.9
TOTAL EMERGY USED (U) without L&S						411.9	520.9	476.1

[a] By definition. [b] Odum (2000) [46]. [c] Odum (2000) [47]. [d] Brown et al (2011) [48]. [e] Buranakarn (1998) [49]. [f] NEAD database [50]. [g] This study. Based on input of 20 cm³ peat and 1/774 l diesel per seedling [31]. [h] Coppola (2009) [51]. [i] This study, based on total M-High emergy use (less emergy for distribution phase and seed potato), allocated based on the yields and share of cultivated area used for main crop potato. [j] Buenfil (1998) [52]. [k] Odum (1996) [23]. [l] Brandt-Williams (2002) [53]. [m] Kamp (2011) [54]. [n] Bargigli (2003) [55]. [o] This study. Weight of vehicle: 1500 kg, lifetime 500.000 km and same transformity as for tractors. [p] This study. Transformity per tkm calculated based on Pulselli (2008) [56]. [q] M-Low and M-High have identical distribution system and need the same L&S calculated based on the consumer prices.

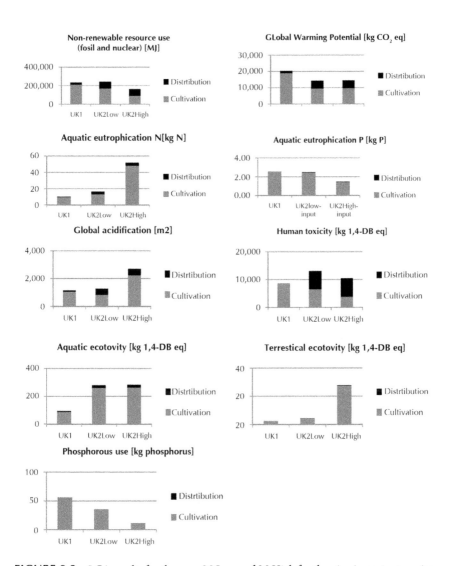

FIGURE 9.2 LCA results for the case, M-Low and M-High for the nine impact categories considered. Impacts per functional unit are divided into distribution phase and cultivation phase.

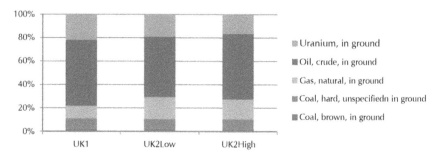

FIGURE 9.3 Contribution of raw materials to the overall result for the Life Cycle Impact category non-renewable resource use (fossil and nuclear) per functional unit for the case, M-Low and M-High.

9.5.2 BENCHMARKING AGAINST MODEL SYSTEMS

An important difference between the food supply system of the case study and the model systems is the amount of food lost. The long chain in the model systems generates a high percentage of food loss, up to 29% for root vegetables [42]. The direct marketing of the case system implies that the crop loss is smaller due to higher acceptance of less-than-perfect crops. Therefore, the case farm does not need to produce as much to provide the same amount of vegetables at the consumer's door as the model production system.

Land required for providing the food service using standard practices vary between 4.29 ha for the high-yielding model system to 9.02 ha for the low-yielding model system (Table 9.2). The area required by the case farm (6.36 ha) is within this range. The land use efficiency at the system level may be calculated as the food energy provided to the consumer per hectare of cultivated area (Table 9.2, vegetables + green manure). This value is 13.3 GJ/ha for the case farm and varies between 9.4 GJ/ha (M-Low) and 19.8 GJ/ha (M-High) for the model systems. This indicates that the case farm has yields within the range of the standard practices.

The consumer price of total output from the case system is £86,800. This is significantly lower than the consumer price for the model systems' output, which is £147,300 (Table 9.4). That the case farmer is able to sell the products at a significantly lower price may be explained by the fact that the full revenue goes directly to the farm (Figure 9.1A) whereas in the modeled systems the supermarkets, freight companies and regional distribution centers (RDC) need to make a profit as well (Figure 9.1B).

9.5.2.1 BENCHMARKING BASED ON EMERGY USE

The emergy use for purchased materials in the model systems is very similar in total but is differently distributed among the different components, e.g., M-Low has twice as much input in the cultivation phase whereas M-High has five times higher input for soil fertility enhancement. By definition, the emergy use in the distribution phase and for L&S is identical for the two systems. The L&S constitute by far the biggest contribution for both M-Low and M-High with about 92% in both systems. The total emergy used for L&S is 5.9 × 1017 seJ for the model systems (Table 9.4), which is 70% more than for the case system. This directly reflects that consumer price for the vegetables are 70% higher in the supermarket than in the direct marketing scheme.

When disregarding L&S, the case system uses less emergy to produce the total amount of vegetables sold compared to both model systems (Figure 9.4). This is especially due to a reduced emergy need for purchased seedlings and seed potato as compared to M-Low and a reduced emergy use for soil enhancements as compared to M-High. In addition, the case distribution system only use one third of the emergy used by the model supply chain. However, the case has a substantial higher consumption of on-farm fuel use and needs more emergy for water for irrigation (Figure 9.4).

The case-study farm uses significantly more diesel in the cultivation phase (Table 9.4). This may partially be explained by the tractors being less efficient than those assumed for the model systems. Another factor is that the diesel use per area is more or less independent of the yields, which means that high

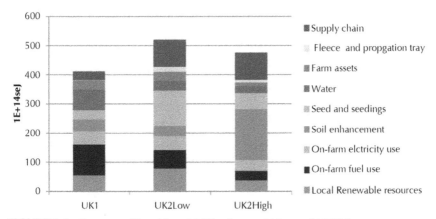

FIGURE 9.4   Emergy profiles without L&S for the case, M-Low and M-High.

yielding crops tend to use less fuel per unit output. This is clearly reflected in the comparison of M-Low and M-High, but does not explain why M-Low uses less diesel than the case system.

On-farm electricity use (Figure 9.4) consists of electricity use for on-farm production of seedlings and offices (only for the case-study) as well as for irrigation (only model farms). Disregarding electricity for irrigation, the case uses significantly more electricity than the model systems. However, the electricity consumption of 4350 kWh is still relatively small as it corresponds to the average UK household (4391 kWh, [57]). The emergy needed for electricity in the case system (4.5×1015 seJ) is partly compensated by the emergy needed for purchased seedlings in M-Low (3.1×1015 seJ) and M-High (1.7×1015 seJ) (Table 9.4). The fact that 30% of the seed potatoes are farm-saved in the case system results in a considerable emergy saving as compared to both model systems. M-Low is particularly bad in this respect since it needs a larger area (Table 9.2) and thus more seed potatoes to produce the required amount of potatoes (Table 9.4).

M-High has the lowest emergy use for irrigation with M-Low using 50% more and the case using twice as much. The latter is in the first place a consequence of that the case uses more water (3.6×109 g groundwater and 1.4×109 g tap water) (Table 9.4). As the annual variation in precipitation is not considered in the model systems, the higher use of water for irrigation in the case system may reflect that the studied period, 2009–2010, was relatively dry, and for instance in 2008 the water use was 70% less. In addition, tap water, which accounts for 28% of total water used, has an UEV value twice as high as ground water due to the extra work that is needed for pumping and treatment (Table 9.4).

Emergy used for soil enhancement is the biggest input to M-High (Table 9.4). With 1.7×1016 seJ it is more than four times higher than the other systems. This reflects that fertilizer is a valuable resource, and that reducing the import of fertilizer is a key element in reducing emergy use in agricultural systems.

The model supply chain needs a total of 805 L diesel (calculated from Table 9.3) for HGV-transport. In addition, 932 L of gasoline is used for the 58% of the shopping trips done by car and 12.6 L of diesel for the 8% of the trips done by bus. The total use of liquid fuels in the model supply chain (4.5×1010 J) is thus bigger than on-farm use of diesel in the cultivation phase in all three systems (Table 9.4). The total emergy use for the distribution system is three times higher for the model systems than for the case system.

M-Low has the largest contribution of local renewable flows as these are calculated directly from the size of the farm (Table 9.4). Further, the emergy indices (Table 9.5) reveal that, disregarding L&S, M-High has the smallest share of local renewable inputs (8%). The renewable resources contribute only with 0.08 seJ per seJ invested from society (EYR = 1.08). M-low is in this respect a bit better than the case. The case on the other hand provides the vegetables with the highest resource efficiency (lowest UEV or transformity) and is as such overall more efficient than both model systems (Table 9.5). This is especially true when also considering L&S in which case the transformity of the case is 39% lower than for M-High.

9.5.2.2   BENCHMARKING BASED ON LCA

The distribution phase has an important contribution to the environmental impacts of the model systems and in particular for the impact categories non-renewable resource use, Global Warming Potential (GWP) and human toxicity (Figure 9.2). The use of non-renewable resources in the case system is similar to M-Low, while the impact of M-High is around 30% lower (Figure 9.2). The GWP of the case is about 40% higher than both model systems. The difference in GWP between M-Low and the case was related to differences in management processes. The on-farm production of seedlings and composting of woodchips, respectively, may not be as efficient as centralized production of seedlings and use of only green manure and rock phosphate for nutrient supply. The impact category Phosphorus use was calculated to be higher in the case study as compared to the model systems due to the use of vermiculite, but it is necessary to bear in mind that the levels of phosphorus use were relatively low for all three analyzed systems. The case system and

TABLE 9.5: Emergy indices for the case system, M-Low and M-High with and without labor and service.

	With labor and service			Without labor and service		
	Case	M-Low	M-High	Case	M-Low	M-High
Emergy Yield Ratio (U/F)	1.01	1.01	1.01	1.15	1.18	1.08
Renewability (R/U)	1%	1%	1%	13%	15%	8%
Solar transformity (seJ/J)	5.20×10^6	8.58×10^6	8.52×10^6	5.54×10^5	7.01×10^5	6.40×10^5

M-Low have significantly lower aquatic eutrophication N potential, terrestrial ecotoxicity and aquatic ecotoxicity than M-High. This is because these impact categories are more dependent on the applied fertilization and irrigation levels rather than on capital goods and on-farm diesel and electricity. Aquatic ecotoxicity and human toxicity effects of the case were also lower than both model systems. Aquatic ecotoxicity levels were found to be similar for model systems while human toxicity of M-Low was shown to be slightly higher than M-High.

In addition, for model systems, environmental impacts for all impact categories are clearly dominated by agricultural cultivation (Figure 9.2). As for the case-study farm, more than 50% of the non-renewable resource use in the model systems is from use of crude oil (Figure 9.4). Nearly 20% of non-renewable resource use in model systems comes from natural gas, while for the case it is only 10%. Uranium ore has a relatively high contribution, nearly 20% of the result for all systems, as the electricity mix in the UK includes nuclear energy. Hard coal contributes to around 9% of the resource use and the remaining raw materials play a minor role (5% or less).

Assessment of environmental impacts exclusively from the distribution phase reveals that the local distribution system provides significantly lower environmental impacts per functional unit for all of the impact categories considered (Table 9.6). The relative advantage of the case system compared to the model system reached from 69% for the non-renewable resource use up to 98% in the case of human toxicity potential.

TABLE 9.6: Environmental impacts per functional unit exclusively for the distribution phase (the same for M-Low and M-High).

Impact Category	Unit	Case Distribution	Model Distribution	Relative Advantage of Case System (%)
Non-renewable, fossil and nuclear	MJ eq	23,783	75,923	69
GWP 100	kg CO_2 eq	1629	4890	67
Acidification, GLO	m^2	92	469	80
Eutrophication aq. N, GLO	kgN	0.73	3.74	81
Eutrophication aq. P, GLO	kgP	0.00	0.05	91
Human toxicity	HTP	122	6646	98
Terrestrial ecotoxicity	TEP	2	21	92
Aquatic ecotoxicity	AEP	513	1445	64
Phosphorus	kg	0	0	71

9.6 DISCUSSION

9.6.1   LCA VERSUS EMERGY ASSESSMENT—HANDLING OF CO-PRODUCTS

The assessment of sustainability of the organic low-input vegetable supply system using emergy accounting and LCA has shown that the two methods lead to the same conclusion regarding the supply chain but differ to some extent in the assessment of the production systems. The sometimes contradictory results of the emergy and LCA results are to a large part due to differences in how co-products, e.g., manure, are accounted for. In emergy accounting, the focus is on the provision of resources, and a key principle in emergy algebra is that all emergy used in a process should be assigned to all co-products as long as they are considered in separate analyses [23]. As manure cannot be produced without producing meat and milk, the entire input to livestock production should be assigned to each of the three products. We have used this approach despite its disadvantages when comparing systems with or without inputs of manure [58]. As a proxy for the UEV of manure, we have combined the UEVs of mineral N, P and K. In the LCA approach, all environmental impacts from animal production were assigned to the main products of animal production being meat and milk. As a result, only emissions associated with transportation, storage and application are considered and the principle of no import of manure in the case system is only partially reflected in the LCA results. Further, this approach has lead to the counter-intuitive result that M-Low has higher phosphorus use than M-High (Figure 9.2) even though the latter system imports twice as much phosphorus as the former (Table 9.A2). The assumption that manure is a waste may not reflect the actual situation for many organic growers who experience that the supply of N is often a limiting factor for maintaining productivity [59].

9.6.2 POTENTIALS FOR REDUCING RESOURCE USE IN THE CASE SYSTEM

Even though the case has a strong focus on minimizing use of purchased resources (M), their contribution is still more than six times larger than the contribution from local renewable resources (R) (Table 9.4). Disregarding L&S, then the largest potential for improving percentage of renewability is to reduce the amount of used fuels (Figure 9.2). However, to substitute fossil fuelled machinery with more labor intensive practices such as draft animals or manual labor, would under current socio-economic conditions increase overall resource consumption due to the high emergy flow associated with labor. In addition, draft animals would require that a considerable amount of land should be used for feed production.

Ground and tap water used for irrigation constitutes 17% of the total emergy use. Due to the differences in UEV between tap and ground water, the emergy use could be substantially reduced by using only ground water. Producing the woodchips, which accounts for 10% of the total emergy used, within the geographical boundaries of the farm would improve renewability. Currently they are residuals supplied from a local gardener who prunes and trims local gardens. In a larger perspective, there are, thus, few environmental benefits from becoming self-sufficient with wood chips in the case system. According to the LCA analysis, the composting process accounts for 30% of greenhouse gas emissions, and using less wood chip compost would reduce the overall global warming potential.

Electricity, which is primarily used for heating and lighting in the production of seedlings, constitutes 11% of the emergy used. It is no doubt convenient to use electricity for heating, but substituting the electricity with a firewood based system would largely reduce the emergy. Alternatively, it may be worthwhile to consider harvesting excess heat from the composting process to heat the green house.

As for the distribution phase, the case has the potential for decimating fossil fuel consumption by replacing the current customers with some of the many households located within few kilometers of the farm. This could dramatically reduce the 70 km round trip each week. The current way of organizing the distribution, however, is extremely efficient when compared to the alternative where customers would go by car each week and pick up the produce. The latter solution would require up to 1,000,000 car-km per year based on the case farmer's calculation. With a fuel efficiency of 15 km/L this translates to 66,666 liter of fuel. This is almost 40 times the fuel consumption for the model system (1737 L).

9.6.3   OUTLOOK FOR EMERGY USE FOR L&S

In a foreseeable future with increasing constraints on the non-renewable resources [6,60], which currently are powering the society with very high EYR-values [23], it is desirable or even necessary that agricultural systems become net-emergy providers, i.e., that more emergy is returned to society from local renewable resources than the society has invested in the production [61]. This requirement means that the contribution from R has to be bigger than F. Bearing in mind that R cannot be increased as the local renewable flows are flow limited, then achieving this can alone be achieved by reducing the emergy currently invested from society, F (3808.2 × 1014 seJ) to less than R (54.8 × 1014 seJ), i.e., by a factor of 70. Such an improvement seems out of reach without transforming the food supply system. Some improvements can be made on the farm as indicated, but the largest change will need to be in the society which determines the emergy use per unit labor. It is important to note that for the standard practices represented by the model system much larger reduction would be required.

Emergy used for L&S accounts for 89% of total emergy flow and constitutes by far the biggest potential for improvements. The L&S-component reflects the emergy used to support people directly employed on the farm and people employed in the bigger economy to manufacture and provide the purchased inputs. Due to the high average material living standard in UK with emergy use per capita being 8.99 × 1016 seJ/year [50], labor is highly resource intensive.

Emergy used for L&S can be reduced by reducing the revenue, but this is highly undesirable. Nevertheless, the employees already have a relatively low salary, which they accept because of the benefits enjoyed, e.g., free access to vegetables, cheap accommodation on the farm perimeter and in their opinion a meaningful job close to nature. Thus, the case system attracts people with a Spartan lifestyle with few expenses and thus below average emergy use.

In the future, it is almost certain that the nation-wide emergy use per capita will be reduced. A likely future scenario for the UK is that the indigenous extraction of non-renewable resources continues to decline (down 23% from 2.4 × 1024 seJ in 2000 to 1.8 × 1024 seJ in 2008 [50]). This is a result of the oil extraction plunging from 2.6 to 1.5 million barrels per day (mbd) from 2000 to 2008 (in 2010 further down to 1.1 mbd) [62]. In the same period the extraction of natural gas dropped from 97.5 to 62.7 million tonnes oil equivalent (and to 40.7 by 2010) [62]. This decline has been compensated by increasing imports of fuels from 57 × 1022 seJ in 2000 to 95 × 1022 seJ by 2008 [50]. The UK has been able to maintain a high level of emergy use per capita by gradually substituting the decline in oil and gas production with imported fuels and services. Such a substitution may continue for some years, but in a longer time perspective a decline in global production of oil, gas and other non-renewable resource is inevitable. Coupled with an increased competition from a growing global population increasing in affluence, it is likely that the import of such resources will eventually decline for the UK as well as other industrialized nations [63].

Such a future scenario imply that the resource consumption per capita will be reduced and thereby that the emergy needed for supporting labor is reduced. However, it may also bring along transformations that are more substantial in the organization of the national economy and all its sub-systems, not least the mass food supply system, which at present uses 9% of UK's petroleum products. In this perspective, the capacity of a system to adapt to changes is a crucial part of its sustainability, and this characteristic is not directly reflected in the quantitative indicators of emergy assessment and LCA.

9.6.4   LOCAL BASED BOX-SCHEME VERSUS NATIONAL-WIDE SUPERMARKET DISTRIBUTION—RESILIENCE

The supermarket based distribution system has during the previous decades been redesigned according to principles of Just-In-Time delivery (JIT). These principles aim at reducing the storage need and storage capacity at every link in a production chain, such that a minimum of capital investment is idle or in excess at any time. Less idle capital means fewer costs and fewer environmental impacts. While JIT may decrease environmental impacts per unit of produce for the particular system, as long as everything is running smoothly, it may compromise the system's resilience as it becomes more vulnerable to disturbance and systemic risks. Systemic risks include disruption in infrastructure supplying money, energy, fuel, power, communications and IT or transport as well as pandemics and climate change [64,65,66]. As can be imagined any disturbance caused by such events may quickly spread throughout the tightly connected network [65,67]. A loss of IT and communication would make it impossible for a national-wide JIT supply system to coordinate supplies [64]. A loss of money would make it impossible to conduct transactions with customers. A loss of fuel for transportation would results in large bulks of produce being stranded. A loss of power in the RDC would stop the entire chain and retailers would run out of products in a few days.

When benchmarking the case study against the national wide supermarket system, the former may be more resilient than the latter as it is in a better position to handle infrastructure failures. Due to the higher degree of autonomy and fewer actors involved, the case system would be able to work around many events, which could bring the supermarket-based system to a halt. For instance, a loss of money supply could be handled by delaying payments until the system recovers. Loss of fuel would be difficult to overcome, but produce could still be collected on bike or by public transport.

However, from the consumer's point of view, the risks have a different nature. Consumers in the case system are vulnerable to a poor or failed harvest (e.g., caused by flooding, unusual weather conditions or pests). The supermarket supply system would be unaffected by a failed harvest at a single farm because of the large number of producers feeding into the system. However, the crop diversity of the case study minimizes the risk of a complete harvest failure.

9.6.5   LIMITATIONS OF STUDY—VALIDITY OF MODEL SYSTEMS

The vegetables produced from the case-study farm have determined the design of the cultivation phase of the two model systems. It is very likely that other model systems would be developed if the systems were optimized for producing any mix of vegetables of certain food energy content but such analyses are outside the scope of this study. However, as the model systems are scale independent the results can be considered as representing the production from larger areas where different farms are producing different crops. Although in many cases the yields will vary from those presented, the high and low yield scenario should capture the range of possible outcomes.

For both emergy assessment and LCA the on-farm use of fuel and irrigation resulted in higher impacts for the case study than for the model systems. This is to a large part due that the model systems are based on standard data for field operations and that annual variation in rainfall is not reflected. This implies that any additional driving of tractors and machinery, that may occur for various reasons in a real farming system are not included. This may result in an underestimation of the actual resource consumption, which real UK organic farms would have needed in the same years under the same conditions.

9.7   CONCLUSIONS

The results of the emergy analysis showed that the case study is more resource efficient than the modeled standard practices, and with the identified potential for further reducing the emergy use, the case-study farm can become substantially better. This is especially true when also considering emergy used to support labor and service. The results of the LCA for the cultivation phase were less conclusive as the case had neither consistently more nor consistently less environmental impacts compared to the model systems. However, for the distribution phase, both the emergy assessment and LCA evaluated the case to perform substantially better than model systems. In addition, we have argued that the case may be in a better position to cope with likely future scenario of reduced access to domestic and imported fossil fuels and other non-renewable resources.

The real value of the case study is that it points out that there are alternative ways of organizing the production and distribution of organic vegetables, which are more resource efficient and potentially more resilient. The case study shows that it is possible to efficiently manage a highly diverse organic vegetable production system independently of external input of nutrients through animal manure, whilst remaining economically competitive. The success of the case system is to a large part due to management based on a clear vision of bringing down external inputs. This vision is generic but the specific practices of the case study may not always be the most appropriate for a farm to improve its resource efficiency and resilience. For systems in other societal contexts, e.g., farms with livestock and crop production or farms in remote locations, other strategies will be needed.

APPENDIX

TABLE 9.A1: Case system output as average of 2009 and 2010 and classes of crops used for establishment of model systems.

	Avg. output per year (kg)	Food energy density (KJ/100 g) [a]	Food energy at consumers (MJ)	
			Individual crops	Groups of crops
STORABLE CROPS				
Potatoes, main crop	8589	298	25,597	25,597
Potatoes, early	2863	298	8532	8532
Carrots	3575	109	3897	
Carrot bunches	678	109	739	
Carrots total				*4635*
STORABLE CROPS				
Beetroot	1446	175	2531	
Beetroot bunches	994	175	1740	
Beetroot total				*4277*

TABLE 9.A1: Continued

	Avg. output per year (kg)	Food energy density (KJ/100 g) [a]	Food energy at consumers (MJ)	
			Individual crops	Groups of crops
Onions	2927	116	3396	
Spring onions	204	116	237	
Onion bunches	48	116	55	
Onions, total				*3688*
Parsnips	1439	247	3555	3555
Leeks	2791	104	2902	2902
Squashes	2474	109	2697	2697
Sprout tops	1661	151	2508	
White cabbage	2086	105	2190	
Kale	140	155	216	
Green cabbage	176	105	185	
Red cabbage	181	92	166	
Black cabbage	75	155	117	
Pak Choi	13	58	7	
Cabbage total				*5390*
Cauliflower	890	92	819	
Swede/rutabaga	476	123	586	
Garlic	91	590	536	
Sprouting broccoli	345	117	403	
Celeriac	489	77	377	
Turnips	184	104	191	
Kohlrabi	153	104	159	
Mooli radish	126	104	131	
Savoy	99	92	91	
Fennel	51	82	42	
Daikon	8	104	9	
Cauliflower and other storable crops, total				*3344*
FRESH CROPS				
Sweet corn	811	369	2992	
Courgette	1723	81	1396	
Broad beans	713	139	991	
Beans, French	649	139	902	

TABLE 9.A1: Continued

	Avg. output per year (kg)	Food energy density (KJ/100 g) [a]	Food energy at consumers (MJ)	
			Individual crops	Groups of crops
Tomatoes	988	73	721	
Chard and leaf beet	1050	58	609	
Salad pack	1214	49	595	
Lettuce	825	49	404	
Artichokes	245	93	228	
Cucumber	391	52	203	
Spinach	273	67	183	
Mange tout	81	179	145	
Pepper	154	81	125	
Spring greens	193	49	95	
Asparagus	84	75	63	
Baby spinach	42	67	28	
Salad	45	49	22	
Rocket	30	49	15	
Fresh crops, total				*9718*
Total output	44,736		74,328	74,328

[a] From Souci *et al.* [68].

TABLE 9.A2: Applied nutrients and nutrient balances for M-Low and M-High.

	Unit	M-Low[a]	M-High[b]
Total input			
N-fixation	kg N	474	225
Atmospheric deposits	kg N	198	94
Applied N	kg N	0	261
Applied P	kg P	32	70
Applied K	kg K	241	279
	kg N	242	
Output in produce	kg P	32	
	kg K	241	
	kg N/ha	54	90
Nutrient balance	kg P/ha	0	10
	kg K/ha	0	10

[a] M-Low (7.84 ha) applies 270 kg rock phosphate and 725 kg kali vinasse; [b] M-High (3.76 ha) applies 44.3 t cattle farm yard manure, 8 kg rock phosphate and 111 kg kali vinasse.

TABLE 9.A3: Notes for emergy table (Table 9.4).

Parameter	Case	M-Low	M-High	Unit	Comments	Reference"
1 Sun						
Area	6.34×10^4	9.02×10^4	4.29×10^4	m²		[CD] [SM]
Net radiation	48.7	48.7	48.7	w/m²		[50]
Conversion factor ($60 \times 60 \times 24 \times 365$)	3.15×10^7	3.15×10^7	3.15×10^7	sek/year		
Net radiation	1.54×10^9	1.54×10^9	1.54×10^9	J/year/m²		
Total energy	9.74×10^{13}	1.39×10^{14}	6.59×10^{13}	J/year		
2 Rain						
Rain (average for the region)	0.65	0.65	0.65	m/year		[69]
Evapotranspiration, arable land	74	74	74	%		[70]
Water density	1000	1000	1000	kg/m³		
Total amount = Area × Rain × Water density × Evapotranpiration	3.04×10^{10}	4.32×10^{10}	2.05×10^{10}	g/year		
3 Wind						
Air density	1.3	1.3	1.3	kg/m³		[69]
Drag coefficient	0.001	0.001	0.001			
Wind speed average 1999	5.3	5.3	5.3	m/s		[50]
Time	3.15×10^7	3.15×10^7	3.15×10^7	sek/year		
Energy = Area × Density × Drag coefficient × Wind speed³ × Time	3.87×10^{11}	5.504×10^{11}	2.618×10^{11}	J/year		[23]
4 Geothermal heat						
Heat flow per area	45.0	45.0	45.0	mW/m²		[71]
Heat flow per area	1.42×10^6	1.42×10^6	1.42×10^6	J/m²/year		
Total heat flow	9.00×10^{10}	1.28×10^{11}	6.09×10^{10}	J/year		

TABLE 9.A3: Continued

Parameter	Case	M-Low	M-High	Unit	Comments	Reference"
CULTIVATION PHASE						
5 *Diesel, fields*						
Quantity used (3.86 × 1007 J/1)	4.03 × 10^10	2.70 × 10^10	1.52 × 10^10	J/year		
6 *Lubricant /grease*						
Quantity used (38.6 MJ/1, 43.1 MJ/kg)	1.93 × 10^9	1.13 × 10^9	5.73 × 10^8	J/year		[CD] [SM]
7 *LPG/NG—thermal weeding (flaming)*						
Quantity used (45.6 MJ/kg)	9.12 × 10^8	6.64 × 10^9	2.26 × 10^9	J/year		[CD] [SM]
8 *Fleece and propagation tray (polypropylene)*						
Propagation tray	0	168	89	kg/year		[SM]
Quantity used (10 years lifetime)	520	1197	496	m²/year		[CD] [SM]
Conversion factor	17	17	17	g/m²		[72]
Quantity used	8840	1.88 × 10^5	9.74 × 10^4	g/year		
9 *Electricity (not irrigation)*						
Electricity						
Quantity used	4350	1044	1044	kWh/year		[CD] [SM]
Quantity used (3.6 MJ/kWh)	1.57 × 10^10	3.76 × 10^9	3.76 × 10^9	J/year		
10 *Seedlings*						
Quantity used		322,933	177,577	pieces		[SM]
Sub-table: Seedlings—calculation of UEV						
Peat	20	cm³/seedling				[31]
Fossil fuels (heating)	1	1/774 seedling				
Fossil fuels (heating)	35.86	MJ/774 seedlings				
Conversion factors						

TABLE 9.A3: Continued

Parameter	Case	M-Low	M-High	Unit	Comments	Reference[a]
Peat, dry matter	0.11			g/cm³		[73]
Peat heating value (average of two values)	17,150			J/g		[74]
Input per seedling		Data	UEV	Emergy		
Peat	J	37,730	31,920	1.204 × 10⁹		
Diesel	J	46,330	181,000	8.386 × 10⁹		
UEV for seedlings				9.59 × 10⁹	sej/one seedling	
11 *Seed*						
Quantity used	20,595	29,295	13,068	g/year		[SM]
12 *Potatoe seed*	1,050,000	3,123,437	1,301,432	g/year		[CD] [SM]
M-High emergy flow (less L&S)						
- less supply chain and seed potato (to avoid circular reference)		4.388 × 10¹⁶	3.446 × 10¹⁶			
Area grown with main crop potato			0.3	ha		
Yield (40 t/ha)			12.5	t/year		
Allocation factor (share of area used for maincrop potatoes—0.31/3.76)			0.083			
UEV seed potatoes			**2.863 × 10⁹**	**sej/g**		
IRRIGATION						
13 *Diesel*						
Quantity used (38.6 MJ/l)	1.53 × 10¹⁰	6.55 × 10⁸	4.59 × 10⁸	J/year		[CD] [SM]
14 *Electricity*						
Electricity—irrigation		3540	2480	kWh/year		[SM]
Electricity—irrigation		1.27 × 10¹⁰	8.93 × 10⁹	J/year		
15 *Ground water*						
Quantity used	3.63 × 10⁹	3.02 × 10⁹	2.11 × 10⁹	g/year		[CD] [SM]

Sustainable Agriculture and Food Supply

TABLE 9.A3: Continued

	Parameter	Case	M-Low	M-High	Unit	Comments	Reference[a]
16	*Tap water*						
	Quantity used	1.35×10^9			g/year		[CD]
17	Woodchips						
	Quantity used	100			m³/year		
	Density	3.88×10^5			g/m³		[75]
	Quantity used (fresh chips, 50% DM)	3.88×10^7			g/year		
	Dry matter	1.94×10^7			g DM		
	Wood energy density	19.0			GJ/t		[76]
	Wood energy density	1.90×10^4			J/g		
	Quantity used	3.69×10^{11}			J/year		
18	Lime						
	Quantity used	2.00×10^4			g		[CD]
19	Nitrogen						
	Quantity used		0.00	2.61×10^5	g N		[CD] [SM]
20	Phosphorus						
	Imported in fertilizer		32.0	70.0	kg P		[CD] [SM]
	Converions factor (kg P_2O_5 per kg P)		2.29	2.29	kg/kg		
	Quantity used		73.0	160	kg P_2O_5		
	Quantity used		7.33×10^4	1.60×10^5	g P_2O_5		
21	Potash						
	Imported in fertilizer		242	279	kg K		Appendix 2
	Conversion factor (kg K_2O per kg K)		1.20	1.20	kg/kg		
	Quantity used		292	336	kg K_2O		
	Quantity used		2.92×10^5	3.36×10^5	g K_2O		

TABLE 9.A3: Continued

	Parameter	Case	M-Low	M-High	Unit	Comments	Reference*
22	Tractors						
	Weight loss per year	1.43×10^5	2.03×10^5	6.78×10^4	g/year	Weight of the case's tractors and lifetime. Model systems scaled to cropping area	
23	*Other machinery*						
	Weight loss per year	1.51×10^5	2.14×10^5	1.02×10^5		Weight of the case's non-motorized machinery divided lifetime min 20 years. Model systems scaled to cropped area.	
24	*Irrigation pipe (HDPE)*						
	Total weight of irrigation pipe	2.23×10^5			g		[CD]
	Estimate lifetime	25.0			years		
	Weight loss per year	8920	8920	8920	g/year		
25	*Wood*						
	Storage building, general, wood construction, non-insulated, at farm	170			m³		[CD]
	Lumber needed per m³ house	200			kg/m³		
	Lumber used for construction of office and storage area	3.40×10^4			kg		
	conversion factor	2.04×10^7			J/kg		
	Estimated lifetime of building	75.0			year		
	Mass depreciation per year	9.23×10^9	9.23×10^9	9.23×10^9	J/year		
26	*Glass*						
	Green house 2 (1970—second hand glass—plastic roof) 7*30 m	210			m²		[CD]
	Glass						

TABLE 9.A3: Continued

Parameter	Case	M-Low	M-High	Unit	Comments	Reference[a]
Thickness	5000			m		
Density	2400			kg/m³		[77]
Glass	2520			kg		
Mass depreciation (estimated glass lifetime 50 year)	5.04×10^{4}			g/year		
Green house 1 (1890—second hand glass—plastic roof) 8 × 10 m						
Glass	107			m²		[CD]
Thickness	5000			m		
Density	2400			kg/m³		
Glass	1280			kg		
Mass depreciation (estimated glass lifetime 50 year)	2.57×10^{4}			g/year		
27 *Plastic*						
Plastic (roof of green houses)	290			m²		[CD]
Thickness	5000			m		
density	950			kg/m³		
Plastic	1380			kg		
Mass depreciation	2.76×10^{4}			g/year	(glass lifetime 50 year)	
Polyethylene covers (polytunnels)						
Size of covers	214			m²	Lifetime 7 years. This is average use per year	
Thickness	1.83×10^{4}			m		

TABLE 9.A3: Continued

Parameter	Case	M-Low	M-High	Unit	Comments	Reference"
Density	950			kg/m³		
Quantity used	3.73 × 10⁴			g/year		
28 *Steel*						
Steel part of polytunnels						
steel arches	84.0			pcs		
Steel per arch	50.0			kg		
total steel	2500			kg		
Lifetime	100			years	Estimated values	
Mass depreciation per year	2.50 × 10⁴			g/year		
DISTRIBUTION PHASE						
29 *Diesel*						
Average haulage distance	70.0			Miles/week	[CD]	
Conversion factor	1.61			km/M		
Distance per year	5750			km/year		
Mileage	8.09			L/100KM		
Mileage	12.0			km/L		
Fuel use	465			L/year		
Conversion factor	3.40 × 10⁷			J/l		
Fuel use	1.58 × 10¹⁰			J/year		
Diesel use		818	818	L/year		Table 3
Diesel use		2.78 × 10¹⁰	2.78 × 10¹⁰	J/year		
30 *Gasoline*						
Gasoline use		466	466	L/year		Table 3

TABLE 9.A3: Continued

	Parameter	Case	M-Low	M-High	Unit	Comments	Reference"
	conversion factor		3.40×10^7	3.40×10^7	J/L		
	Gasoline use		1.58×10^{10}	1.58×10^{10}	J/year		Table 3
31	*Electricity*						
	Electricity use (RDC and Retail)		1001	1001	kWh/year		
	conversion factor		3.60×10^6	3.60×10^6	J/kWh		
	Electricity use		3.61×10^9	3.61×10^9	J/year		
32	*Natural gas*						Table 3
	NG use (Retail)		4.64×10^{09}	4.64×10^{09}	J/year		
33	*Machinery (van)*						
	Weight	1.50×10^{06}			g		
	UEV for lorry	8.20×10^{09}			seJ/g	Same as for tractors	
	Lifetime in km	5.00×10^{05}			km	Estimated values	
	UEV per km	2.46×10^{10}			seJ/km		
34	*Machinery (Truck)*						
	Total ton-km by truck		1.53×10^4	1.53×10^4	tkm	Calculated from Table 3	
	Emergy per truck		8.48×10^{16}	8.48×10^{16}	seJ/truck		[56]
	Truck lifetime in tkm		2.07×10^7	2.07×10^7	tkm		[56]
	Emergy per tkm		4.10×10^9	4.10×10^9	seJ/tkm		
35	*Labor and service*						
	Consumer price	86,804.5					[CD]

TABLE 9.A4: Revenue for model system based on supermarket prices.

Model systems-Labor and service	12 month avg. price (March 2012 to March 2013) [43] (£/kg)	Sold to consumers (t)	Total consumer price (£)	Reference [a]
Potatoes, main crop (3/4 of potatoes)	£ 1.11	8.6	9491	
Potatoes. Early (1/4 of potatoes)	£ 1.39	2.9	3986	
Carrots	£ 1.34	4.3	5705	
Cabbages	£ 1.72	5.1	8850	
Cauliflower	£ 2.25	3.6	8173	
Parsnips	£ 2.88	1.4	4145	
Beetroots	£ 3.46	2.4	8444	
Onions	£ 1.32	3.2	4196	
Leeks	£ 4.95	2.8	13,813	
Squash	£ 2.72	2.5	6730	
Courgettes	£ 2.00	6.0	12,003	
Lettuce (little gem)	£ 7.45	9.9	73,872	
Total/Total price index corrected	£		159,409/147.264	
Price index adjustment	2012	2009	2010	avg 2009–2010
01.1.7 Vegetables including potatoes and tubers (2005 = 1)	141.1	128.5	132.2	130.35
Em$ratio (2008 data)			2.10 × 10^{12} Sej/$	[50]
Conversion factor			0.53 USD/GBP	[78]
EM£ratio			**3.98 × 10^{12}** Sej/£	

[a] Numbers in parentheses refer to references in the reference list, [CD] = Collected data, and [SM] = Supplementary Material.

REFERENCES

1. Pelletier, N.; Audsley, E.; Brodt, S.; Garnett, T.; Henriksson, P.; Kendall, A.; Kramer, K.J.; Murphy, D.; Nemecek, T.; Troell, M. Energy Intensity of Agriculture and Food Systems. Annu. Rev. Environ. Resour. 2011, 36, 223–246.

2. Heller, M.C.; Keoleian, G.A. Assessing the Sustainability of the US Food System: A Life Cycle Perspective. Agric. Syst. 2003, 76, 1007–1041.

3. Barnosky, A.D.; Matzke, N.; Tomiya, S.; Wogan, G.O.U.; Swartz, B.; Quental, T.B.; Marshall, C.; McGuire, J.L.; Lindsey, E.L.; Maguire, K.C.; et al. Has the Earth's Sixth Mass Extinction Already Arrived? Nature 2011, 471, 51–57.

4. Rockström, J.; Steffen, W.; Noone, K.; Persson, A.; Chapin, S.; Lambin, E.; Lenton, T.; Scheffer, M.; Folke, C.; Schellnhuber, H.J.; et al. Planetary Boundaries: Exploring the Safe Operating Space for Humanity. Ecol. Soc. 2009, 14, 1–33.

5. Sorrell, S.; Speirs, J.; Bentley, R.; Brandt, A.; Miller, R. Global Oil Depletion: A Review of the Evidence. Energ. Pol. 2010, 38, 5290–5295.

6. Cordell, D.; Drangert, J.; White, S. The Story of Phosphorus: Global Food Security and Food for Thought. Global Environ. Change 2009, 19, 292–305.

7. Smil, V. Nitrogen and Food Production: Proteins for Human Diets. Ambio 2002, 31, 126–131.

8. IAASTD. International Assessment of Agricultural Science and Technology for Development, IAASTD; Island Press: Washington, DC, USA, 2009.

9. Defra. Food Industry Sustainability Strategy; Department for Environment Food and Rural Affairs: London, UK, 2006.

10. Smith, A.; Watkiss, P.; Tweddle, G.; McKinnon, A.; Browne, M.; Hunt, A.; Treleven, C.; Nash, C.; Cross, S. The Validity of Food Miles as an Indicator of Sustainable Development: Final Report; AEA Technology Environment: London, UK, 2005.

11. Coley, D.; Howard, M.; Winter, M. Local Food, Food Miles and Carbon Emissions: A Comparison of Farm Shop and Mass Distribution Approaches. Food Policy 2009, 34, 150–155.

12. Mundler, P.; Rumpus, L. The Energy Efficiency of Local Food Systems: A Comparison between Different Modes of Distribution. Food Policy 2012, 37, 609–615.

13. Cardinale, B.J.; Duffy, J.E.; Gonzalez, A.; Hooper, D.U.; Perrings, C.; Venail, P.; Narwani, A.; Mace, G.M.; Tilman, D.; Wardle, D.A.; et al. Biodiversity Loss and its Impact on Humanity. Nature 2012, 486, 59–67.

14. Østergård, H.; Finckh, M.R.; Fontaine, L.; Goldringer, I.; Hoad, S.P.; Kristensen, K.; Lammerts van Bueren, E.T.; Mascher, F.; Munk, L.; Wolfe, M.S. Time for a Shift in Crop Production: Embracing Complexity through Diversity at all Levels. J. Sci. Food. Agr. 2009, 89, 1439–1445.

15. Bjorklund, J.; Westberg, L.; Geber, U.; Milestad, R.; Ahnstrom, J. Local Selling as a Driving Force for Increased on-Farm Biodiversity. J. Sustain. Agr. 2009, 33, 885–902.

16. Nemecek, T.; Dubois, D.; Huguenin-Elie, O.; Gaillard, G. Life Cycle Assessment of Swiss Farming Systems: I. Integrated and Organic Farming. Agric. Syst. 2011, 104, 217–232.

17. Audsley, E.; Branderm, M.; Chatterton, J.; Murphy-Bokern, D.; Webster, C.; Williams, A. How Low Can We Go? An Assessment of Greenhouse Gas Emissions from the UK Food System and the Scope to Reduce Them by 2050; FCRN-WWF: London, UK, 2009.

18. Cavalett, O.; Ortega, E. Emergy, Nutrients Balance, and Economic Assessment of Soybean Production and Industrialization in Brazil. J. Clean. Prod. 2009, 17, 762–771.

19. Bastianoni, S.; Marchettini, N.; Panzieri, M.; Tiezzi, E. Sustainability Assessment of a Farm in the Chianti Area (Italy). J. Clean. Prod. 2001, 9, 365–373.

20. Rugani, B.; Benetto, E. Improvements to Emergy Evaluations by using Life Cycle Assessment. Environ. Sci. Technol. 2012, 46, 4701–4712.

21. Raugei, M.; Rugani, B.; Benetto, E.; Ingwersen, W.W. Integrating Emergy into LCA: Potential Added Value and Lingering Obstacles. Ecol. Model. 2014, 271, 4–9.

22. Stockfree Organic Services. Available online: http://www.stockfreeorganic.net/ (accessed on 19 June 2013).

23. Odum, H.T. Environmental Accounting: Emergy and Environmental Decision Making; John Wiley & Sons, Inc.: New York, NY, USA, 1996.

24. ISO. ISO 14040: Environmental Management—Life Cycle Assessment—Principles and Framework; 2006.

25. Frischknecht, R.; Jungbluth, N.; Althaus, H.; Bauer, C.; Doka, G.; Dones, R.; Hischier, R.; Hellweg, S.; Humbert, S.; Margni, M.; et al. Implementation of Life Cycle Impact Assessment Methods; Swiss Center for Life Cycle Inventories: Dübendorf, Switzerland, 2007; p. 151.

26. IPCC. Climate Change 2001: The Scientific Basis. In Third Assessment Report of the Intergovernmental Panel on Climate Change (IPCC); Houghton, J.T., Ding, Y., Griggs, D.J., Noguer, M., van der Linden, P.J., Xiaosu, D., Eds.; Cambridge University Press: Cambridge, UK, 2001.

27. Guinée, J.; van Oers, L.; de Koning, A.; Tamis, W. Life Cycle Approaches for Conservation Agriculture; CML Report 171. Leiden University: Leiden, The Netherlands, 2006; p. 156.

28. Nemecek, T.; Freiermuth, R.; Alig, M.; Gaillard, G. The Advantages of Generic LCA Tools for Agriculture: Examples SALCAcrop and SALCAfarm. In Proceedings of the 7th Int. Conference on LCA in the Agri-Food Sector, Bari, Italy, 22–24 September 2010; pp. 433–438.

29. Pré Consultants. SimaPro V 7.3.3, Pré Consultants, Amersfoort, The Netherlands, 2007.

30. Frischknecht, R.; Jungbluth, N.; Althaus, H.; Doka, G.; Dones, G.; Heck, T.; Hellweg, S.; Hischier, R.; Nemecek, T.; Rebitzer, G.; et al. The Ecoinvent Database: Overview and Methodological Framework. Int. J. LCA 2005, 10, 3–9.

31. Stoessel, F.; Juraske, R.; Pfister, S.; Hellweg, S. Life Cycle Inventory and Carbon and Water Foodprint of Fruits and Vegetables: Application to a Swiss Retailer. Environ. Sci. Technol. 2012, 46, 3253–3262.

32. Patel, M.; Bastioli, C.; Marini, L.; Würdinger, E. Life-cycle Assessment of Bio-based Polymers and Natural Fiber Composites. Biopolymers online 2003.

33. Wihersaari, M. Evaluation of Greenhouse Gas Emission Risks from Storage of Wood Residue. Biomass Bioenerg. 2005, 28, 444–453.

34. ISO. ISO 14044: Environmental Management—Life Cycle Assessment—Requirements and Guidelines; 2006.

35. Lampkin, N.; Measures, M.; Padel, S. 2011/12 Organic Farm Management Handbook, 9th ed.; The Organic Research Center: Hamstead Marshall, UK, 2011.

36. Gerrard, C.L.; Smith, L.; Pearce, B.; Padel, S.; Hitchings, R.; Measures, M. Public goods and farming. In Farming for Food and Water Security; Lichtfouse, E., Ed.; Springer: London, UK, 2012; pp. 1–22.

37. Smith, L.; Padel, S.; Pearce, B.; Lampkin, N.; Gerrard, C.L.; Woodward, L.; Fowler, S.; Measures, M. Assessing the public goods provided by organic agriculture: Lessons learned from practice. In Organic is Life-Knowledge for Tomorrow, Proceedings of the Third Scientific Conference of ISOFAR, Namyangju, Korea, 28 September–1 October 2011; Volume 2, pp. 59–63.

38. Nemecek, T.; Kägi, T. Life Cycle Inventories of Swiss and European Agricultural Production Systems; Ecoinvent report No. 15a; Swiss Centre for Life Cycle Inventories: Zürich and Dübendorf, Switzerland, December 2007.

39. Williams, A.; Audsley, E.; Sandars, D. Determining the Environmental Burdens and Resource Use in the Production of Agricultural and Horticultural Commodities; Final Report to Defra on Project IS0205; Department for Envronment Food and Rural Affairs: London, UK, 2006.

40. Milà i Canals, L. LCA Methodology and Modelling Considerations for Vegetable Production and Consumption; Working Papers 02/07; Center for Environmental Strategy (CES), University of Surrey: Guildford, UK, 2007.

41. Tassou, S.A.; De-Lille, G.; Ge, Y.T. Greenhouse Gas Impacts of Food Retailing; Project code: FO405; Defra: London, UK, 2009.

42. Terry, L.A.; Mena, C.; Williams, A.; Jenney, N.; Whitehead, P. Fruit and Vegetable Resource Maps: Mapping Fruit and Vegetable Waste through the Retail and Wholesale Supply Chain; Project code: RSC-008; WRAP: Cranfield, UK, 2011.

43. Mysupermarket. Available online: http://www.mysupermarket.co.uk/ (accessed on 3 June 2013).

44. UK National Statistics. Price Indices and Inflation: UK National Statistics Publication Hub. Available online: http://www.statistics.gov.uk/hub/economy/prices-output-and-produc-tivity/price-indices-and-inflation/index.html (accessed on 30 May 2013).

45. World Food Programme. What is hunger? Available online: http://www.wfp.org/hunger/what-is (accessed on 10 September 2012).

46. Odum, H.T. Folio #2 Emergy of Global Processes. In Handbook of Emergy Evaluation: A Compendium of Data for Emergy Computation; Center for Environmental Policy, University of Florida: Gainesville, FL, USA, 2000.

47. Odum, H.; Mark, B.; Brandt-Williams, S. Folio #1 Introduction and Global Budget. In Handbook of Emergy Evaluation: A Compendium of Data for Emergy Computation; Center for Environmental Policy, University of Florida: Gainesville, FL, USA, 2000.

48. Brown, M.T.; Protano, G.; Ulgiati, S. Assessing Geobiosphere Work of Generating Global Reserves of Coal, Crude Oil, and Natural Gas. Ecol. Model. 2011, 222, 879–887.

49. Buranakarn, V. Evaluation of Recycling and Reuse of Building Materials using the Emergy Analysis Method. Ph.D. Thesis, University of Florida, Gainesville, FL, USA, 1998.

50. National Environmental Accounting Database. Available online: http://cep.ees.ufl.edu/nead/ (accessed on 3 June 2013).

51. Coppola, F.; Bastianoni, S..; Østergård, H. Sustainability of Bioethanol Production from Wheat with Recycled Residues as Evaluated by Emergy Assessment. Biomass Bioenerg. 2009, 33, 1626–1642.

52. Buenfil, A. Emergy Evaluation of Water. Ph.D. Thesis, University of Florida, Ganiesville, FL, USA, 1998.

53. Brandt-Williams, S. Folio #4 Emergy of Florida Agriculture. In Handbook of Emergy Evaluation: A Compendium of Data for Emergy Computation; Center for Environmental Policy, University of Florida: Gainesville, FL, USA, 2002.

54. Kamp, A.; Østergård, H.; Gylling, M. Sustainability Assessment of Growing and Using Willow for CHP Production. In Proceedings from the 19th European Biomass Conference and Exhibition, Berlin, Germany, 6–10 June 2011; pp. 2645–2656.
55. Bargigli, S.; Ulgiati, S. Emergy and Life-Cycle Assessment of Steel Production in Europe. In Emergy Synthesis 2: Theory and Application of the Emergy Methodology, Proceedings from the Second Biennial Emergy Analysis Research Conference, Gainesville, FL, USA, 20–22 September 2001; pp. 141–156.
56. Pulselli, R.M.; Simoncini, E.; Ridolfi, R.; Bastianoni, S. Specific Emergy of Cement and Concrete: An Energy-Based Appraisal of Building Materials and their Transport. Ecol. Ind. 2008, 8, 647–656.
57. Regional and Local Authority Electricity Consumption Statistics: 2005 to 2011. Available online: https://www.gov.uk/government/statistical-data-sets/regional-and-local-authority-electricity-consumption-statistics-2005-to-2011 (accessed on 3 June 2013).
58. Kamp, A.; Østergård, H. How to Manage Co-Product Inputs in Emergy Accounting Exemplified by Willow Production for Bioenergy. Ecol. Model. 2013, 253, 70–78.
59. Berry, P.M.; Sylvester-Bradley, R.; Philipps, L.; Hatch, D.J.; Cuttle, S.P.; Rayns, F.W.; Gosling, P. Is the Productivity of Organic Farms Restricted by the Supply of Available Nitrogen? Soil Use Manag. 2002, 18, 248–255.
60. Brown, M.T.; Ulgiati, S. Understanding the Global Economic Crisis: A Biophysical Perspective. Ecol. Model. 2011, 223, 4–13.
61. Odum, H.T.; Odum, E.C. The Prosperous Way Down. Energy 2006, 31, 21–32.
62. Statistical Review of World Energy 2012. Available online: http://www.bp.com./bp_internet/globalbp/globalbp_uk_english/reports_and_publications/statistical_energy_review_2011/STAGING/local_assets/pdf/statistical_review_of_world_energy_full_report_2012.pdf (accessed on 3 June 2013).
63. UK Ministry of Defense. Regional Survey: South Asia Out to 2040. Strategic Trends Programme. 2012. Available online: https://www.Gov.uk/government/uploads/system/uploads/attachment_data/file/49954/20121129_dcdc_gst_regions_sasia.pdf (accessed on 3 June 2013).
64. Peck, H. Resilience in the Food Chain: A Study of Business Continuity Management in the Food and Drink Industry: Final Report to the Department for Environment, Food and Rural Affairs; Defra: Shrivenham, UK, 2006; p. 171.
65. Goldin, I.; Vogel, T. Global Governance and Systemic Risk in the 21st Century: Lessons from the Financial Crisis. Global Pol. 2010, 1, 4–15.
66. Fantazzini, D.; Höök, M.; Angelantoni, A. Global Oil Risks in the Early 21st Century. Energ. Pol. 2011, 39, 7865–7873.
67. Korhonen, J.; Seager, T.P. Beyond Eco-Efficiency: A Resilience Perspective. Bus. Strat. Environ. 2008, 17, 411–419.
68. Souci, S.; Fachmann, W.; Kraut, H. Food Composition and Nutrition Tables, 6th ed.; CRC Press: Stuttgart, Germany, 2000.
69. Regional Climates: Southern England. Available online: http://www.metoffice.gov.uk/climate/uk/so/ (accessed on 28 February 2014).
70. Dunn, S.M.; Mackay, R. Spatial Variation in Evapotranspiration and the Influence of Land use on Catchment Hydrology. J. Hydrol. 1995, 171, 49–73.
71. British Geological Survey. Available online: http://www.bgs.ac.uk/research/energy/geothermal/ (accessed on 28 February 2014).

72. Shawbury Garden Center. Gardening Supplies. Available online: http://www.shawburygar-dencentre.co.uk/ (accessed on 28 February 2014).
73. Brown, M.; Bardi, E. Folio #3 Emergy of Ecosystems. Handbook of Emergy Evaluation: A Compendium of Data for Emergy Computation; Center for Environmental Policy, University of Florida: Gainesville, FL, USA, 2001.
74. Engineeringtoolbox. Fuels-Higher Calorific Values. Available online: http://www.engineeringtoolbox.com/fuels-higher-calorific-values-d_169.html (accessed on 28 February 2014).
75. Francescato, V.; Antonini, E.; Luca, B. Wood Fuels Handbook; Italian Agriforestry Energy Association: Legnaro, Italy, 2008; p. 79.
76. Skov-og Naturstyrelsen. Træ som brændsel. Available online: http://skov-info.dk/haefte/18/kap07.htm (accessed on 28 February 2014). (In Danish).
77. The Physics Factbook. Density of Glass. Available online: http://hypertextbook.com/facts/2004/ShayeStorm.shtml (accessed on 28 February 2014).
78. OANDA. Historical Exchange Rates. Available online: http://www.oanda.com/currency/historical-rates/ (accessed on 28 February 2014).

The Sustainability of Organic Grain Production on the Canadian Prairies—A Review

Crystal Snyder and Dean Spaner

10.1 INTRODUCTION

Organic agriculture is described by the International Federation of Organic Agriculture Movements (IFOAM) as "a whole system approach based upon a set of processes resulting in a sustainable ecosystem, safe food, good nutrition, animal welfare and social justice" [1]. Organic production systems operate according to standards which, among other things, aim to promote ecosystem health, while discouraging the use of many non-organic inputs, such as synthetic fertilizers, pesticides, and certain veterinary drugs. Interest in organic production and organic food products has been increasing rapidly in recent years, due to a number of factors, including concerns about environmental sustainability, human health, and rising input costs of conventional agriculture.

Globally, the market for organic food products doubled between 2002 and 2007, to more than $46 billion (USD) [2,3], with North America representing one of the fastest growing markets in the sector. Canadian sales of organic products exceeded an estimated $1 billion in 2006 [4]. In 2009, Canada enacted new federal regulations for organic production, requiring mandatory certification to a revised national standard for all products represented as organic in inter-provincial or international trade. These regulations replace a previously volun-tary certification process and address issues of regulatory equivalency between major trading partners [5].

The number of certified organic farms in Canada has also been on the rise, increasing 60% between 2001 and 2006. In 2006, there were about 3,500 cer-tified organic farms, representing 1.5% of all farms in Canada [6]. Nearly half (45%) of these farms are situated in the Prairie Provinces, with Saskatchewan accounting for about one-third of the nationwide total. Like their conventional counterparts, most (95%) organic producers on the Prairies are engaged in the production of hay or field crops, primarily wheat and barley, but also including a variety of other grains, pulses, and oilseeds [6].

Despite the steady growth in the organic sector in recent years, it remains a fledgling research area, particularly in Western Canada. Most of the informa-tion on the benefits and impacts of organic agriculture is based on research from Europe, and there has been comparatively little research focused on the contribution of organic production to sustainable agriculture in the Canadian context. While many recognize the intuitive appeal of organic agriculture as a low-input, holistic alternative to conventional production systems, serious ques-tions remain about its long-term sustainability. In the Canadian Prairies, there is particular concern about the depletion of soil phosphorous from organic grain production [7], and the long-term impacts of tillage practices employed by organic producers [8]. Grain yields under organic management are, on average, lower than under conventional management, and it has been suggested that the yield deficit is more severe on the Canadian Prairies than some other regions [9]. Even where yields are similar, reliance on rotational strategies over syn-thetic fertilizers to maintain soil nutrients may place a further constraint on the overall productivity of organic cash crops [10]. Conversely, some studies have suggested that organic production on the Prairies requires less overall energy and contributes less to greenhouse gas emissions than conventional production, largely owing to its rejection of synthetic nitrogen fertilizers [11,12]. From a consumer's perspective, besides the environmental impacts, there are questions about food quality, safety and affordability.

The contribution of organic production to sustainable agriculture, then, in large part depends on how sustainability is defined and evaluated. Agriculture and Agri-Food Canada's Sustainable Development Strategy suggests that sus-tainable agriculture: (1) "protects the natural resource base; prevents degrada-tion of soil, water, and air quality; and conserves biodiversity," (2) "contributes to the economic and social well-being of all Canadians," (3) "ensures a safe and high-quality supply of agricultural products," and (4) "safeguards the livelihood and well-being of agricultural and agri-food businesses, workers and their families" [13]. Many proponents of organic agriculture accept it as a system that is by definition sustainable. For example, the Rodale Institute describes organic food as food produced by "tried and true sustainable methods that are as close to nature as possible" [14]. IFOAM has integrated the concept of sustainability into its official definition as well as its four overarching principles of organic agriculture—health, ecology, fairness, and care [15]. Other advocates of sustainable agriculture have more clearly delineated differences between sustainable agriculture as a general concept and organic agriculture as a specific example of a sustainable production system; inherent in this separation is a recognition that not all organic systems are necessarily sustainable [16].

In this review, we will summarize Western Canadian research on organic grain production and evaluate the sustainability of organic grain production on the Canadian Prairies in relation to its agronomic, environmental, and socioeconomic aspects.

10.2 AGRONOMIC ASPECTS OF ORGANIC GRAIN PRODUCTION ON THE CANADIAN PRAIRIES

A dichotomy exists between the extensive nature of conventional grain farming (average farm size = 424 ha; [17,18]) and the more intensive nature of organic grain production on the Prairies (average farm size = 132 ha; [18]). Organic grain producers rely on many non-chemical agronomic techniques to remain viable, and agronomic issues were consistently ranked as major priorities in recent research needs surveys. In Saskatchewan, Manitoba and Alberta, three of the top four overall production concerns related to field crops, and specifically called for research into weed management, crop rotations, and managing soil fertility/soil quality [19-21]. This is not surprising in light of the reduced yields, increased weed pressure, and reliance on non-chemical approaches for weed control and soil fertility management typical on organic farms. Most Prairie producers are relative newcomers to organic production, with 50–86% of respondents reporting less than 10 years of experience in organic management. The greatest yield reductions are often experienced in the transitional and early years of organic production [22], and this is reflected in the priorities identified by these surveys. In the following section, we review the state of Canadian research into organic weed control and soil fertility management, and comment on their potential impact on the sustainability of organic grain production on the Prairies.

10.2.1 WEED MANAGEMENT

Competition from weeds is known to reduce grain yields in both conventional and organic systems, but is often a particular challenge for organic producers due to the greater weed abundance and diversity on organically managed lands [23]. Organic producers employ a variety of methods to manage weeds, including increased seeding rates, mechanical weeding, crop rotations that disrupt the growth habit of problem weeds, and selection of cultivars that are highly competitive against weeds. Canadian organic standards also permit the use of acetic acid and plant extracts (i.e., pine oil) for weed control, but these may not be economical on a large scale [24]. Biological weed controls such as the fungus *Phoma macrostoma*, have shown promise against a variety of broadleaf weeds (including annual sow thistle and wild mustard) in preliminary research trials, but have not yet been released for widespread agricultural use [25].

Mechanical weeding methods, particularly pre-seeding tillage, are common on organic farms, but have been criticized as a primary method of weed control due to their disruption of soil structure, leading to increased erosion risk. The widespread adoption of zero-tillage practices on the Canadian Prairies has been considered a major advancement in the sustainability of conventional systems, due in large part to the reduced erosion risk and increased retention of soil moisture [26]. An assessment of management practices in the United Kingdom, where more long-term data on organic systems is available, concluded that conventional zero-tillage is environmentally superior to organic systems employing intensive tillage practices, based on a number of criteria [27]. A nine year study from the United States, on the other hand, found that organic management with minimum tillage could provide greater long-term benefits to soil quality than conventional zero-tillage [28]; however, the authors concede that reduced tillage under organic management may not provide satisfactory weed control. Weed populations on the Canadian Prairies have been shown to be responsive to different tillage intensities, with many biennial and perennial weeds prevalent under reduced tillage and annual weeds more strongly associated with conventional tillage systems [29]. A survey of Canadian organic and conventional farmers indicated that around 60% of organic farmers had reduced tillage practices on their farms [30]. Conventional farmers were more

likely to use zero-tillage and/or direct seeding systems, while organic producers relied on other forms of conservation tillage which aim to minimize the
amount of soil disturbance. In Canada, there have been few studies specifically comparing erosion risk on organic and conventional farms, but one study
comparing soil samples from organic and conventional farms in the Canadian
Prairies suggested that crop rotation had a much larger influence than the type
of production system on erosion risk [8].

There are a number of other practices that can be used in conjunction with
mechanical methods to manage weeds and reduce soil erosion risk. The use of
perennial forage crops such as alfalfa, in crop rotations, has been reported to
markedly reduce weeds in the following year [31]. Cover cropping (planting
generally leguminous crops in lieu of fallow), underseeding (planting nurse
leguminous crops with grains) and the use of green manures (plowing in cover
crops) are cropping strategies of potential value for organic grain production,
as they represent non-chemical methods for controlling weeds and improving
soil quality [32]. Nitrogen (N) recovery from green manures is generally much
higher (70–90%) than from synthetic fertilizers [32]. Fast growing leguminous
species grown as cover crops and harvested as silage (or plowed under as green
manure) have potential as a weed control strategy in organic systems. There is,
however, little scientific literature on these strategies for organic systems on the
Canadian prairies.

Wiens et al. [33] reported that in the wetter eastern regions of the prairies,
alfalfa mulch derived from strip farming in association with wheat could suppress weeds in the wheat crop. They also reported higher N uptake with alfalfa
mulch treatments than with synthetic fertilizers in the wheat and second-year
oat crop, and the oat crop also had a higher grain yield. Malhi et al. [34] reported
that organic cropping systems employing some form of fallow, or green manure
partial-fallow, tend to accumulate more nitrate-N in the rooting zone than high
input systems. They further suggested that fallow systems employing a green
manure limited leaching because they temporarily stored available nitrate-
N, while using soil water that could drive leaching, compared with fallow that
excluded vegetation.

There have been a number of integrated weed management studies in south-
central Alberta incorporating cover crops, underseeding and green manures
[35-39]. While all of these studies included some form of chemical management
in the protocol, all related their work to potential for organic systems. Sweet clover green manure used in lieu of fallow in dryland systems strongly suppressed
weeds whether harvested as hay, left on the surface, or incorporated [35,39].

The authors that some of the weed suppression effect of sweet clover may have been due to allelopathic compounds. Alfalfa, red clover, or Austrian winter pea were grown as spring or winter planted cover crops in dryland systems of the western Prairies [37]. Spring planted legumes exhibited limited growth, and there were some problems associated with winter kill, crop yield suppression and/or weed control, with all cover crops except alfalfa.

In general, while the theoretical benefits of cover cropping, underseeding and the use of green manures are evident, many have not been tested in the diverse growing conditions represented by organic management systems of the Canadian prairies. Anecdotally, however, our research group has collaborated for many years with a large-scale (600 ha) organic grain producer in Alberta who plows in leguminous grain mixtures every second year for weed control and nutrient management. He thus profitably sacrifices economic yield in every second year. In addition, this farmer incorporates crop fields where weeds become too prevalent prior to seed set, as a matter of course. The long-term effect on soil as a result of this extensive use of tillage has not been studied.

Optimization of seeding rates for organic production may also be beneficial for yield maintenance and weed control, provided the increased input costs are not prohibitive. Increasing seeding rates has been shown to be an effective strategy for enhancing crop competitiveness in integrated weed management systems [39,40], or other reduced input systems aiming to decrease herbicide use [41]. O'Donovan et al. [42], found that increasing barley crop densities enhanced the effectiveness of the herbicide tralkoxydim on wild oats, allowing for reduced application rates. Increasing seeding rates in a wheat-canola rotation reduced weed biomass and the weed seedbank after four years, with no reduction in crop yield [43]. The same authors found that when the increased seeding rates were used, herbicide application at 50% of the recommended rate was often as effective as the recommended rate. In canola, cultivar selection and increasing seeding rates were major factors in reducing dockage [44]. Economic analyses of barley-field pea and wheat-canola rotations in an integrated weed management system have demonstrated such practices to be cost-effective, particularly in the case of wheat and barley where the increased seed costs are readily offset by the agronomic gains [45]. Recognition of these benefits has led many farmers to increase their seeding rate by 50% in the past five years, with many organic farmers doubling or tripling their seeding rate [46].

In organically managed wheat and barley, doubling the seeding rate enhanced weed suppression and increased grain yields by about 10% on average [47]. This effect was not cultivar specific, and the estimated net economic returns were generally positive. A farm-scale, Canada-wide trial of different seeding rates in organically managed spring wheat suggested that a 1.25x seeding rate was nearly as effective as 1.5x or 2x seeding rates for increasing grain yield [48], and would likely make the economic return even more favourable. In organically managed pulses in Saskatchewan, increasing the seeding rate substantially above the conventional recommendation led to weed biomass reductions of up to 59% and 68% for lentil and field pea, respectively [49,50]. In lentil, economic returns were positive at the highest recommended seeding rate of 375 viable seeds m^{-2} [49], while in pea, an intermediate seeding rate (200 seeds m^{-2}) provided the best compromise between weed biomass reductions, yield gains, and input costs [50].

Crop mixtures have been considered as an agronomic approach to reducing weed pressure, protecting against pests and diseases, and enhancing yield stability [51,52]. Mixtures of Park wheat and Manny barley, for example, were shown to have equal or greater yields than monoculture wheat under organic conditions, which may be partly attributed to the weed suppressive ability of Manny barley [51]. Mixtures of AC Superb and AC Intrepid (1:1 or 1:2) wheat were found to have greater stability than AC Superb alone [53]. Pridham et al. [52] found that mixtures of wheat did not provide a yield advantage, but helped stabilize yields in the presence of disease susceptible cultivars. Further evaluation of intercropping wheat with other cereals and several noncereal crops, however, did not demonstrate a clear benefit over monoculture wheat [54].

A number of studies have suggested it may be possible to develop more competitive wheat cultivars for organic management through breeding [23]. Conventional breeding programs have largely focused on maximizing the yield potential of grains and oilseeds, with less emphasis on selection for competitive traits, due to the widespread use of synthetic herbicides in conventional agriculture. In some cases, selection for increased yield may have resulted in the loss of certain competitive traits. For example, modern semidwarf wheat cultivars have increased grain yield at the expense of plant height, which has been associated with weed competitiveness [23]. This has led some to suggest that cultivars developed before the advent of modern, high-input agriculture may be better suited to organic production. A comparison of 63 historic and modern spring wheat cultivars under low-input conditions generally supported the trend toward higher yield in modern cultivars, coupled with a reduction in weed suppression ability [55]. In another study, 27 wheat cultivars spanning more than a century of Canadian wheat breeding were compared, and it was found that certain traits were associated with increased grain yield and/or reduced weed biomass under organic management [56]. Based on this, the authors proposed an idiotype for organic wheat that included early flowering and maturity, increased tillering capacity, and increased plant height. In another study, they further compared nine wheat cultivars differing in height, tillering capacity and maturity on organic and conventional lands with different degrees of natural and simulated weed pressure [57]. Under high weed pressure, plant height, early heading and maturity were associated with increased grain yield. Tillering capacity was important at medium and low weed pressure, but was not associated with increased grain yield under high weed pressure, suggesting that the contribution of different traits to overall competitive ability depends at least in part on the degree of weed pressure. Stability analyses indicated that older cultivars (released between 1890 and 1963) were generally more yield-stable across environments, and the cultivar Park (1963), a medium height, high tillering, early maturing cultivar, may be particularly suitable for low-input management [57]. Despite the differences in competitive traits observed under different levels of weed pressure, Reid et al. [58] found that heritability estimates were similar for conventionally grown wheat under weed-free versus simulated-weedy environments. In a direct comparison of organically managed versus conventionally managed wheat, however, heritability estimates were significantly different for several traits, suggesting that cultivars for organic management should be bred under organic conditions [59]. Murphy et al. [60] also found evidence supporting the need for breeding programs specifically tailored for organic and low-input systems. In their study of 35 different soft white winter wheat breeding lines, they found that direct selection within organic systems resulted in yields 5–31% higher than indirect selection in conventional systems [60]. Reid et al. (unpublished data) corroborated this apparent need for different breeding programs but did report that of the eight highest yielding (10%) wheat lines from a recombinant inbred population tested in multi-site organic trials, five were in the top 15% in multi-site conventional trials.

10.2.2   MANAGING SOIL FERTILITY/QUALITY

According to the Canadian organic production standards, soil fertility should be managed using practices that "maintain or increase soil humus levels, that promote an optimum balance and supply of nutrients, and that stimulate biological activity within the soil" [61]. Effective management of soil fertility in organic systems requires an awareness of various interdependent factors, including choice of crop rotation, soil chemistry (i.e., pH, salinity), soil structure, and soil microbial communities whose composition and diversity can influence nutrient cycling and availability.

On the Canadian Prairies, depletion of soil phosphorous (P) under long-term organic management appears to be a significant problem. Entz et al. [7] tested soil nutrient levels on several organic farms across the Prairies and found that, while nitrogen (N), sulphur (S), and potassium (K) were generally sufficient, several farms were P deficient. A broader survey of organic farms on the Canadian Prairies confirmed low phosphorous levels, particularly on farms under long-term organic management [62]. Long-term rotational studies at Scott, Saskatchewan have also reported lower soil extractable P under organic management [34,63].

Management of soil P can be a challenge because much of the total soil P occurs in forms unavailable to plants. While it is believed that mycorrhizal colonization of plant roots can enhance P availability by making recalcitrant forms of P more accessible to plants, mycorrhiza populations are particularly sensitive to management practices. For example, higher levels of active hyphae were found in clay soil treated with manure than in soils treated with inorganic fertilizers [64]. Manure processing has also been shown to have an impact on mycorrhiza, with greater colonization under composted manure compared to raw manure or inorganic fertilizer [65], and reduced colonization when using sterile versus unsterile manure [66]. This may be attributable to greater nutrient availability and has also been reported with inorganic phosphorous fertilizers [67].

Increased tillage intensity, common in organic systems, disrupts soil microbial communities and can also have a negative impact on mycorrhizal colonization due to the destruction of the mycelial network [68]. Such disruption may exacerbate the P depletion problem. In general, soil microbial diversity and biological fertility is best encouraged by management systems with minimal tillage, increased above-ground biodiversity (i.e., diverse crop rotations or crop mixtures), and reduced synthetic inputs [68,69]. It has been suggested that a well-managed, reduced-input, zero-tillage conventional system could compete favourably against organic systems with regard to maintaining soil biological fertility [27,68].

Crop rotations may also have a major influence on P availability. For example, forage-grain rotations were shown to deplete available P more rapidly than recalcitrant forms could be mobilized [70]. Organic grain-only rotations, on the other hand, did not deplete available P as quickly, but suffered substantially reduced yields compared to both conventional grain-only and organic or conventional forage-grain rotations [70]. Conversely, Malhi et al. [34] did not observe a consistent effect of crop diversity on extractable P under organic management, even though P tended to be lower under organic management than under reduced or high input conventional management. Despite the more rapid P depletion under forage-grain rotations, there are a number of potential benefits of including forage crops in rotation, such as increased grain yield following the forage crop, enhanced weed suppression, nitrogen fixation, and carbon sequestration [31]. Such studies highlight the challenges of balancing rotational strategies for maintaining soil quality with overall productivity and grain yield.

There are few options available for organic management of soil phosphorous through soil amendments. Rock phosphate, while permitted by organic standards, is non-renewable and may contain unacceptable levels of heavy metals. Composted livestock manure can be applied, but sources of organic livestock manure are limited, particularly on the Prairies where organic farms are primarily engaged in crop production. The use of manure from conventional sources is permitted by Canadian organic standards provided no organic source is available and it meets certain conditions [61], but critics have voiced concerns about the presence of antibiotics and other contaminants from conventionally-raised livestock [71]. Recently, there has been renewed interest in integrated crop-livestock systems [72], which could help mitigate the P depletion issue on organically managed land while maximizing the rotational benefits of forages for both grazing and subsequent grain production [31]. In fact, it has been suggested that such an integrated approach may be key to the long-term sustainability of organic grain production on the Canadian Prairies [73].

10.3   ENVIRONMENTAL ASPECTS OF ORGANIC GRAIN PRODUCTION ON THE CANADIAN PRAIRIES

The influences of organic management on soil fertility often represent the most direct and immediate environmental impact of organic agriculture, and is often a key factor in producers' decisions to adopt organic practices. Proponents of organic agriculture have argued that the environmental benefits extend further to include reductions in greenhouse gas emissions and improvements in energy use efficiency, water quality and plant and wildlife diversity [74]. To date, however, most of the long-term research into the environmental impacts of organic agriculture has been conducted in Europe and only a few studies have examined the potential impacts of organic systems on the Canadian prairies.

Modeling of a hypothetical transition to organic production in Canada suggested that a total transition of Canadian canola, corn, soy and wheat production to organic management would reduce overall national energy consumption by 0.8%, global warming emissions by 0.6% and acidifying emissions by 1% [12]. Despite slightly higher fuel-related energy consumption in organic systems, the average cumulative energy demand for organic systems was estimated to be about 39% that of conventional management, mainly due to the energy-intensiveness of synthetic fertilizer and pesticide production for conventional systems. These estimates, however, are based on a number of assumptions which may not be broadly applicable to the Canadian Prairies. The study assumes yield reductions of only 5–10% under organic management, which may not be realistic, especially during and immediately following the transition to organic management [9]. Second, while the study may be useful for best-case illustrative purposes, a complete national transition from conventional to organic production is probably impractical, particularly for canola, which has already been polluted by genetically modified varieties (>95% of all varieties grown), to the extent that organic canola can no longer be grown in Canada due to outcrossing.

Field studies of wheat-pea cropping systems in Manitoba under various conventional management regimes demonstrated that nitrogen fertilizer had the greatest impact on farm energy use and greenhouse gas emissions, and was associated with reduced economic returns at application rates above 20 kg N/ha [11]. A twelve year comparison of grain-based and integrated crop rotations under organic and conventional management in Manitoba concluded that integrated rotations under organic management were the most energy efficient [75]. The authors caution, however, that soil phosphorous levels were lower in the integrated rotations than in the grain-based rotations after 12 years, and were lower under organic than conventional management. It is unclear whether any apparent near-term energy savings would remain significant once the energy costs associated with long-term phosphorous management are accounted for. In his review of a more extensive body of European research, Trewavas [27] argued that continued reliance on conventionally derived animal manures in part nullifies the perceived energy savings associated with organic production.

One long-term North American study found that although there were significant environmental benefits to organic management, adoption of some organic technologies in conventional systems would ameliorate some of the negative environmental impacts associated with conventional systems [10]. This again reinforces the importance of management quality; it may be that a well-managed conventional system could be as good as a typical organic system. Others have also sought more of an ideological and practical middle ground, suggesting that agricultural and environmental sustainability might best be advanced through a combination of organic and conventional practices, even suggesting that organic producers should adopt transgenic crops [76,77]. This is rather unlikely given that the exclusion of genetically modified organisms is one of the central tenets of organic agriculture, but it would nevertheless be short-sighted to neglect the potential for either system to be improved through the ideological or techno-logical contributions of the other.

10.4   SOCIO-ECONOMIC ASPECTS OF ORGANIC GRAIN PRODUCTION ON THE CANADIAN PRAIRIES

10.4.1   *FACTORS INFLUENCING CONSUMER PREFERENCE FOR ORGANIC PRODUCTS*

The rapid expansion of the organic food industry in North America has been attributed to consumer perceptions that organic food products are healthier and more environmentally friendly than those produced under conventional management. A number of environmental and socio-economic problems have been associated with conventional, high-input cropping systems, and although organic production systems are often believed to have fewer negative impacts, many of the perceived benefits cannot be directly measured and necessitate faith on the part of the consumer.

A global online survey by AC Nielsen found that in North America, nearly 80% of respondents chose organic foods based on a perception that they repre-sented a healthier option, while 11% cited the environmental benefits as their major motivation for choosing organic [78]. This is in contrast to the situa-tion in Europe, where a greater proportion of respondents cited environmen-tal benefits (20%) and animal welfare (12%) as reasons for choosing organic. Interestingly, a Canada-wide survey of consumers' attitudes and willingness-to-pay for foods with enhanced health benefits reported that while a large propor-tion of Canadians were willing to pay a premium for the health benefit, when controlled for price, most consumers would choose conventional food products over genetically-modified (GM) or organic products [79]. The same study also found that less than 5% of Canadians were able to correctly answer six knowl-edge questions about conventional, organic and GM food production practices, which could indicate the preference for conventional food is one based on familiarity. The distribution of consumer valuation of organic foods was broader than for GM foods, consistent with the idea of organic food occupying a niche market in Canada [79]. In an investigation of the role of sensory, health, and environmental information on Canadians' willingness-to-pay for organic wheat bread, Annett et al. [80] reported that willingness-to-pay was greater when health information was coupled with sensory evaluation. Overall, sensory evaluations revealed that organic bread was preferred in both blind and fully labeled tests [80], despite the fact that a trained sensory panel detected no differences in color, flavour, or aroma [81].

A few studies have assessed the bread-making quality of organically grown Canadian wheat. Mason et al. [82] compared the bread-making quality of several Canadian Western Hard Red Spring wheat cultivars grown under organic and conventional management, and found that despite differences in soil nitrogen availability between the management systems, grain protein content was high enough for bread-making under both organic and conventional management. They also reported a significant management x cultivar interaction for some traits, suggesting it may be possible to breed for high-quality organic wheat. Gelinas et al. [83] compared several wheat cultivars under organic management, and concluded that both cultivar and environment played an important role in bread-making characteristics. Both Gelinas et al. [83] and Annett et al. [81] reported reduced loaf volume in organic wheat bread, which was consistent with observation by a trained sensory panel that organic wheat bread was more "dense" than conventional bread [81].

Turmel et al. [84] reported that crop rotation and management system both played a role in the mineral nutrient content of wheat produced under organic and conventional management, but no direct comparison of bread-making or nutritional quality was made. In a comparison of five Canadian spring wheat cultivars, Nelson et al. [85] reported higher grain Zn, Fe, Mg and K levels in organically produced grain. Turmel et al. [84] also reported increased Zn content in organically managed wheat, but there was an interactive effect between management system and crop rotation. The various interactions between management system and crop rotation [84], environmental conditions [83] and cultivars [82], highlight the potential complications inherent in making valid nutritional comparisons between organic and conventional food. Such complexities have also been recognized by other authors attempting to review the larger body of international literature comparing the nutritional and sensory attributes of organic vs. conventional food [27,86]. Bourn and Prescott [86] examined a variety of nutritional, sensory, and food safety studies covering a wide range of organic and conventionally produced food products, and concluded that overall, there was little evidence to support the perception that organic foods are nutritionally superior. Might this be cause for concern about the sustainability of the health and nutrition-driven North American organic marketplace?

Organic agriculture is a process, and its standards only dictate what is acceptable in relation to the production process, not the end product itself. No testing is required, for instance, to verify that the end product meets the consumer's perception that it is indeed nutritionally superior and untainted by pesticides or genetically modified organisms. Given the difficulty of truly isolating an organic system from its conventional surroundings, and the likely ongoing dependence of organic production systems on some conventional by-products (i.e., manure; [71]), it is questionable whether process standards alone will be sufficient to sustain consumer confidence in organic food products over the long-term. As consumer awareness about organic agriculture and its standards increases, it is possible that consumers will increasingly demand the implementation of product standards on organic food, which is subject to price premiums based on the (perhaps unjustified) perception that it is superior to its conventional counterparts. Cranfield et al. [87] evaluated Canadian consumer preferences of production standards for organic apples, and found that respondents preferred an organic standard that required testing of apples for pesticide residue, in contrast to the current Canadian organic process standard which only prohibits the use of pesticides on organic farms. Such product standards would undoubtedly have consequences on the price of organic food and could impact the affordability for at least some of the current market share.

10.4.2 FACTORS INFLUENCING THE ECONOMIC SUSTAINABILITY OF ORGANIC PRODUCERS

For producers, the profitability and financial stability of their operation is of paramount concern and is often a driving influence in management decisions. Although the reduction in yield under organic management is often a concern, several other factors work in favour of increased profitability of crops under organic management. Overall input costs are generally lower for organic systems, in spite of increased seed and equipment costs associated with cultural and mechanical weed control [9,10,88]. Such gains are not unique to organic systems, however, as it has been shown that reduced inputs, particularly of nitrogen, can also increase economic margins under conventional management [11,89].

Price premiums are a major factor in determining the profitability of organic systems in general, and specifically in relation to comparable conventional rotations. For example, Smith et al. [88] found that the relative profitability of several organic and conventional crop rotations was heavily dependent on the value of the price premium for the organic product. The net returns for the most profitable organic rotation tested (wheat-peas-oilseed-sweet clover) only exceeded that of the most profitable conventional rotation (continuous wheat) when price premiums on the organic product were high (50–60%). Long term economic analyses of the Rodale Institute Farming Systems Trial in the United States suggested that although net returns for an organic corn-soybean system were lower than a conventional corn-soybean system when all explicit, transitional and labour costs were taken into account, the premium required to offset this difference was only about 10%, much lower than the typical premium of 65–140% for organic grains [10].

While some may question whether such high premiums can be sustained, others have argued that organic food prices better reflect the range of production, processing, distribution and environmental costs that remain externalized in conventional systems and artificially deflate the price of conventional food [9]. Nevertheless, it seems likely that as more producers enter the organic market, increasing supply will force a reduction in some production premiums. Furthermore, as marketing of organic food products increasingly moves from direct sales (i.e., farmer's markets, community supported agriculture) into supermarkets, other players in the food distribution chain will likely capture a share of the premiums. Currently in Canada, sales of organic products in supermarkets account for about 40% of the value of the organic market [4], and more than two-thirds of each consumer dollar is captured by the food distribution and retail system [9]. Thus, the trend toward more mainstream marketing of organic food products may result in a shift of the economic benefits from the producer to the retail sector, while at the same time, increased production resulting from the mass-market demand may lead to a reduction in production premiums. On the other hand, many organic producers have expressed concern; suggesting the lack of developed distribution and marketing infrastructure for organic products represent a major constraint on the industry [19-21,90].

10.5   CONCLUSIONS

Despite the tremendous growth in demand for organic food products in the North American marketplace and a widespread perception that organic agriculture represents a more sustainable alternative to conventional production systems, questions remain about the long-term sustainability of organic grain production on the Canadian Prairies. Cropping system comparisons are inherently challenging for reductionist science, since both organic and conventional systems are characterized by a range of management practices which vary according to site-specific requirements and farmer choice. For example, although the absence of synthetic fertilizers and pesticides is a defining characteristic common to all organic systems, there is considerable diversity in crop choice, rotation, and other management practices, the sum of which determine the placement of farms along a spectrum of "organic production systems." While such diversity makes generalizations difficult, there are a number of practices commonly different between organic and conventional systems which nevertheless make such comparisons valuable.

Considerable strides have been made toward addressing the agronomic challenges inherent in organic systems, including weed control and soil fertility management, but more work is needed to ensure that production is sustainable over the long-term. Further research is needed to fully understand the impacts of long-term organic management on soil phosphorous availability, and to optimize cropping systems and management standards accordingly. Integrated crop-livestock systems [72] may play an important role in maintaining soil nutrients on organic farms and more research will be needed to determine the best practices for organic systems on the Canadian Prairies.

Concerns about soil conservation still need to be addressed through the development of methods to further reduce soil disturbance from tillage. The benefits of zero-tillage have long been recognized in conventional systems [26,69], and although adoption of zero-tillage in conventional systems has been greatly assisted by the use of herbicides for weed control, high-input costs are supporting a shift toward reduced input systems. In terms of long-term sustainability, such well-managed conventional systems may rival some organically managed systems.

The development of more competitive cultivars suitable for organic production would likely also benefit such reduced-input conventional systems. Some authors have argued that the focus on genetic engineering as a technological paradigm has in fact hindered agroecological innovations which are vital to the sustainability of agricultural systems [91]. There is some merit in the suggestion that certain agricultural research policies and funding priorities do greatly favour biotechnological approaches, but there may be some room for an ideological middle ground and a willingness for both organic and conventional systems to adopt innovations that are mutually beneficial. Conventional systems may benefit greatly from adoption of low-input agronomic strategies borrowed from organic systems, allowing for a reduced input system which can realize many of the environmental benefits of organic systems, such as increased energy efficiency and reduced greenhouse gas emissions.

From the perspective of advancing overall agricultural sustainability and productivity, this would seem to be a prudent approach, but for organic systems in particular, this may be difficult to achieve while preserving the ||identity|| of organic agriculture as something recognizably distinct from conventional systems. Given the importance of price premiums for ensuring the economic viability of organic producers, preservation of this high-value niche market will be important for the ongoing sustainability of organic production. For the same reason, the organic sector may need to address the issue of relying solely on process standards in its certification requirements [87].

There is also a need for greater consumer education on agricultural production systems. This has been recognized by both organic producers [19-21] and market researchers [79]. While there is growing awareness of both health and environmental issues associated with agricultural production, many Canadians are unaware of the differences between different production systems [79], and there is little recognition of the large externalized costs of conventional systems [9].

A full accounting of the costs associated with high-input conventional systems must consider the range of negative impacts, including reduced ground and surface water quality, crop pest problems, soil erosion, energy use, high input costs and compromised farm economic resilience. If we consider sustainable agriculture to include systems which permit indefinite future use without causing irrecoverable degradation of resources and biological integrity [92], it is clear that conventional systems relying on synthetic inputs are not sustainable over the long-term. Organic production systems offer a good alternative, but the extensive nature and commodity-driven reality of Prairie grain production may limit its widespread adoption.

REFERENCES

1. IFOAM. The IFOAM Basic Norms for Organic Production and Processing Version 2005; Available online: http://www.ifoam.org/about_ifoam/standards/norms/norm_documents_library/ Norms_ENG_V4_20090113.pdf (access on 16 January 2010).
2. Willer, H. The world of organic agriculture 2009: Summary. In The World of Organic Agriculture: Statistics and Emerging Trends 2009; Willer, H., Klicher, L., Eds.; IFOAM: Bonn, Germany; FiBL: Frick, Switzerland; ITC: Geneva, Switzerland, 2009; pp. 19-24.

3. Sahota, A. Overview of the Global Market for Organic Food and Drink. In The World of Organic Agriculture: Statistics and Emerging Trends 2004; Willer, H., Yussefi, M., Eds.; International Federation of Organic Agriculture Movements: Bonn, Germany, 2004; pp. 21-26.

4. Macey, A. Retail Sales of Certified Organic Food Products in Canada in 2006; Available online: http://www.organicagcentre.ca/Docs/RetailSalesOrganic_Canada2006.pdf (access on 30 December 2009).

5. Canadian Food Inspection Agency. Organic Products; Available online: http://www.inspection. gc.ca/english/fssa/orgbio/orgbioe.shtml (access on 19 December 2009).

6. Kendrick, J. Organic: From Niche to Mainstream (Statistics Canada: Canadian Agriculture at a Glance); Available online: http://www.statcan.gc.ca/bsolc/olc-cel/olc-cel?lang=eng&catno=96-325-X200700010529 (access on 17 January 2010).

7. Entz, M.H.; Guilford, R.; Gulden, R. Crop yield and soil nutrient status on 14 organic farms in the eastern portion of the northern Great Plains. Can. J. Plant Sci. 2001, 81, 351-354.

8. Nelson, A.; Froese, J.; Beavers, R.L. Lowering Soil Erosion Risk in Organic Cropping Systems; Final Research Report W2006-09; Available online: http://www.organicag-centre.ca/ Docs/OACC_bulletins06/OACC_Bulletin9_erosion_risk.pdf (access on 26 February 2010).

9. Macrae, R.J.; Frick, B.; Martin, R.C. Economic and social impacts of organic production systems. Can. J. Plant Sci. 2007, 87, 1037-1044.

10. Pimentel, D.; Hepperly, P.; Hanson, J.; Douds, D.; Seidel, R. Environmental, energetic, and economic comparisons of organic and conventional farming systems. Bioscience 2005, 55, 573-582.

11. Khakbazan, M.; Mohr, R.M.; Derksen, D.A.; Monreal, M.A.; Grant, C.A.; Zentner, R.P.; Moulin, A.P.; McLaren, D.L.; Irvine, R.B.; Nagy, C.N. Effects of alternative management practices on the economics, energy and GHG emissions of a wheat-pea cropping system in the Canadian prairies. Soil Till. Res. 2009, 104, 30-38.

12. Pelletier, N.; Arsenault, N.; Tyedmers, P. Scenario Modeling Potential Eco-Efficiency Gains from a Transition to Organic Agriculture: Life Cycle Perspectives on Canadian Canola, Corn, Soy, and Wheat Production. Environ. Manage. 2008, 42, 989-1001.

13. Agriculture and Agri-Food Canada. Sustainable Agriculture: Our Path Forward; Available online: http://www4.agr.gc.ca/AAFC-AAC/display-afficher.do?id=1175533355176&lang= eng (access on 30 December 2009).

14. Rodale Institute. Organic or "Natural"; Available online: http://www.rodaleinstitute.org/ organic_or_natural (access on 7 February 2010).

15. IFOAM. Principles of Organic Agriculture; Available online: http://www.ifoam.org/ about_ifoam/ principles/index.html (access on 16 January 2010).

16. Sustainable Table. The Issues: Organic; Available online: http://www.sustainabletable.org/ issues/organic/ (access on 7 February 2010).

17. Statistics Canada. Total Area of Farms, Land Tenure and Land in Crops, by Province; Available online: http://www40.statcan.ca/l01/cst01/agrc25a.htm (access on 26 February 2010).

18. Macey, A. Certified Organic Production in Canada, 2004; Available online: http:// www.cog.ca/ documents/certified_organic_production_2004_report.pdf (access on 26 February 2010).

19. Frick, B.; Beavers, R.L.; Hammermeister, A.M.; Thiessen-Martens, J.R. Research Needs Assessment of Saskatchewan Organic Farmers; Available online: http://www.organicag-centre.ca/ Docs/Saskatchewan%20Research%20Needs%20Survey%20with%20cover.pdf (access on 13 December 2009).

20. Organic Agriculture Centre of Canada. Research Needs Assessment of Manitoba Organic Farmers; Available online: http://oacc.info/Docs/Manitoba%20Research%20Needs%20 Survey%20Final%20Report_dec08.pdf (access on 13 December 2009).

21. Organic Agriculture Centre of Canada. Research Needs Assessment of Alberta Organic Farmers; Available online: http://www.organicagcentre.ca/Docs/Alberta%20survey%20 Nov12.pdf (access on 13 December 2009).

22. Canadian Organic Growers Economics of Organic Farming. In Organic Field Crop Handbook; Wallace, J., Ed.; Canadian Organic Growers: Ottawa, ON, Canada, 2001; pp. 8-10.

23. Mason, H.E.; Spaner, D. Competitive ability of wheat in conventional and organic management systems: A review of the literature. Can. J. Plant Sci. 2006, 86, 333-343.

24. Johnson, E.; Wolf, T.; Caldwell, B.; Barbour, R.; Holm, R.; Sapsford, K. Efficacy of vinegar (acetic acid) as an organic herbicide (ADF Project # 20020202, AAFC Project # A03637); Available online: http://www.agr.gov.sk.ca/apps/adf/adf_admin/reports/20020202.pdf (access on 31 December 2009).

25. Bailey, K.; Johnson, E.; Kutcher, R.; Braaten, C. An Organic Option for Broadleaved Weed Control in Cereals Using a Microbial Herbicide; Interim Report; Organic Sector Market Development Initiative (OSMDI), Canadian Wheat Board: Manitoba, Canada, 2009; Available online: http://www.organicagcentre.ca/Docs/OSMDI%20Oct%202009%20 Bailey%20Interim%20 Report.pdf (access on 31 December 2009).

26. Lafond, G.P.; Derksen, D.A. Long-term potential of conservation tillage on the Canadian prairies. Can. J. Plant Pathol. 1996, 18, 151-158.

27. Trewavas, A. A critical assessment of organic farming-and-food assertions with particular respect to the UK and the potential environmental benefits of no-till agriculture. Crop Prot. 2004, 23, 757-781.

28. Teasdale, J.R.; Coffman, C.B.; Mangum, R.W. Potential Long-Term Benefits of No-Tillage and Organic Cropping Systems for Grain Production and Soil Improvement. Agron. J. 2007, 99, 1297-1305.

29. Blackshaw, R.E. Tillage intensity affects weed communities in agroecosystems. In Invasive Plants: Ecological and Agricultural Aspects; Inderjit, S., Ed.; Birkhauser Verlag: Basel, Switzerland, 2005; pp. 209-221.

30. Nelson, A. Soil Erosion Risk and Mitigation through Crop Rotation on Organic and Conventional Cropping Systems; M.Sc. Thesis; University of Manitoba: Winnipeg, MB, Canada, 2005.

31. Entz, M.H.; Baron, V.S.; Carr, P.M.; Meyer, D.W.; Smith, S.R.; McCaughey, W.P. Potential of forages to diversify cropping systems in the northern Great Plains. Agron. J. 2002, 94, 240-250.

32. Smil, V. Feeding the World: A Challenge for the Twenty-First Century; The MIT Press: Cambridge, MA, USA, 2000.

33. Wiens, M.J.; Entz, M.H.; Martin, R.C.; Hammermeister, A.M. Agronomic benefits of alfalfa mulch applied to organically managed spring wheat. Can. J. Plant Sci. 2006, 86, 121-131.

34. Malhi, S.S.; Brandt, S.A.; Lemke, R.; Moulin, A.P.; Zentner, R.P. Effects of input level and crop diversity on soil nitrate-N, extractable P, aggregation, organic C and N, and nutrient balance in the Canadian Prairie. Nutr. Cycl. Agroecosyst. 2009, 84, 1-22.

35. Blackshaw, R.E.; Moyer, J.R.; Doram, R.C.; Boswell, A.L. Yellow sweetclover, green manure, and its residues effectively suppress weeds during fallow. Weed Sci. 2001, 49, 406-413.

36. Blackshaw, R.E.; Moyer, J.R.; Doram, R.C.; Boswall, A.L.; Smith, E.G. Suitability of under-sown sweetclover as a fallow replacement in semiarid cropping systems. Agron. J. 2001, 93, 863-868.

37. Blackshaw, R.E.; Molnar, L.J.; Moyer, J.R. Suitability of legume cover crop-winter wheat intercrops on the semi-arid Canadian Prairies. Can. J. Plant Sci. 2010, (in press).

38. Moyer, J.R.; Blackshaw, R.E.; Huang, H.C. Effect of sweetclover cultivars and manage-ment practices on following weed infestations and wheat yield. Can. J. Plant Sci. 2007, 87, 973-983.

39. O'Donovan, J.T.; Blackshaw, R.E.; Harker, K.N.; Clayton, G.W.; Moyer, J.R.; Dosdall, L.M.; Maurice, D.C.; Turkington, T.K. Integrated approaches to managing weeds in spring-sown crops in western Canada. Crop Prot. 2007, 26, 390-398.

40. O'Donovan, J.T.; Blackshaw, R.E.; Harker, K.N.; Clayton, G.W.; McKenzie, R. Variable crop plant establishment contributes to differences in competitiveness with wild oat among cereal varieties. Can. J. Plant Sci. 2005, 85, 771-776.

41. Nazarko, O.M.; Van Acker, R.C.; Entz, M.H. Strategies and tactics for herbicide use reduc-tion in field crops in Canada: A review. Can. J. Plant Sci. 2005, 85, 457-479.

42. O'Donovan, J.T.; Harker, K.N.; Clayton, G.W.; Newman, J.C.; Robinson, D.; Hall, L.M. Barley seeding rate influences the effects of variable herbicide rates on wild oat. Weed Sci. 2001, 49, 746-754.

43. Blackshaw, R.E.; Beckie, H.J.; Molnar, L.J.; Entz, T.; Moyer, J.R. Combining agronomic practices and herbicides improves weed management in wheat-canola rotations within zero-tillage production systems. Weed Sci. 2005, 53, 528-535.

44. Harker, K.N.; Clayton, G.W.; Blackshaw, R.E.; O'Donovan, J.T.; Stevenson, F.C. Seeding rate, herbicide timing and competitive hybrids contribute to integrated weed management in canola (Brassica napus). Can. J. Plant Sci. 2003, 83, 433-440.

45. Smith, E.G.; Upadhyay, B.M.; Blackshaw, R.E.; Beckie, H.J.; Harker, K.N.; Clayton, G.W. Economic benefits of integrated weed management systems for field crops in the Dark Brown and Black soil zones of western Canada. Can. J. Plant Sci. 2006, 86, 1273-1279.

46. Blackshaw, R.E.; Harker, K.N.; O'Donovan, J.T.; Beckie, H.J.; Smith, E.G. Ongoing develop-ment of integrated weed management systems on the Canadian prairies. Weed Sci. 2008, 56, 146-150.

47. Mason, H.; Navabi, A.; Frick, B.; O'Donovan, J.; Spaner, D. Cultivar and seeding rate effects on the competitive ability of spring cereals grown under organic production in northern Canada. Agron. J. 2007, 99, 1199-1207.

48. Beavers, R.L.; Hammermeister, A.M.; Frick, B.; Astatkie, T.; Martin, R.C. Spring wheat yield response to variable seeding rates in organic farming systems at different fertility regimes. Can. J. Plant Sci. 2008, 88, 43-52.

49. Baird, J.M.; Shirtliffe, S.J.; Walley, F.L. Optimal seeding rate for organic production of lentil in the northern Great Plains. Can. J. Plant Sci. 2009, 89, 1089-1097.

50. Baird, J.M.; Walley, F.L.; Shirtliffe, S.J. Optimal seeding rate for organic production of field pea in the northern Great Plains. Can. J. Plant Sci. 2009, 89, 455-464.

51. Kaut, A.H.E.E.; Mason, H.E.; Navabi, A.; O'Donovan, J.T.; Spaner, D. Organic and conven-tional management of mixtures of wheat and spring cereals. Agron. Sustain. Dev. 2008, 28, 363-371.

52. Pridham, J.C.; Entz, M.H.; Martin, R.C.; Hucl, R.J. Weed, disease and grain yield effects of cultivar mixtures in organically managed spring wheat. Can. J. Plant Sci. 2007, 87, 855-859.

53. Kaut, A.H.E.E.; Mason, H.E.; Navabi, A.; O'Donovan, J.T.; Spaner, D. Performance and stability of performance of spring wheat variety mixtures in organic and conventional management systems in western Canada. J. Agr. Sci. 2009, 147, 141-153.

54. Pridham, J.C.; Entz, M.H. Intercropping spring wheat with cereal grains, legumes, and oilseeds fails to improve productivity under organic management. Agron. J. 2008, 100, 1436-1442.

55. Murphy, K.M.; Dawson, J.C.; Jones, S.S. Relationship among phenotypic growth traits, yield and weed suppression in spring wheat landraces and modern cultivars. Field Crop Res. 2008, 105, 107-115.

56. Mason, H.E.; Navabi, A.; Frick, B.L.; O'Donovan, J.T.; Spaner, D.M. The weed-competitive ability of Canada western red spring wheat cultivars grown under organic management. Crop Sci. 2007, 47, 1167-1176.

57. Mason, H.; Goonewardene, L.; Spaner, D. Competitive traits and the stability of wheat cultivars in differing natural weed environments on the northern Canadian Prairies. J. Agr. Sci. 2008, 146, 21-33.

58. Reid, T.A.; Navabi, A.; Cahill, J.C.; Salmon, D.; Spaner, D. A genetic analysis of weed competitive ability in spring wheat. Can. J. Plant Sci. 2009, 89, 591-599.

59. Reid, T.A.; Yang, R.C.; Salmon, D.F.; Spaner, D. Should spring wheat breeding for organically managed systems be conducted on organically managed land? Euphytica 2009, 169, 239-252.

60. Murphy, K.M.; Campbell, K.G.; Lyon, S.R.; Jones, S.S. Evidence of varietal adaptation to organic farming systems. Field Crop Res. 2007, 102, 172-177.

61. Canadian General Standards Board. Organic Production Systems General Principles and Management Standards; Available online: http://www.organicagcentre.ca/Docs/Cdn_ Stds_ Principles2006_e.pdf (access on 16 January 2010).

62. Martin, R.C.; Lynch, D.; Frick, B.; van Straaten, P. Phosphorous status on Canadian organic farms. J. Sci. Food Agric. 2007, 87, 2737-2740.

63. Malhi, S.S.; Brandt, S.A.; Ulrich, D.; Lemke, R.; Gill, K.S. Accumulation and distribution of nitrate-nitrogen and extractable phosphorous in the soil profile under various alternative cropping systems. J. Plant Nutr. 2002, 25, 2499-2520.

64. Kabir, Z.; OHalloran, I.P.; Fyles, J.W.; Hamel, C. Seasonal changes of arbuscular mycorrhizal fungi as affected by tillage practices and fertilization: Hyphal density and mycorrhizal root colonization. Plant Soil 1997, 192, 285-293.

65. Douds, D.D.; Galvez, L.; Franke-Snyder, M.; Reider, C.; Drinkwater, L.E. Effect of compost addition and crop rotation point upon VAM fungi. Agr. Ecosyst. Environ. 1997, 65, 257-266.

66. Brechelt, A. Effect of Different Organic Manures on the Efficiency of Va Mycorrhiza. Agr. Ecosyst. Environ. 1990, 29, 55-58.

67. Hamel, C.; Strullu, D.G. Arbuscular mycorrhizal fungi in field crop production: Potential and new direction. Can. J. Plant Sci. 2006, 86, 941-950.

68. Nelson, A.; Spaner, D. Cropping systems management, soil microbial communities, and soil biological fertility: A review. In Genetic Engineering, Biofertilisation, Soil Quality and Organic Farming, Sustainable Agriculture Reviews 4; Lichtfouse, E., Ed.; Springer Science+Business Media B.V.: Dordrecht, The Netherlands, 2010.

69. Clapperton, M.J.; Yin Chan, K.; Larney, F.J. Managing the soil habitat for enhanced biological fertility. In Soil Biological Fertility—A Key to Sustainable Land Use in Agriculture; Abbott, L.K., Murphy, D.V., Eds.; Springer: Dordrecht, The Netherlands, 2007; pp. 203-222.

70. Welsh, C.; Tenuta, M.; Flaten, D.N.; Thiessen-Martens, J.R.; Entz, M.H. High Yielding Organic Crop Management Decreases Plant-Available but Not Recalcitrant Soil Phosphorus. Agron. J. 2009, 101, 1027-1035.

71. Duval, J. Co-dependency between Organic and Conventional Agriculture: Transient or Long-lasting? Available online: http://www.organicagcentre.ca/Docs/DiscussionPapers09/Codependency%20final%20version.pdf (access on 18 January 2010).

72. Russelle, M.P.; Entz, M.H.; Franzluebbers, A.J. Reconsidering Integrated Crop-Livestock Systems in North America. Agron. J. 2007, 99, 325-334.

73. Entz, M.H.; Hoeppner, J.W.; Wilson, L.; Tenuta, M.; Bamford, K.C.; Holliday, N. Influence of organic management with different crop rotations on selected productivity parameters in a long-term Canadian field study. In Researching Sustainable Systems, Proceedings of the International Scientific Conference on Organic Agriculture, Adeledaide, Australia, 21–23 September 2005.

74 Lynch, D. Environmental impacts of organic agriculture: A Canadian perspective. Can. J. Plant Sci. 2009, 89, 621-628.

75. Hoeppner, J.W.; Entz, M.H.; McConkey, B.G.; Zentner, R.P.; Nagy, C.N. Energy use and efficiency in two Canadian organic and conventional crop production systems. Renew. Agr. Food Syst. 2006, 21, 60-67.

76. Ammann, K. Why farming with high tech methods should integrate elements of organic agriculture. New Biotechnol. 2009, 25, 378-388.

77. Ammann, K. Integrated farming: why organic farmers should use transgenic crops. New Biotechnol. 2008, 25, 101-107.

78. AC Nielsen. Functional Foods and Organics: A Global AC Nielsen Online Survey on Consumer Behavior and Attitudes; Available online: http://it.nielsen.com/trends/2005_cc_functional_ organics.pdf.pdf (access on 17 January 2010).

79. West, G.E.; Gendron, C.; Larue, B.; Lambert, R. Consumers' valuation of functional properties of foods: Results from a Canada-wide survey. Can. J. Agr. Econ. 2002, 50, 541-558.

80. Annett, L.E.; Muralidharan, V.; Boxall, P.C.; Cash, S.B.; Wismer, W.V. Influence of health and environmental information on hedonic evaluation of organic and conventional bread. J. Food Sci. 2008, 73, H50-H57.

81. Annett, L.E.; Spaner, D.; Wismer, W.V. Sensory profiles of bread made from paired samples of organic and conventionally grown wheat grain. J. Food Sci. 2007, 72, S254-S260.

82. Mason, H.; Navabi, A.; Frick, B.; O'Donovan, J.; Niziol, D.; Spaner, D. Does growing Canadian Western Hard Red Spring wheat under organic management alter its breadmaking quality? Renew. Agr. Food Syst. 2007, 22, 157-167.

83. Gelinas, P.; Morin, C.; Reid, J.F.; Lachance, P. Wheat cultivars grown under organic agriculture and the bread making performance of stone-ground whole wheat flour. Int. J. Food Sci. Technol. 2009, 44, 525-530.

84. Turmel, M.S.; Entz, M.H.; Bamford, K.C.; Martens, J.R.T. The influence of crop rotation on the mineral nutrient content of organic vs. conventionally produced wheat grain: Preliminary results from a long-term field study. Can. J. Plant Sci. 2009, 89, 915-919.

85. Nelson, A.; Quideau, S.; Frick, B.; Hucl, P.; Thavarajah, D.; Clapperton, J.; Spaner, D. The soil microbial community and grain micronutrient content of wheat grown organically and conventionally. Can. J. Plant Sci. 2010, (submitted).

86. Bourn, D.; Prescott, J. A comparison of the nutritional value, sensory qualities, and food safety of organically and conventionally produced foods. Crit. Rev. Food Sci. 2002, 42, 1-34.

87. Cranfield, J.; Deaton, B.J.; Shellikeri, S. Evaluating Consumer Preferences for Organic Food Production Standards. Can J. Agr. Econ. 2009, 57, 99-117.

88. Smith, E.G.; Clapperton, M.J.; Blackshaw, R.E. Profitability and risk of organic production systems in the northern Great Plains. Rene. Agr. Food Syst. 2004, 19, 152-158.

89. Khakbazan, M.; Grant, C.A.; Irvine, R.B.; Mohr, R.M.; McLaren, D.L.; Monreal, M. Influence of alternative management methods on the economics of flax production in the Black Soil Zone. Can. J. Plant Sci. 2009, 89, 903-913.

90. Degenhardt, R.; Martin, R.; Spaner, D. Organic farming in Central Alberta: Current trends, production constraints and research needs. J. Sustain. Agr. 2005, 27, 153-173.

91. Vanloqueren, G.; Baret, P.V. How agricultural research systems shape a technological regime that develops genetic engineering but locks out agroecological innovations. Res. Policy 2009, 38, 971-983.

92. Love, B.; Spaner, D. Agrobiodiversity: Its value, measurement, and conservation in the context of sustainable agriculture. J. Sustain. Agr. 2007, 31, 53-82.

PART 5
The Future of Sustainable Agriculture

What Do We Need to Know to Enhance the Environmental Sustainability of Agricultural Production? A Prioritisation of Knowledge Needs for the UK Food System

Lynn V. Dicks, Richard D. Bardgett, Jenny Bell,
Tim G. Benton, Angela Booth, Jan Bouwman,
Chris Brown, Ann Bruce, Paul J. Burgess,
Simon J. Butler, Ian Crute, Frances Dixon,
Caroline Drummond, Robert P. Freckleton,
Maggie Gill, Andrea Graham, Rosie S. Hails,
James Hallett, Beth Hart, Jon G. Hillier,
John M. Holland, Jonathan N. Huxley,
John S. I. Ingram, Vanessa King, Tom Macmillan,
Daniel F. Mcgonigle, Carmel Mcquaid,
Tim Nevard, Steve Norman, Ken Norris,
Catherine Pazderka, Inder Poonaji,
Claire H. Quinn, Stephen J. Ramsden,
Duncan Sinclair, Gavin M. Siriwardena,
Juliet A. Vickery, Andrew P. Whitmore,
William Wolmer, and William J. Sutherland

11.1   INTRODUCTION

The sustainability of food production has become a strong focus of attention in recent years, in policy, within the food industry [1] and in research [2]. This is partly in response to emerging risks to food production from global environmental change, particularly climate change [3], risks to food security from increasing global population and changing dietary habits [4] and the rising prominence of the sustainability agenda amongst consumers and in corporate governance [5].

One aspect of this 'sustainability' is environmental sustainability. Environmentally sustainable food production can be defined as food production that makes efficient use of natural resources and does not degrade the environmental systems that underpin it, or deplete natural capital stocks. In a recent working paper [6], the United Nations Sustainable Development Solutions Network proposed four post-2015 environmental development goals for agriculture. These can be summarised as: (1) slow the expansion of agriculture into sensitive natural ecosystems; (2) increase the efficiency of resource use, (3) stop unsustainable withdrawal of water and degradation of soil and (4) protect biodiversity and other ecosystem services in farmland. These proposed goals sit alongside (and will have to be reconciled with) proposed food security, economic and social development goals. The latter include the suggested requirement to increase global food supply through a combination of increased productivity and less waste of food products both pre- and post-harvest, whilst minimising the use of food crops for bio-energy.

As part of the focus on sustainable food production, many organizations within and linked to the food and farming industry, are actively supporting or developing more environmentally sustainable agriculture. Government, industry and third sector organizations use a number of levers to influence farming practices to this end. They include regulation to impose minimum environmental standards; advice, guidance or voluntary approaches to encourage the uptake of good practice; investment in research and development [7] and economic incentives and disincentives through agri-environment or farm assurance schemes, including business-to-business schemes such as GlobalGAP, and consumer-facing certification schemes such as organic, Rainforest Alliance and LEAF Marque [8]. These organizations commonly express interest in ensuring that they have access to and use the best available science, to increase their likelihood of improving environmental sustainability.

Scientific knowledge about the environmental sustainability of agriculture is derived from a wide range of disciplines including: agronomy, livestock science, ecology, hydrology, climate science, plant and animal pathology, entomology and economics. Some aspects of environmental sustainability in agriculture have been well researched, such as management methods to supply floral resources for pollinators on farmland [9]. In these cases, there is typically a need for knowledge exchange, as the knowledge is rarely synthesized and users of research in the food and farming sector do not have full access to the scientific literature, or the time to synthesize it and develop an integrated understanding. Other aspects of agricultural sustainability, such as means to reduce greenhouse gas emissions from crop and livestock production systems, are of relatively recent interest [10] and even less well integrated with the full range of environmental concerns for example, [11]. The type of integrated, multidisciplinary research required to develop novel production systems that are better for the environment overall is rarely conducted. Little has been done since the projects on integrated farming were conducted in the 1990s [12].

With increased appreciation of the importance of sustainable food production, new funding streams are being developed in the UK and elsewhere to bridge these key research and development gaps. It is now important to identify the most pressing knowledge needs, from both scientist and practitioner perspectives. Three of the authors (LVD, RPF and WJS) have been involved in several previous exercises to identify questions of importance to policymakers and practitioners [13]. These have generated substantial interest and been used to shape science policy. For example, in the UK Government's Marine Science Strategy [14], the research questions were acknowledged as being based on an exercise to identify 100 ecological questions of high policy relevance [15]. In addition an exercise to identify the top questions in agriculture [1] was subsequently used as the basis for a workshop that informed the initial priorities of the UK Global Food Security Research Programme. Both examples demonstrate the importance of bringing a diverse interest group together to identify knowledge needs, and the demand for studies of this kind.

Improving the environmental sustainability of agriculture will inevitably involve changes in farming practices, aided by the development and adoption of new products and technologies [16]. There are a number of interacting drivers of farming practice [17], with farmers' attitudes and beliefs expected to be key determinants of the management actions ultimately undertaken [18,19]. In the UK, farmers receive advice from a complex array of sources that include government, industry and non-governmental organizations [20]. A growing body of evidence highlights farmers' confidence in different sources of advice. Other farmers and family members have long been considered as valued sources of advice [21], and research has shown that this advice is more highly valued than information from commercial, Government or other organisations, which might be viewed as having vested interests [18,22]. Farmers also value trusted relationships with professional advisers, including agronomists, feed advisers, land agents and vets, who focus on improving farm business performance and resource efficiency to deliver economic as well as environmental outcomes [21]. If the research community is to engage the farming community in developing environmentally sustainable agricultural systems, it must be a collaborative process, carried out in partnerships between researchers and farmer-led organizations, or organizations trusted and respected by the farming industry.

The exercise described here uses such a collaborative approach to identify and rank knowledge needs which, if addressed, would facilitate change and enhance the environmental sustainability of agriculture in some way. It goes beyond identifying research questions, as it includes cases where the scientific knowledge already exists but needs to be communicated or adapted for use. Often the challenge comes in translating a large body of existing scientific information to address an applied problem. In common with a recent exercise on conservation of wild insect pollinators [23], the organisers made a particular effort to include the full range of interest groups directly linked to the production aspects of farming, including large private sector commercial interests such as supermarkets, agrochemical companies and food manufacturers that play a significant role in the agri-food supply chain, along with policymakers and non-Governmental organisations.

11.1.1 SCOPE

This exercise is about enhancing the environmental sustainability of agricultural production within the UK or for the UK market. It includes agricultural areas around the world that provide food for UK markets, as well as all UK farming. It does not include issues surrounding consumption and demand for food, or human health and nutrition.

11.2 METHODS

Our process of defining priority knowledge needs for enhancing sustainable agriculture involved collaborative development of an initial long list, followed by three stages of voting or scoring. Online surveys were designed and conducted using the online survey tool Qualtrics [24].

11.2.1   WHO WAS INVOLVED?

Forty-six people participated in one or more stages of the prioritisation process. They comprised 17 research scientists, 20 from businesses or trade associations involved in agriculture, food production or retail, five from government departments or agencies and four from charitable organizations with a strong focus on sustainable agriculture. Of these 46, eight participated in only the early stages of prioritisation, by suggesting knowledge needs or taking part in the initial round of voting. The remaining 38 (those included here as authors) participated in all stages.

The research scientists were either from UK higher education institutions, Government-funded or independent research institutes, including the British Trust for Ornithology and the UK Centre for Ecology and Hydrology. They were all either leading academics, selected by the Natural Environment Research Council (NERC) Knowledge Exchange Programme on Sustainable Food Production, or Knowledge Exchange Fellows working on aspects of sustainable agriculture, funded by NERC. Leading scientists from the following research areas pertinent to sustainable farming were selected, based on relevant research grants awarded and publications in the scientific literature: soil science, crop ecology, weed ecology, biogeochemistry, farmland biodiversity and ecosystem services, plant disease, livestock disease and farm management. Of the 14 research scientists who took part in the full process, three had expertise in soils, three in biodiversity, three in crop ecology or health and two in whole farming systems, while one had expertise in livestock health (specifically diseases of dairy cattle). The remaining two researchers were Knowledge Exchange Fellows working with large food sector businesses.

We use the term 'practitioners' for all the non-academics, or end-users of research in the process. This encompasses people engaged in promoting or practising sustainable agriculture at a wide range of levels, from corporate sustainability strategies and large-scale international procurement decisions to management of individual farms. The selected participants were all directly linked with the production aspects of farming. We did not seek organisations whose main focus was food processing, waste management, packaging, distribution, consumer choice or health and nutrition. We invited representatives from the seven largest UK supermarkets by market share [25], three of the four largest agrochemical companies globally, seven food or feed manufacturers playing leadership roles in corporate sustainability in the UK, either through involvement with the NERC's Knowledge Exchange Programme on Sustainable Food Production or the Cambridge Programme for Sustainability Leadership, seven major trade associations representing UK food retail or farming, nine charitable organizations well-known for campaigning on issues related to UK farming or farm wildlife, and nine UK or devolved national government departments or agencies responsible for agricultural, land management and environmental policy. Of the 41 practitioner organizations invited, twenty-nine participated. Twenty-four of the 38 people involved in every stage of the process (those listed as authors) were classed as practitioners.

11.2.2   THE PROCESS OF DEFINING PRIORITY KNOWLEDGE NEEDS

The initial long list of potential knowledge needs was developed by all participants submitting up to 10 possible knowledge or evidence needs for consideration by the group. Participants were invited to select issues that they considered important for making agriculture more sustainable. The objective was for suggestions to be as specific as possible, but for the list as a whole to consider all aspects of the natural environment—air, soil, water, biodiversity, climate, weather and natural hazards, disease, pollution and human health. Participants consulted their colleagues and told them that the outcomes from the process would be used to shape future science investment. Thirteen example knowledge needs were provided. These were either gathered from previous or concurrent exercises that identified knowledge needs or research priorities, such as the Feeding the Future project [26] and recent workshops on soil carbon [27] or nitrogen [28], or they had been identified in individual meetings with businesses as part of the NERC Knowledge Exchange Programme on Sustainable Food Production. The examples were carried through into the subsequent stage of the process alongside new suggestions generated by the current exercise.

All the submitted knowledge needs were compiled into a list by the first author (LVD) and categorized by subject area. Nine categories were selected to provide groups of knowledge needs as similar in size as possible. Each knowledge need appeared in only one category. Where possible, the categories were chosen and named to reflect priority research areas for sustainable agriculture recently identified for new funding by the Department of Environment Food and Rural Affairs (Defra) and two UK Research Councils (NERC and the Biotechnology and Biological Sciences Research Council, BBSRC). This matching was a deliberate effort to enhance the linkage between the identified knowledge needs and future research investment, but no matching category was created if there were insufficient suggested knowledge needs to make a group of at least 15.

In the first voting stage, all members of the group anonymously voted on the long list of knowledge needs, using an online survey. Participants were asked to select between three and five items from each of the nine categories that represented the most important current knowledge needs for sustainable agriculture either within the UK, or serving UK food markets. For groups of similar knowledge needs, participants were asked to select one that most matched their concern, and told there would be a chance either to amalgamate similar needs at the workshop or to identify the most useful need. Participants had the opportunity to make comments or suggest alternative wording for each knowledge need.

The final prioritisation of knowledge needs took place at a one-day workshop held in Cambridge on 8 January 2013. In the second stage, each item on the long list was discussed during a 90-minute session dedicated to each section of the list (see Table 11.1). Three sessions ran in parallel, so each person was involved in discussions for three of the nine categories, and was able to choose which they attended. The only rule was that each group had to have between three and seven research scientists. This was imposed to ensure that the groups always had a mix of scientists and practitioners.

During the discussion sessions, all participants could see the anonymous comments or alternative wording others had suggested during the first voting stage along with the number of votes for each knowledge need. Similar knowledge needs were placed together, shaded in colour and then placed in the rank according to the one in the group that received the most votes during Stage 1. However, the original wording of each need was always retained, to maintain transparency in the process and avoid misinterpretation of individual knowledge needs.

Participants were told to identify knowledge needs that, if met, would allow their organizations to take action, change practice or enhance agricultural sustainability. They were guided by session chairs to discuss, for each knowledge need, what could be done if this was known, and to prioritise needs that would catalyse or facilitate change and could be delivered in a reasonable timeframe of less than five years.

TABLE 11.1: Structure of the initial long list of knowledge needs for environmentally sustainable agriculture.

Category	Label	Description	Number of proposed knowledge needs
Crops	C	Crop plants, including crop selection, rotation, plant breeding, yields and agroforestry, but not including pests and diseases. Future impacts of climate change on crop suitability or cropping patterns were included here.	36
Livestock	Li	Management of livestock, including livestock health and interactions between environmental sustainability and animal welfare.	31
Nutrients	N	Includes the use of non-mineral fertilizers, nutrient-related emissions to air and water.	16
Soil	S	Managing and understanding soil health and fertility, including soil carbon.	38
Pest control	P	Sustainable strategies to deal with crop pests (including plant diseases) and weeds, but not livestock disease.	23
Farm-scale management of biodiversity and ecosystem services	F	Management of biodiversity, habitats & ecosystem services at farm scale (except soil carbon and water quality issues associated with nutrients, which are in soils and nutrients respectively).	37
Landscape-scale planning for sustainable agriculture	La	Balancing ecosystem services and food production at landscape scale (beyond the individual farm). Includes governance and decision-making at this scale.	20
Markets and Drivers	Ma	Understanding the wider drivers for decision-making on farms, including influences of global commodity markets, global environmental change (expect impacts of climate change on cropping patterns) and developing markets for ecosystem services or biofuels. Issues related to farm economics and farmer behaviour were included here, including questions about implementing precision agriculture and other new technologies.	45
Monitoring	Mo	Understanding the status of the farmed environment through monitoring or application of existing datasets.	18
TOTAL NUMBER OF PROPOSED KNOWLEDGE NEEDS			264

In general, knowledge needs with more votes were given more discussion time, but there was ample opportunity to speak up for needs that had no votes, or few votes. Some knowledge needs were re-worded or amalgamated with others at this stage, by consensus. It was also possible to add additional needs at this stage. Knowledge needs were first eliminated to create a short list, then voting by show of hands during each session was used to produce a shorter list of knowledge needs under each section. We aimed to emerge from each session with four, six or eight knowledge needs, depending on the number of knowledge needs in the initial list for each category. The thresholds were set so that approximately 20% of the initially suggested knowledge needs from each category could proceed to the next stage. The selected knowledge needs in each category were also ranked, by counting votes or by consensus.

In a final plenary session, 53 knowledge needs drawn from all categories of the long list were presented to all participants, each showing the category it came from and the ranking it achieved. The list was ordered so that high-ranking knowledge needs from all categories appeared first, the low-ranking knowledge needs were at the end and the categories were dispersed evenly through the list. Each knowledge need was briefly discussed by the whole group (largely for the benefit of those who had not been in the relevant sessions). Then all participants privately scored each knowledge need between 0 and 10, with 10 being of highest importance, using hand-held electronic voting devices. The workshop facilitators (WJS and LVD) did not vote or score the knowledge needs at any stage. An initial ranking of all knowledge needs was presented to participants at the end of the one-day workshop.

As the final scores are ordinal data, we used medians to rank them. This also reduced the influence of unusually high or low scores from single individuals. The list of priority knowledge needs was initially drawn up as the top 20 knowledge needs according to scoring by practitioners. Taking a cut at the top 20, allowing for ties, included 23 knowledge needs scored by practitioners with a median of 7 or higher. To give scientists an equal hearing, we added to the priority list any knowledge needs scored with a median of 7 or higher by scientists, even if scored lower by practitioners.

During the workshop, all participants were asked to record how many people they had consulted to suggest knowledge needs or to help them with the initial voting stage. At least 293 people were consulted in identifying and voting on the knowledge needs prior to the workshop, in addition to the 46 involved in the process, giving 339 consultees in total.

We used a Friedman test to identify whether any of the knowledge needs were scored significantly differently from others. We used a Multiple Factor Analysis, using the R Package FactomineR [29], to look for differences in scoring patterns between scorers, for all 53 knowledge needs scored in the final session. We also used a Spearman rank correlation test to assess the correlation between practitioner and scientist median scores for each knowledge need. All statistical analyses were carried out using R version 2.15.0 [30].

11.3   RESULTS AND DISCUSSION

Our initial long list comprised 253 knowledge needs. Three were discarded because they were too general or out of scope, and 14 were added after the initial voting stage, due to late submissions. In total 264 knowledge needs fed into the meeting where the final stages of voting and scoring took place.

The knowledge needs were categorized into groups as shown in Table 11.1. Five of these categories could be matched to already emerging research priorities at Defra, NERC and BBSRC. These were: nutrients, pest control, landscape-scale planning for sustainable agriculture, farm-scale management of biodiversity and ecosystem services, and markets and drivers. One area identified for research investment by the two research councils as part of a new Sustainable Agriculture Research and Innovation initiative—water—did not emerge as a category of knowledge needs in its own right from this group. However, nine suggested knowledge needs were specific to water, and these were placed in 'crops,' 'nutrients' or 'farm-scale management of biodiversity and ecosystem services.'

Similarly, twelve questions about adoption of practices by farmers, on-farm economics, use of evidence/decision support and precision agriculture were placed in 'markets and drivers,' because there were not quite enough of them to warrant their own section, which might have been called 'farmer adoption.' This made 'markets and drivers' the largest section.

We identified 26 priority knowledge needs that met the scoring criteria identified above. These are shown in Table 11.2.

A Friedman test found that there were significant differences between the scores given by scientists and practitioners for the 53 different knowledge needs scored at the end of the workshop (Friedman test statistic M = 238.01, p = 2.2 × 10−16). We do not present the results of post-hoc tests to identify where these significant differences lie, because the high number of pair-wise tests required with 53 knowledge needs makes it difficult to assign significance to any differences.

TABLE 11.2: The 26 highest scoring knowledge needs, ranked according to median score (0 = not a priority, 10 = high priority) from practitioners (n = 24). The median scores according to scientists (n = 14) are also given and the overall median from scores across both groups. This list includes the knowledge needs scored with a median of 7.0 or more by either practitioners or scientists. Cutting off at 7.0 gives the top 23 practitioner knowledge needs and the top 14 as scored by scientists. When practitioner medians were equal, knowledge needs are ordered according to overall median, then scientist median score. The sections of the list to which each knowledge need belongs are described in Table 1: C = Crops, Li = Livestock, N = Nutrients, S = Soil, P = Pest control, F = Farm-scale management of biodiversity and ecosystem services, La = Landscape-scale planning for sustainable agriculture, Ma = Markets and Drivers, Mo = Monitoring.

Rank	Number	Knowledge need	List section	Median practitioner score	Median scientist score	Overall median (interquartile range)
1	1	How can we develop a sustainable animal feed strategy?	Li	8.5	7	8 (3.75)
2	2	What are the trade-offs between delivering different ecosystem services (including biodiversity and crop production)?	F	8	8.5	8 (1.75)
3	3	How can phosphorus be recycled effectively for farming systems?	N	8	7.5	8(3)
4	4	How can we develop 'multi-functional' land management options to maximise both agricultural productivity and environmental benefits?	La	8	7.5	8(2)
5	5	What is the smallest set of metrics to evaluate the sustainability (economic, social and environmental) of agricultural systems and interventions at farm and landscape scales?	Mo	8	6	8(3)
6	6	Can integrated control strategies protect crop yield and quality as the number of available plant protection products falls?	P	8	7	7(2)
7	7	What metrics define "soil health" and how can we measure this?	S	8	5	7(3)
8=	8	What is the relationship between soil biodiversity and agricultural production?	S	7.5	7	7(2)
8=	9	What measures might be adopted to deliver more effective means of marketising ecosystem services (such as auctions) and rewarding land managers for their delivery?	Ma	7.5	7	7(3.5)

TABLE 11.2: Continued

Rank	Number	Knowledge need	List section	Median practitioner score	Median scientist score	Overall median (interquartile range)
10	10	Assuming a substantial increase in the demand for livestock products, what systems of production, and in which locations, have the least adverse effects?	Li	7.5	6	7(3)
11	11	Why is there an increasing gap between observed yields and maximum attainable yields in arable systems, and how can this be closed?	C	7	8	7(2)
12	12	How will climate change affect the suitability, yields and management of crops on which the UK is currently or could become reliant?	C	7	7.5	7(4)
13=	13	How much further can we increase potential yield and quality of the crops on which the UK is reliant, via whatever technology?	C	7	7	7(3)
13=	14	How do we best make crop production more water efficient?	C	7	7	7(2)
15	15	How can we optimise nitrogen inputs for different agricultural systems whilst minimising nitrous oxide emissions?	N	7	6.5	7(3)

Figure 11.1 shows the results of our multiple factor analysis. This would reveal differences between individuals in scoring patterns, according to whether they were a scientist or a practitioner, and according to the nine knowledge need categories, which were analysed separately as groups of variables in the analysis. Scorers are plotted according to the first two dimensions generated by the analysis (top panel). This showed no strong difference in scoring pattern between scientists (open circles) and practitioners (closed circles). The percentages of variation explained by each of the first two axes are shown in Figure 11.1, with a cumulative total of 26.08% explained by these two dimensions. The third dimension, not shown, explained a further 10.62% of the variation. Analyses of variance showed no significant differences between scores of scientists and practitioners on the first, second or third axes ($p = 0.123$, 0.252 and 0.505 for dimensions 1, 2 and 3 respectively).

The importance of each category of knowledge need to the classification is plotted in the bottom panel of Figure 11.1. Monitoring is important to both dimension 1 and dimension 2. Crops, soil, and nutrients are also important in the first dimension, while farm-scale management, landscape-scale planning, markets and drivers were also important in the second dimension. This means there was some evidence that people in the group could be divided primarily according to how they weighted (1) the importance of monitoring, (2) agronomic considerations in agriculture, such as soil condition, plant nutrition and which crops to grow, and (3) issues of biodiversity, ecosystem services and wider drivers of sustainability. It should be noted, however, that most of the variation in scores (74%) is not represented in the first two dimensions, so numerous other factors clearly also influenced the scorers' priorities.

Interestingly, the type of scorer (scientist or practitioner) was relatively unimportant in the first two dimensions. We also ran the multiple factor analysis with the scorers re-categorised as 'business,' 'government or NGO' and 'scientist,' to see if the practitioner groups separated out. Results were very similar, again with no clear differences between groups.

Figure 11.2 shows how the practitioner and scientist median scores for all 53 scored knowledge needs were positively correlated (Spearman rank correlation test: $rs = 0.488$, $p = 0.00021$).

11.3.1   WHAT DO THE PRIORITIES TELL US?

Our analysis demonstrates that the practitioners and scientists in this exercise generally agreed on what knowledge is needed, despite coming from a wide variety of backgrounds and knowledge bases.

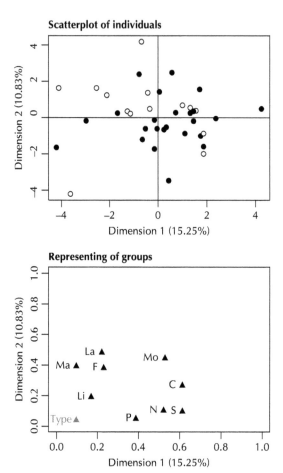

FIGURE 11.1   Results of Multiple Factor Analysis. Upper panel: Individual scorers plotted in multivariate space according to the first two dimensions. The percentage of variance explained by each dimension is given in brackets. Closed circles (●) = practitioners, open circles (o) = scientists. Lower panel: Groups of knowledge needs corresponding to relevant categories of the list (described in Table 11.1), each represented as a single point: C = Crops, Li = Livestock, N = Nutrients, S = Soil, P = Pest control, F = Farm-scale management of biodiversity and ecosystem services, La = Landscape-scale planning for sustainable agriculture, Ma = Markets and Drivers, Mo = Monitoring. The grey triangle shows the representation of scorer type (practitioner or scientist).

Many of the knowledge needs (particularly numbers 2, 4, 5, 6, 8, 9, 10, 17, 21, 24 and 26) demand integration of strands of knowledge from different disciplines to inform policy and practice. As argued above, much existing evidence on the environmental sustainability of agriculture focuses on individual

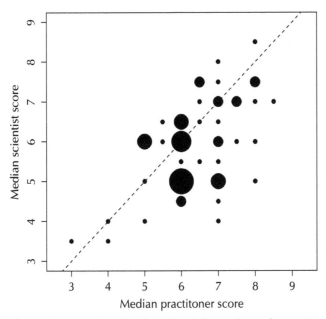

FIGURE 11.2   Median scores for each of the 53 knowledge needs given by practitioners (n=24) and scientists (n=14); 0 = not a priority, 10 = high priority. Spearman rank correlation coefficient rs = 0.488, p < 0.001. Points are sized according to the number of knowledge needs with each combination of scores. The largest circles represent five knowledge needs, the smallest just one. The dashed line shows the identity line where y = x.

environmental outcomes or aspects of land management. The challenge now is to integrate this evidence in a coherent way to inform farm and landscape management. Our exercise demonstrates that this challenge is coming from practitioners in business and government who are trying to implement sustainable agriculture in the real world. There is a need for integrated sets of metrics, understanding of how different aspects of agricultural systems and different environmental services interact, and adequately substantiated modelling tools that allow the results of management decisions to be visualized easily.

The sustainability of livestock systems in a global market is prominent in the priority list. Four knowledge needs relate to it (number 1, 10, 18 and 23 in Table 11.2), including the highest priority of all. Three of them are concerned with how livestock are fed and where their feed comes from (1, 18 and 23), whilst the other (10) is concerned with where in the world livestock production systems have the least environmental impact. Livestock management is a particularly prominent issue for agricultural sustainability when viewed on a global scale. Global demand for livestock products is increasing, and expected to continue increasing for at least the next three decades [31]. Both demand and production increases in livestock products are expected to derive largely from the developing world, where eating habits and wealth are changing rapidly. The implications of using imported feed products in the UK are also related to land-use change and habitat loss elsewhere in the world.

Seven out of the top 14 priorities had a specific crop- or animal-feed focus (numbers 1, 6, 8, 11, 12, 13 and 14). In general the animal component of livestock systems was not highly ranked, featuring in only knowledge needs 10 and 23. Livestock feed, rather than the animal component, was also the focus of knowledge need 18. The relatively low prioritisation of knowledge needs related to the animal component of livestock systems is perhaps surprising, seeing that the value of the output of the UK livestock sector is about 40% greater than the value of the UK crop sector [32]. In addition, the most recent environmental accounts for UK agriculture [33] are dominated by the negative effects of methane and ammonia emissions, which are largely associated with livestock. There is however, a bias towards crop production systems in the community of publicly-funded research scientists working on environmental aspects of agriculture. Biodiversity and ecosystem services are more thoroughly studied in arable systems [34]. Of the 14 research scientists who took part in the full process, three had specific expertise in crop systems and one in livestock farming, although several had broad interests across farming systems. At least five of the practitioners were primarily interested in the sustainability of livestock, rather than arable or all farming.

Practitioners tended to consider metrics to measure sustainability as more important than did scientists. Two knowledge needs rated with relatively high priority by practitioners (numbers 5 and 7 in Table 11.2) are concerned with sets of metrics to measure sustainability and soil health respectively. Both were scored lower by scientists. The need for metrics to monitor ecosystem services is widely acknowledged [35,36], and development of globally standardized sustainability metrics for agriculture is currently being discussed [37,38,39]. Metrics are needed for national-scale reporting, monitoring the success of policy interventions, environmental sustainability programmes in the food and drinks industry and for self-assessment of performance by farmers. Clearly there is specific support for this work from those involved in moving towards more sustainable agriculture within the UK food system. There was good understanding amongst the group that the number of metrics is likely to be reasonably high, given the complexity of agroecosystems. For example, the recent Linking

Environment And Farming (LEAF) report, focused on sustainable farming includes 24 objectives [40]. This is why knowledge need 5 refers specifically to identifying the minimum number of metrics needed.

Phosphorus is a greater concern than nitrogen in this set of priorities. Knowledge needs 3 and 15 are the only two from the nutrients category that made it to the priority list. The priority need related to phosphorus is not 'Where are the supplies going to come from?' or 'How much do we have left globally,' or even 'How can we reduce dependence on phosphorus?,' all of which have been raised in several fora recently [1,41,42,43]. This group prioritised the knowledge need for practical approaches to recycle phosphorus more effectively in farming systems. One specific example of this—recovering phosphorus from manures—was also raised as a researchable issue by the Feeding for the Future project [26] in the context of new models of integrated mixed farming.

Three of the four priority knowledge needs relating to soil (knowledge needs 7, 16 and 20), are about monitoring and maintaining its natural integrity and fertility, often referred to in policy as 'soil health' [44] and its resilience to environmental change.

Soil is the only feature of agro-ecosystems for which this group wanted to know more about the role biodiversity plays in delivering ecosystem services (knowledge need 8). In a previous exercise focused on the conservation of wild insect pollinators [23], understanding the relationship between biodiversity and delivery of the pollination service was the highest priority knowledge need. This was felt to reflect a desire within the group to reconcile new ecosystem service objectives with more traditional objectives to conserve the diversity of species and habitats for their own sake. In the broader context of environmentally sustainable agriculture, protection of farmland biodiversity is often stated as an aim [6], but it seems that the role this biodiversity plays in delivering farm productivity through services other than those related to soil fertility, such as pest control, pollination and water quality, is less of a concern.

There are three knowledge needs at the end of the list which would not have appeared in the top 20 according to scoring by practitioners, but which were scored with medians of seven or higher by scientists. One is about the appropriate scale of management of environmental services (knowledge need 24). Whilst both groups saw a strong need for knowledge about the implications of managing for several different services at once (knowledge needs 2 and 4), scientists saw scale as a stronger component of this than practitioners. Scale is much discussed in the academic and policy communities and known to be very important for understanding and managing trade-offs between ecosystem services [36,45,46]. For example, scale was identified as an important consideration for policy development by a recent stakeholder-led exercise to balance food production and environmental objectives [47].

The opportunities for re-integrating arable and livestock farming at the farm-scale was an issue represented in three separate proposed knowledge needs in the initial long list, but none made it through even to the final scoring session of our workshop. This may reflect the increasingly global nature of food production systems and the economic advantages of specialisation in farm businesses, processing and logistics. The priority questions on livestock feed (1, 18 and 23) are perhaps indicative of the interdependencies between specialist arable and livestock farms on a larger scale than individual farms.

Of the 26 priority knowledge needs in Table 11.2, six can be matched to one (or in some cases two) of the 100 questions of importance to the future of global agriculture identified by Pretty et al. [1]. These are knowledge need numbers 5, 10, 12, 14, 17 and 26. The overlap is small, with less than 25% of our knowledge needs matching the 100 questions. The global agriculture exercise identified questions that, "if addressed, would have a significant impact on global agricultural practices worldwide." Here we prioritised knowledge needs that, if addressed, would enable a move towards more sustainable agriculture for the UK food system. There was a stronger focus on the large-scale commercial production enterprises that dominate the UK food system in the current exercise, as opposed to the low-income small-scale agricultural enterprises that dominate global agriculture in terms of numbers of people. The priorities here also reflect the European policy context of the UK food system, with several of our knowledge needs relating specifically to EU policies such as agri-environment schemes (number 21), regulation of plant protection products (number 6) and prominence of the ecosystem services agenda in the European Union (numbers 2, 4, 9, 21 and 24) [45].

11.3.2 NEXT STEPS

In the next stage of this process, the same group of scientists and practitioners will consider how the priority knowledge needs they have identified can be addressed. We will follow similar methods, working collaboratively with iterated discussions to specify, in detail, what is already known in each area, where the relevant knowledge lies and what steps can be taken within or outside the group to meet the knowledge need cost-effectively. Our aim will be to ensure that knowledge and data from all sectors are taken into account.

The list of knowledge needs includes a range of different types of question and levels of information. Some are scientific questions that require large research programmes. These may need to be unpacked into smaller, more manageable scientific questions.

Aspects of many of the knowledge needs have already been tackled, or are in the process of being answered by existing projects in the UK or internationally. Trade-offs between ecosystem services, for example (knowledge need number 2), are the subject of active on-going research [36]. Here, the need is to review and synthesize existing and emerging knowledge, and make it accessible to an array of users. Other priority knowledge needs may require new standardised data collection, or stakeholder-driven policy development.

Clearly, the priorities that emerge from a process like this depend to an extent on the participants involved. As in previous similar exercises, we made every effort to be as inclusive as possible, and to involve representatives from all sectors, so we suggest that these results reflect a broad range of interests relevant to the implementation of environmentally sustainable agriculture for UK food systems.

REFERENCES

1. Pretty, J.; Sutherland, W.J.; Ashby, J.; Auburn, J.; Baulcombe, D.; Bell, M.; Bentley, J.; Bickersteth, S.; Brown, K.; Burke, J.; et al. The top 100 questions of importance to the future of global agriculture. Int. J. Agric. Sustain. 2010, 8, 219–236.
2. Godfray, H.C.J.; Beddington, J.R.; Crute, I.R.; Haddad, L.; Lawrence, D.; Muir, J.F.; Pretty, J.; Robinson, S.; Thomas, S.M.; Toulmin, C. Food security: The challenge of feeding 9 billion people. Science 2010, 327, 812–818.
3. Beddington, J.R.; Asaduzzaman, M.; Clark, M.E.; Bremauntz, A.F.; Guillou, M.D.; Howlett, D.J.B.; Jahn, M.M.; Lin, E.; Mamo, T.; Negra, C.; et al. What next for agriculture after Durban? Science 2012, 335, 289–290.
4. Duchin, F. Sustainable consumption of food—a framework for analyzing scenarios about changes in diets. J. Ind. Ecol. 2005, 9, 99–114.
5. Lockwood, M.; Davidson, J.; Curtis, A.; Stratford, E.; Griffith, R. Governance principles for natural resource management. Soc. Nat. Resour. 2010, 23, 986–1001.
6. Dobermann, A.; Nelson, R. Opportunities and solutions for sustainable food production; Sustainable Development Solutions Network: Paris, France, 2013.
7. McGonigle, D.F.; Harris, R.C.; McCamphill, C.; Kirk, S.; Dils, R.; Macdonald, J.; Bailey, S. Towards a more strategic approach to research to support catchment-based policy approaches to mitigate agricultural water pollution: A UK case-study. Environ. Sc. Policy 2012, 24, 4–14.
8. Tallontire, A.; Nelson, V.; Dixon, J.; Benton, T.G. A Review of the Literature and Knowledge of Standards and Certification Systems in Agricultural Production and Farming Systems; University of Greenwich: London, UK, 2012.

9. Dicks, L.V.; Ashpole, J.E.; Danhardt, J.; James, K.; Jönsson, A.; Randall, N.; Showler, D.A.; Smith, R.K.; Turpie, S.; Williams, D.; Sutherland, W.J. Farmland conservation synopsis. Available online: http://www.conservationevidence.com/data/index?synopsis_id[ ]=9 (accessed on 20 March 2013).

10. Smith, P.; Martino, D.; Cai, Z.; Gwary, D.; Janzen, H.; Kumar, P.; McCarl, B.; Ogle, S.; O'Mara, F.; Rice, C.; et al. Agriculture. In Climate change 2007: Mitigation. Contribution of Working Group III to the Fourth Assessment Report of the Intergovernmental Panel on Climate Change; Metz, B., Davidson, O.R., Bosch, P.R., Dave, R., Meyer, L.A., Eds.; Cambridge University Press: New York, NY, USA, 2007.

11. Garnett, T. Where are the best opportunities for reducing greenhouse gas emissions in the food system (including the food chain)? Food Policy 2011, 36 (Supplement 1), S23–S32.

12. Holland, J.M.; Frampton, G.K.; Cilgi, T.; Wratten, S.D. Arable acronyms analyzed - a review of integrated arable farming systems research in western-europe. Ann. Appl. Biol. 1994, 125, 399–438.

13. Sutherland, W.J.; Fleishman, E.; Mascia, M.B.; Pretty, J.; Rudd, M.A. Methods for collaboratively identifying research priorities and emerging issues in science and policy. Methods Ecol. Evol. 2011, 2, 238–247.

14. Defra. UK Marine Science Strategy; Department for Environment, Food and Rural Affairs, on behalf of the Marine Science Co-ordination Committee: London, UK, 2010.

15. Sutherland, W.J.; Armstrong-Brown, S.; Armsworth, P.R.; Brereton, T.; Brickland, J.; Campbell, C.D.; Chamberlain, D.E.; Cooke, A.I.; Dulvy, N.K.; Dusic, N.R.; et al. The identification of 100 ecological questions of high policy relevance in the UK. J. Appl. Ecol. 2006, 43, 617–627.

16. Burgess, P.J.; Morris, J. Agricultural technology and land use futures: The UK case. Land Use Policy 2009, 26, S222–S229.

17. Defra. Understanding Behaviours in a Farming Context: Bringing Theoretical and Applied Evidence Together from Across Defra and Highlighting Policy Relevance and Implications for Future Research; Defra: London, UK, 2008.

18. Elliott, J.; Sneddon, J.; Lee, J.A.; Blache, D. Producers have a positive attitude toward improving lamb survival rates but may be influenced by enterprise factors and perceptions of control. Livest. Sci. 2011, 140, 103–110.

19. Cooke, I.R.; Mattison, E.H.A.; Audsley, E.; Bailey, A.P.; Freckleton, R.P.; Graves, A.R.; Morris, J.; Queenborough, S.A.; Sandars, D.L.; Siriwardena, G.M.; et al. Empirical test of an agricultural landscape model: The importance of farmer preference for risk-aversion and crop complexity. SAGE Open 2013.

20. Defra. Review of Environmental Advice, Incentives and Partnership Approaches for the Farming Sector in England. Available online: https://www.gov.uk/government/uploads/system/uploads/attachment_data/file/181835/pb13900-review-incentives-partnership-approaches.pdf.pdf (accessed on 30 April 2013).

21. Agricultural Industries Confederation, The Value of Advice Report; AIC: Peterborough, UK, 2013.

22. Garforth, C.; McKemey, K.; Rehman, T.; Tranter, R.; Cooke, R.; Park, J.; Dorward, P.; Yates, C. Farmers' attitudes towards techniques for improving oestrus detection in dairy herds in south west england. Livest. Sci. 2006, 103, 158–168.

23. Dicks, L.V.; Abrahams, A.; Atkinson, J.; Biesmeijer, J.; Bourn, N.; Brown, C.; Brown, M.J.F.; Carvell, C.; Connolly, C.; Cresswell, J.E.; et al. Identifying key knowledge needs for

evidence-based conservation of wild insect pollinators: A collaborative cross-sectoral exercise. Insect Conserv. Diver. 2012, 3, 435–446.

24. Qualtrics. Available online: http://www.Qualtrics.com (accessed on 4 July 2012).

25. Defra. Food Statistic Pocketbook 2012; Defra: London, UK, 2012.

26. Pollock, C. Feeding the Future - Innovation Requirements for Primary Food Production in the UK to 2030, 2012. Available online: http://feedingthefutureblog.files.wordpress.com/2012/11/feedingthefuture2013-web.pdf (accessed on 12 July 2013).

27. Rosser, A. How Can Incentives for Soil Carbon Management Contribute to Food Security and Biodiversity Conservation? Cambridge Conservation Initiative: Cambridge, UK, 2012.

28. Centre for Ecology and Hydrology, Taking a Market Lead to Tackle the Nitrogen Problem. Centre for Ecology and Hydrology: Edinburgh, UK, 2012.

29. Husson, F.; Josse, J.; Le, S.; Mazet, J. Factominer : Multivariate Exploratory Data Analysis and Data Mining with R. Available online: http://cran.r-project.org/web/packages/FactoMineR/index.html (accessed on 6 July 2012).

30. R Development Core Team, R: A Language and Environment for Statistical Computing; R Foundation for Statistical Computing: Vienna, Austria, 2010.

31. Thornton, P.K. Livestock production: Recent trends, future prospects. Phil. Trans. Roy. Soc. B 2010, 365, 2853–2867.

32. Defra. Statistical Data Set Agriculture in the United Kingdom. Available online: https://www.gov.uk/government/statistical-data-sets/agriculture-in-the-united-kingdom (accessed on 30 April 2013).

33. Department for Environment; Food and Rural Affairs; Welsh Assembly Government; Scottish Government; Department of Agriculture and Rural Development (Northern Ireland). Environmental Accounts for Agriculture; Project SFS0601 Final Report; Defra: London, UK, 2008.

34. Defra. Potential for Enhancing Biodiversity on Intensive Livestock Farms (PEBIL); BD1444; Defra: London, UK, 2007.

35. Boyd, J.; Banzhaf, S. What are ecosystem services? The need for standardized environmental accounting units. Ecol. Econ. 2007, 63, 616–626.

36. Burkhard, B.; de Groot, R.; Costanza, R.; Seppelt, R.; Jorgensen, S.E.; Potschin, M. Solutions for sustaining natural capital and ecosystem services. Ecol. Indic. 2012, 21, 1–6.

37. Lindenmayer, D.B.; Likens, G.E. Effective monitoring of agriculture. J. Environ. Monitor. 2011, 13, 1559–1563.

38. Sachs, J.; Remans, R.; Smukler, S.; Winowiecki, L.; Andelman, S.J.; Cassman, K.G.; Castle, D.; DeFries, R.; Denning, G.; Fanzo, J.; et al. Monitoring the world's agriculture. Nature 2010, 466, 558–560.

39. Sachs, J.D.; Remans, R.; Smukler, S.M.; Winowiecki, L.; Andelman, S.J.; Cassman, K.G.; Castle, D.; DeFries, R.; Denning, G.; Fanzo, J.; et al. Effective monitoring of agriculture: A response. J. Environ. Monitor. 2012, 14, 738–742.

40. LEAF, LEAF-Driving Sustainability. A Review of our Impact, Achievements and Challenges 2013; Linking Environment and Farming: Warwickshire, UK, 2013.

41. Cordell, D.; Neset, T.S.S.; Prior, T. The phosphorus mass balance: Identifying 'hotspots' in the food system as a roadmp to phosphorus security. Curr. Opin. Biotechnol. 2012, 23, 839–845.

42. Elser, J.J. Phosphorus: A limiting nutrient for humanity? Curr. Opin. Biotechnol. 2012, 23, 833–838.

43. Schroder, J.J.; Cordell, D.; Smit, A.L.; Rosemarin, A. Sustainable Use of Phosphorus; Plant Research International, Wageningen UR: Wageningen, The Netherlands, 2010.

44. Kibblewhite, M.G.; Ritz, K.; Swift, M.J. Soil health in agricultural systems. Phil. Trans. Roy. Soc. B 2008, 363, 685–701.

45. Hauck, J.; Görg, C.; Varjopuro, R.; Ratamäki, O.; Jax, K. Benefits and limitations of the ecosystem services concept in environmental policy and decision making: Some stakeholder perspectives. Environ. Sci. Policy 2013, 25, 13–21.

46. Nelson, E.; Mendoza, G.; Regetz, J.; Polasky, S.; Tallis, H.; Cameron, D.R.; Chan, K.M.A.; Daily, G.C.; Goldstein, J.; Kareiva, P.M.; et al. Modeling multiple ecosystem services, biodiversity conservation, commodity production, and tradeoffs at landscape scales. Front. Ecol. Environ. 2009, 7, 4–11.

47. Defra. Green Food Project Conclusions. Available online: http://www.defra.gov.uk/publications/files/pb13794-greenfoodproject-report.pdf (accessed on 3 April 2013).

CHAPTER 12

The Role of Biotechnology in Sustainable Agriculture: Views and Perceptions Among Key Actors in the Swedish Food Supply Chain

Karin Edvardsson Björnberg, Elisabeth Jonas, Håkan Marstorp, and Pernilla Tidåker

12.1 INTRODUCTION

Researchers have put forward agricultural biotechnology, that is "any technique that uses living organisms or substances from these organisms to make or modify a product" [1] (p. 8), as a tool for increasing food production, while, at the same time, making agriculture more sustainable from an environmental point of view [2]. Research suggests that genetic engineering can be used to develop crop varieties that cope better with drought and salinity [3,4]; are more disease resistant [5,6]; and use nutrients more efficiently. These features are particularly desirable in a changing climate where the population grows and competition over arable land increases. However, agricultural biotechnology is a controversial topic, and not everyone is convinced that the net benefits of genetically-modified (GM) varieties will be positive overall. Critics point to the ecological and health risks involved and to the negative impacts of GM varieties on small-scale traditional farming, especially in the global South [7,8,9]. Thus, the current debate over agricultural biotechnology, especially GMOs (genetically-modified organisms), is framed in strongly dichotomous terms: biotechnology is considered either an important part of or a severe threat to the effort to create sustainable agricultural production systems.

To be able to assess the potential sustainability implications of biotechnology, a closer look into the sustainability concept and its definition and application is necessary. According to the received view, agricultural sustainability deals with the maintenance of agricultural production systems over time [10,11], but further detailing of the concept has been widely discussed [12,13,14,15,16]. No single agreed-on definition of agricultural sustainability exists today; instead, there is a variety of definitions in academic and policy discussions [17,18].

The diversity of meanings attached to the concept of agricultural sustainability has led some authors to argue that it is an essentially contested concept [19,20,21,22]. This means that it is a normative concept with two levels of meaning. The first level expresses a number of core ideas, which are substantive and non-redundant in the sense that even if actors have very different views on how the concept should be interpreted, they can still agree that in some situations, the conditions for agricultural sustainability are not present [23]. At the second level of meaning are a number of different conceptions, that is there are, "legitimate, yet incompatible and contested, interpretations of how the concept should be put into practice" [24] (p. 262), [25,26]. This means that, even if people agree about the core of the concept, there is considerable disagreement concerning how the concept ought to be implemented. One such point of disagreement concerns the role of biotechnology in creating sustainable agricultural production systems.

In this paper, it is investigated how key actors in the Swedish food supply chain understand and operationalize the concept of agricultural sustainability (sustainable agricultural food production systems) and what influence this may have on their views of agricultural biotechnology and its role in creating sustainable agricultural production systems. Five actors participate in the study: The Federation of Swedish Farmers (*Lantbrukarnas riksförbund*, LRF), Lantmännen, ICA, Axfood and Coop. Based on policy documents and semi-structured interviews with representatives from the selected organizations, an attempt is made to answer the following questions:

- How does the organization perceive the concept of agricultural sustainability (sustainable agricultural production systems)?
- Who/what influences the organization's sustainability policies, including the organization's standpoint on agricultural biotechnology?
- What are the organization's views and perceptions of biotechnology and its possible role in creating agricultural sustainability?

Based on the empirical data, it is argued that, although there is no single agreed-on definition of agricultural sustainability, there is a shared understanding among key actors in the Swedish food supply chain of what the key constituents of agricultural sustainability are. At the same time, however, there is less explicit consensus on how the concept should be put into practice or what role biotechnology has in creating sustainable agricultural production systems. The open-ended character of the concept of agricultural sustainability provides an opportunity for various actors to make their favored sustainability discourse the dominant one in the general sustainability debate. The interview data suggest that the current Swedish agricultural sustainability discourse has been influenced by consumer opinion and the views of strong environmental organizations.

Section 12.2 describes the methods used. Section 12.3 briefly summarizes the results of the desk-based review of the selected organizations' sustainability policies. In Section 12.4, Section 12.5 and Section 12.6, the three aforementioned research questions are discussed based on the studied sustainability policies and the interviews with the representatives of the selected organizations. Section 12.7 and Section 12.8 comprises some conclusions and suggestions for further research.

From here on, the terms "agricultural sustainability" and "sustainable agricultural production systems" will be used interchangeably. However, it should be noted that "agricultural sustainability" is the more commonly-used term, and even though it targets the sustainability of the production systems, it may also be used in a wider sense, including additional elements within the food production chain.

12.2   METHODS

The study is a desk-based review of policy documents of key actors in the Swedish food supply chain, that is actors that play a role in the organization of producing, processing and retailing food and that consequently have a significant impact on the actual (un)availability of genetically-modified foodstuff on the Swedish market. At an initial stage of the research, two groups of actors were selected from a preliminary survey of the organizations active in the Swedish food supply chain: (i) organizations either owned by Swedish farmers or with the primary objective of representing Swedish farmers' interests; and (ii) food retailers. The inclusion of these two groups of actors enabled a review of policies related to food production in Sweden, as well as policies related to the import of food [27].

Three farming organizations were identified: The Federation of Swedish Farmers (LRF), Lantmännen and Ekologiska lantbrukarna (Association of Organic Farmers). Since Ekologiska lantbrukarna has organic farming as their main objective, which rules out the use of biotechnology, they were not selected for this study. LRF is a politically independent interest and business organization with around 170,000 individual members representing some 90,000 enterprises [28], including the larger part of both organic and conventional farms (http://www.lrf.se). The organization's overall mission is to improve farming and forestry businesses and enable individual members to achieve their goals in terms of profitability, growth and quality of life. Lantmännen is an agricultural cooperative that is owned by 33,500 Swedish farmers (http://www.lantmannen.se). The organization's core mission is to increase the economic profitability of its members and maximize their capital returns. Lantmännen buys, refines and sells farmers' produce, but it is also a retailer of agricultural commodities and machinery. Lantmännen is also involved in research and development, such as plant breeding.

Three companies were selected from the identified grocery retailers: ICA, Axfood and Coop. Taken together, these three retailers represent approximately 87 percent of the Swedish food retail market [29]. The ICA Group is Sweden's biggest food retailer. It runs around 2400 retail stores in five geographic markets with a 50 percent market share in Sweden (http://www.ica.se). Axfood's retail business mainly consists of entirely owned chains, but it also has franchise agreements (http://www.axfood.se). It represents around 16 percent of the food retail market in Sweden. Coop is a consumer cooperative with 3.3 million members conducting its business via entirely owned chains (http://www.coop.se). It has a market share of around 21 percent.

Published policy documents from the five organizations were downloaded and thoroughly examined. A number of strategies were discussed in order to identify what would be the most informative method to identify the selected organizations' views on the role of biotechnology in furthering agricultural sustainability. Based on the available policy documents and the discussions, semi-structured interviews with those responsible for sustainability issues within the selected organizations (typically the organization's head of sustainability/environmental department) were identified as the most effective method of data collection. Five respondents, one from each organization, were approached and agreed to represent the organizations' views on the identified research questions. Thus, a total of five interviews were conducted.

The interviews were performed between June and November 2013 and followed an interview template structured around three themes: the concept of agricultural sustainability, the role of biotechnology in creating sustainable agricultural production systems and the role of external actors in shaping the organizations' sustainability (including biotechnology) policies (Appendix. The template was sent out to the responsible personnel of each organization beforehand. All interviews took between 45 and 75 min and were transcribed and analyzed through repeated readings of statements.

12.3   SUSTAINABILITY POLICIES OF THE SELECTED ORGANIZATIONS

None of the selected organizations has a policy explicitly dealing with agricultural sustainability. Instead, sustainability aspects related to agriculture are dealt with in the organizations' general sustainability policies/programs, codes of conduct, corporate social responsibility (CSR) policies and/or policies dealing with specific environmental aspects. In general, LRF's and Lantmännen's policies are more explicitly oriented towards the sustainability impacts of food and fodder production than the policies of the grocery retailers. The latter predominantly focus on the sustainability impacts of activities associated with the distribution of food (carbon emissions from transport, energy consumption in food distribution premises, climate impacts from refrigerants, etc.). None of the surveyed policy documents contains any explicit definition of the concept of agricultural sustainability. Only LRF and Lantmännen have adopted policies on agricultural biotechnology. In the next sections, the results of the desk-based review and the interviews are summarized and analyzed, for each of the respondents.

12.3.1   ORGANIZATIONS OWNED BY THE SWEDISH FARMERS

12.3.1.1   LRF

As noted above, none of the selected organizations explicitly defines "agricultural sustainability" or "sustainable agricultural production systems" in their sustainability policy documents. However, in a publication entitled Sustainability in Swedish Agriculture 2012, written by Statistics Sweden (*Statistiska centralby-rån*, SCB) with the contribution of LRF among others, it is stated that:

> Sustainable agriculture integrates three different aspects: environmental health, economic profitability and social and economic equity. Sustainability rests on the principle that we must meet the needs of the present without compromising the ability of future generations to meet their own needs. A sustainable agriculture conserves our natural resources, is adapted to the environment and is environmentally ethical. An economically and socially sustainable development in the countryside requires for instance that agriculture produces high quality food at reasonable prices to the consumer and provides the producers a reasonable income.[30] (p. 8)

These three sustainability aspects are also referred to in LRF's pesticide policy, in which the organization describes how reduced pesticide use and economic competitiveness and growth can be reconciled [31].

LRF has adopted a policy on biotechnology that consists of six core principles concerning sustainability, precaution and ethics, competitiveness, freedom of choice, labelling and transparency and responsibility [32]. LRF has a positive attitude towards using GMOs in agriculture if they contribute to an environmentally and economically sustainable development and do not affect human quality of life negatively [32]. According to LRF, the risks and benefits of GMOs should be assessed on a case-by-case basis, taking into account the long-term benefits for society, humans and human health, non-human animals and the environment. As an example of positive environmental impacts of GMOs, the policy document mentions decreased use of substances that pollute water or impact negatively on biological diversity or human health.

12.3.1.2 Lantmännen

Lantmännen's Code of Conduct [33] contains statements on environmental and social sustainability. The Code specifies that Lantmännen should aim to decrease its emissions to air, land and water and make resource use (including energy use) more efficient. The organization should work to develop transport logistics solutions that optimize resource use, decrease environmental impact and increase safety. The Code also mentions social aspects, such as creating safe and healthy work environments (including decent pay and a ban on child labor) and human rights protection (including a ban on discrimination and harassment).

Lantmännen's Code of Conduct contains a separate section on genetic engineering. Lantmännen has a positive view of biotechnology and explicitly acknowledges technology's potential to contribute to more sustainable societies. At the same time, the organization recognizes that there might be risks associated with biotechnology and that every application has to be thoroughly assessed before it is put to use. The Code stresses that biotechnology assessments should be guided by the precautionary principle and that consideration must be paid to the demands and expectations of customers, as well as to market conditions. Regarding food products on the European market, the Code prescribes that they must not contain any raw materials from genetically modified crops. In relation to fodder, Lantmännen ensures that it "can deliver GMO-free raw materials for feeds and feed products according to customer requirements" [33] (p. 2).

12.3.2 RETAILERS

12.3.2.1 ICA

ICA's different CSR policies, for example the Business Ethics Policy, Quality, Environmental and Social Compliance Policy and Health Policy, cover environmental and social sustainability issues. ICA's most important environmental goals include reducing its greenhouse gas emissions, energy use and waste and safeguarding biological diversity [34]. ICA participates in the UN Global Compact initiative and is thus committed to supporting social sustainability goals, such as the protection of human rights, the freedom of association and the elimination of discrimination in employment.

ICA does not have a cohesive policy on biotechnology. However, the organization has a positive attitude towards new technology that can contribute to better products from a consumer perspective. For ethical and environmental reasons, ICA questions the production and cultivation of genetically modified food and crops that are not produced in a closed environment. GMOs and ingredients should, in ICA's view, be stored separately and be traceable [34]. At present, there are no such products available in ICA's stores. In order for ICA to market such a product in the future, the product must have consumer benefits and be safe for humans and the environment. In line with the UN Global Compact initiative, ICA strives to "encourage the development and diffusion of environmentally friendly technologies" [35] (p. 6). The ICA Group management team makes decisions about which GM products, if any, should be included in ICA's assortment.

12.3.2.2   Axfood

Axfood has adopted a sustainability program that describes goals and strategies for the group's sustainability work [36]. The program outlines goals related to products, transport, energy and use of premises, suppliers, employees and animal welfare. Like ICA, Axfood has no cohesive policy on biotechnology, but from its webpage, one can read, "Genetic modification is a relatively new tool, and Axfood realizes that this technology may be helpful in certain contexts" (http://www.axfood.se/en/Sustainability/How-we-work/Standpoints/). At present, Axfood has no GM products in its stores. If introduced, such products should be labelled in order for customers to be able to make informed choices about whether or not to purchase them.

12.3.2.3   Coop

Coop has adopted a policy on sustainable development that addresses the financial, environmental and social consequences of their business [37]. The policy is generic, but it is clear from Coop's most recent annual sustainability report [38] that organic and Fair-trade labelling play an important role in the implementation of the policy. Other parts of the sustainable development policy concern Coop's own energy consumption, transportation, use of premises, recycling and food wastes. Coop is also engaged in a continuous dialogue on sustainability issues with several NGOs, such as the World Wide Fund for Nature (WWF), Fair-trade, the Swedish Society for Nature Conservation (SSNC) and the Swedish organic labelling organization KRAV. Coop also participates in the UN Global Compact initiative.

Coop has no policy on biotechnology, but according to their website, the organization believes that GMOs might have negative environmental impacts in the long term, not least on biodiversity, and that cultivation of GM varieties could have negative socio-economic impacts on farmers at the mercy of the big multinational companies. Today, Coop does not sell any products with genetically modified additives.

12.4   THE CONCEPT OF AGRICULTURAL SUSTAINABILITY

As noted above, with the possible exception of LRF, none of the selected organizations has adopted a clear definition of "agricultural sustainability" or "sustainable agricultural production systems" to guide their policy work.

However, the empirical data obtained through readings of the policy documents and the interviews show that some themes are recurring in the organizations' sustainability discussions and, thus, could legitimately be said to be part of the core of the concept of agricultural sustainability as understood by key actors in the Swedish food supply chain. Three such themes, or core aspects, can be identified: a clear commitment to environmental protection, in particular emission reductions and increased resource efficiency; a commitment to securing intra- and inter-generational equity, that is meeting at least the basic needs of everyone today and in the future; and a realization that sustainability (or sustainable development) involves several dimensions, or areas of concern, that need to be integrated in all planning and decision-making concerning food production, processing and retailing (cf. [23,24]).

12.4.1 ENVIRONMENTAL PROTECTION

Reductions of emissions and increased resource efficiency are at the core of the selected organizations' sustainability policies. Several of the organizations have adopted targets concerning carbon emissions and energy consumption. One of ICA's key environmental goals is to reduce the group's direct greenhouse gas emissions by 30 percent by 2020 compared to 2006 [34]. Axfood [36] has adopted the goal to reduce its carbon footprint by 75 percent by 2020, and Coop [37] has adopted the target to minimize its direct and indirect climate impacts and to become "climate neutral" in the long run. In addition to emissions reductions goals and goals relating to increased resource efficiency, some of the organizations have adopted goals concerning other aspects of the natural environment, such as biological diversity and animal welfare.

From the policy documents and the interviews, it is clear that the organizations focus on environmental aspects and goals that relate to their own core areas of activity. All of the grocery retailers have adopted goals concerning waste reduction and the reduction of energy consumption in their premises. For example, Axfood has adopted the goal to halve its climate impact from refrigerants by 2015 using 2009 as the base year, and Coop has adopted the goal to minimize its waste and to recycle as much as possible. In contrast, the farming organization LRF has adopted sustainability policies specifically targeting food production activities. Two examples of these policies are LRF's chemical pesticide policy [31] and LRF's policy on the discharge of sewage sludge [39].

12.4.2   EQUITY

When asked about whose interests the organizations are safeguarding or prioritizing in their sustainability work, several of the informants mention spatially-
and/or temporally-distant people. Using resources efficiently is considered
vital in satisfying the needs or interests of the present generation, but also in
order to "be able to deliver raw materials to a growing (future) population"
(Lantmännen), so that in the end, "everyone gets their share" (LRF). Two of
the interviewed retailers specifically emphasize their efforts to support socially
sustainable production in supplier countries. Both ICA and Coop participate in
the UN Global Compact initiative, and Coop is actively working to increase its
sale of Fair-trade products [38].

The interview data suggests that sustainability (sustainable development)
is primarily understood in anthropocentric terms. That is, the main rationale
for protecting the natural environment is to safeguard equal opportunities for
welfare among humans living in different spatial and temporal locations. The
needs and interests of non-human animals are referred to in some of the organizations' policies; however, they are not at the heart of the organizations' sustainability policies.

12.4.3   SUSTAINABILITY DIMENSIONS AND PRIORITIZATIONS

All of the interviewed representatives acknowledge that the concept of agricultural sustainability has several dimensions that need to be integrated in the organizations' sustainability work. This is in line with academic discussions on the
concept of agricultural sustainability that emphasize that agricultural production should not only be economically profitable and environmentally benign,
but should also meet human needs for food and contribute to quality of life
[14,17]. From the organizations' sustainability policy documents, it is clear that
most of the goals that have been included in the policies concern either environmental or social aspects of sustainable development. However, as explained by
the representative of Lantmännen, this is not surprising, since economic goals
are a 'natural' part of the organizations' reason for being and, therefore, do not
have to be included in their sustainability policies.

Although the informants agree that the concept of agricultural sustainability
consists of different aspects (ecological, economic and social) that are all central to the idea of creating sustainable agricultural production systems, they give

slightly different answers when asked to exemplify what falls under the (three) sustainability dimensions. For example, when asked about the social dimension of agricultural sustainability, all informants agree that work environment issues, such as safety, pay and human rights issues, are central. These aspects are also included in the organizations' sustainability policies, but again, refer partly to different parts of the food supply chain. LRF states that making farming a viable means of sustenance is an important social goal, whereas Lantmännen and the retailers also include their own staff and suppliers not directly involved in agricultural production.

A central idea in the sustainability literature is that sustainability is situated at "the intersection of environmental protection, economic growth, and social justice" [40] (p. 72). Thus, it is assumed that the three sustainability dimensions are reconcilable, at least in principle. However, in actual planning and decision-making, compromises and trade-offs often have to be made between goals belonging to different sustainability dimensions. When asked about whether they could think of any conflicts between the various sustainability aspects, all informants answered affirmatively. An example of a goal conflict mentioned by ICA was climate change mitigation versus local production. For example, lifecycle assessments could show that from an emissions reductions perspective, it is better to buy a certain product from another country. However, buying from abroad would conflict with ICA's policy of purchasing products locally.

Even goals belonging to the same sustainability dimension could conflict with each other. The representative from Lantmännen mentions how in the treatment of sewage sludge, recycling of nutrients could conflict with the national environmental objective of a non-toxic environment. The representative from Axfood explains how different environmental goals concerning energy and water use in vegetable production could be in conflict depending on the climatic conditions in the production area.

When asked about whether the organizations prioritize any sustainability aspects or interests, the informants again gave somewhat different answers. ICA performs annual "heat-map" analyses, which structure sustainability questions and issues into four fields according to their perceived importance. Which issues are prioritized can differ from one year to the next. Among the issues ICA currently prioritizes are product safety, climate change and social responsibility (workplace safety, salary, etc.). They also emphasize this in their annual sustainability report [34]. Lantmännen considers all dimensions important, but acknowledges that because of the organization's historical focus on economic profitability at the farm level, the challenge today is to strengthen the environmental and social aspects of agricultural sustainability.

LRF considers all aspects equally important in the short run, but emphasizes that over a longer-term perspective, the ecological aspects are the most important ones. The stated reason for this is that if we "saw off the branch we are sitting on, it does not really matter what we do with the money" (representative from LRF).

12.5 FACTORS INFLUENCING THE ORGANIZATIONS' SUSTAINABILITY POLICIES

According to the interviewed representatives, both internal and external actors influence the organizations' sustainability policies and their present standpoints on biotechnology. Internally, the organizations' members, owners and officers (sustainability departments or departments responsible for CSR issues, etc.) have the opportunity to raise issues and initiate policy reforms, which the organizations' decision-making bodies then discuss. Among the external actors that have an influence on the studied organizations' sustainability policies are regulatory authorities at national and EU levels, consumers and consumer organizations, various NGOs (in particular environmental organizations) and the general public. The exact extent to which these external actors are allowed to influence the policy process could not be measured in this study, but it is clear from the informants' answers that consumers and NGOs have particular influence on the organizations' policies, not least their policies or standpoints on biotechnology. This finding is in line with previous studies on the influence of environmental/consumer groups on EU GMO regulation and, in particular, how NGOs have been able to influence the market behavior of downstream producers in the EU [41].

The interview data show that internal actors, such as the organizations' sustainability departments, predominantly raise some sustainability issues. Axfood mentions the environmental impact of refrigerators as an example of a sustainability issue that the organization deals with internally, without initiation or involvement of any external actors. This is in comparison with, for example, the issue of palm oil production, which many Swedish consumers consider to be a controversial issue, since in many cases it involves indiscriminate forest clearing and gives rise to conflicts with local populations. In relation to palm oil, consumers have exerted significant pressure on the organizations to adopt certain standpoints.

Several of the interviewed representatives admit that their organizations are sensitive to campaigns from NGOs. This is particularly evident in the case of biotechnology. Coop, for example, states that the anti-GMO campaigns of the

Swedish Society for Nature Conservation have been particularly successful. The interviewees also mention other NGOs, such as Greenpeace and various environmental labelling organizations. When asked about what concrete influence the NGOs have had on the organizations' sustainability and/or biotechnology policies/positions, the representative from Lantmännen emphasizes that:

In one way or another, NGOs do affect our policies. Obviously, we do not change any policies overnight as a consequence of different campaigns being directed at us, but the public debate and activities of this kind do affect our policies over time.

NGOs have also been actively involved in the development of some of the organizations' sustainability policies. For example, Coop asked for advice from NGOs when developing their certification policy for wild-caught and cultivated fish. As noted above, Coop is also engaged in a continuous sustainability dialogue with several NGOs, such as WWF, Fair-trade and the Swedish organic labelling organization, KRAV.

In addition to NGOs, all retailers emphasize the importance of the consumers' views in shaping their sustainability policies. ICA and Coop refer to customer polls conducted on a regular basis with the aim to get a better picture of consumer opinions. The results of these polls feed into the policy process in various ways. ICA, for example, uses the results from consumer polls when deciding what sustainability issues to prioritize as part of the organizations' annual "heat-map analysis." At present, biotechnology is not a prioritized policy area. Part of the reason for this is that the issue has received relatively little attention from ICA's customers, compared to other issues, such as palm oil production and consumption. This is in comparison with Coop, which considers biotechnology a topic of importance among both its members and customers, a stance that strongly affects its standpoint on GMOs.

When asked about what they believe are the biggest obstacles to including biotechnology as a part of the organizations' sustainability work, several informants point to public attitudes and consumer opinion in addition to the stringent regulatory framework in force. According to the representative from ICA, "there is a built-in fear and skepticism toward biotechnology among many consumers." In ICA's view, skepticism among Swedish consumers is mainly grounded in environmental thinking and the activities of strong environmental organizations that oppose GMO. This can be contrasted with consumer opposition in the Baltic States (where ICA is also a major retailer), where GMOs are mainly opposed out of fear that large biotech companies and products will outcompete local production [42].

The interviewee from LRF also raised the issue of the public fear that large biotech companies will monopolize the market. He believes that the prevalent negative opinion of GMO is not so much about the technology itself, but rather has to do with the perception of how the big companies have acted and how patent rights have been granted in the past. One possible way of overcoming the present obstacles identified by the LRF representative is to initiate a more nuanced public discussion about the role of GM varieties in creating more sustainable agricultural production systems. This discussion should not be limited to discussions of the risks associated with GM varieties, but should also cover the potential environmental benefits involved.

12.6   THE ROLE OF BIOTECHNOLOGY

During the interviews, questions were asked about the organizations' views on biotechnology in a broad sense, including non-transgenic techniques. However, the discussion often led into analyses of the pros and cons of GMOs specifically; this was especially the case with the retailers. There was a consensus among the interviewees that drawing up policies based on a distinction between GMOs and biotechnology broadly construed might be too complicated for the general public.

None of the representatives mentioned specific examples in which biotechnology could be a way of obtaining a more sustainable agricultural production system. However, LRF and Lantmännen are generally positive towards the possible use of biotechnology if the technology could contribute to more environmentally and economically sustainable agricultural systems. Crop traits should be considered on a case-by-case basis, taking risks and benefits into consideration. This is in line with the current debate on the need for increased, but sustainable, food production to meet the predicted future demands and the challenges imposed by climate change where a number of strategies have been suggested, among those the use of biotechnology [43,44]. Examples of important crop traits targeted with biotechnological methods in the coming decades include resistance to plant fungal and viral diseases, drought and heat tolerance, improved use of plant nutrients and healthier products from a dietary perspective [43].

Despite its potential, biotechnology is a highly controversial and politicized issue. It is either viewed as an important part of sustainable agricultural systems or a severe threat to such systems. Both LRF and Lantmännen explain that environmental organizations and other NGOs affect a company greatly via their publicity campaigns by communicating their concerns about biotechnology to the broader community, thus restricting some opportunities players in the food chain may take. At present, those organizations (environmental organizations and other NGOs) still view GMOs and related technologies negatively. ICA and Axfood acknowledge that biotechnology could be a useful tool in obtaining more sustainable food production. However, none of the companies have or plan to have GM products in their stores. The representative from Coop states that she does not predict a change in Coop's view on biotechnology in the near future. This is due to NGOs' influence and negative public attitude toward GMOs. All retailers concur that they do not anticipate marketing GM products in the current situation, as the risk of a general non-acceptance from consumers remains too high.

12.7 DISCUSSION

The interview data support the argument that the concept of agricultural sustainability is essentially contested. From the interviews, it is clear that, although "agricultural sustainability" lacks a formal definition, key actors in the Swedish food production chain have a shared understanding of the concept's core constituents. Through the document review and the interviews, three such constituents could be identified: commitment to environmental protection, commitment to securing intra- and inter-generational equity and realization that agricultural sustainability involves several areas of concern that need to be integrated in decision-making concerning food production, processing and retailing.

Biotechnology can affect the sustainability of agricultural production systems depending on how the concept of agricultural sustainability is put into practice. According to some writers, biotechnology, including transgenic varieties, could make the world's agricultural production systems less sustainable from an environmental point of view through, for example, gene spread or increased invasiveness. It could also make the world's agricultural production systems socially less sustainable if the regulatory system that develops in parallel with the introduction of GM varieties prevents socioeconomic development among certain segments of the population [7]. However, biotechnology could also contribute to making our agricultural production systems more sustainable, for example by making nutrient use more efficient or by reducing the land area needed for agriculture [45]. These are very important aspects as competition over natural resources, including land, increases due to population growth and changes in climate. Arguably, there is nothing inherently or manifestly unsustainable about biotechnology; it all boils down to particular applications and the environmental, social and economic risks that those applications involve, as pointed out by some of the informants in this study. There is growing scientific evidence that, if put to use wisely, biotechnology can indeed yield significant environmental benefits [2].

The lack of precise action guidance provided by the concept of agricultural sustainability, for example in relation to the use of biotechnology, does not mean that the concept has no policy relevance at all [46]. However, it does make the concept vulnerable to 'hi-jacking' by actors who have an interest in instantiating the concept to correspond to their own political agendas, as discussed by Aerni [47] and Gunnarsson Östling et al. [48]. How the concept is put into practice and which sustainability discourse is prevalent at a particular point in time is largely the result of a struggle between different actors over the second-level meaning of the sustainability concept. The actors who are strong in the debate also have the opportunity to make their favored sustainability discourse the dominant one in planning, decision-making and the public debate.

The influence of external actors on the policy process and the resulting conceptualization of the sustainability concept are noticeable in our study. In Sweden, the dominant current discourse says that biotechnology is not part of sustainable agriculture, at least not when it comes to food for human consumption. This is clear from the policies of the organizations participating in this study. Although a majority of the interviewed organizations claim to have a positive attitude towards new technologies in general and admit that genetically engineered crop traits ought to be assessed on a case-by-case basis, they categorically reject food products containing GM-varieties in their present assortments. Thus, the perceived role of biotechnology in creating sustainable agricultural production systems is somewhat ambiguous.

The interview data suggest that the prevalent agricultural sustainability discourse has been largely shaped by consumer attitudes and pressure from strong environmental organizations. The sensitivity of anti-GM campaigns generally increases as one moves further down in the food supply chain from production to retailing. That is, among the organizations that participated in our study, LRF appeared to be the least sensitive and the food retailers the most sensitive to anti-GM campaigns. Lantmännen positioned themselves somewhere in between LRF and the food retailers. This may be because of how susceptible the organizations are to changes in consumer behavior (choice). Consumer behavior can change easily and rapidly, sometimes overnight, as a result of political campaigns and media coverage. Although they affect all actors in the food supply chain, these changes in consumer behavior have a much more direct impact on food retailers than on an organization like LRF.

Researchers tend to believe that consumer's attitudes concerning GMOs and the use of biotechnological methods in plant breeding are more negative in Europe than in the rest of the world. However, a recent meta-study of 214 studies [49] suggests that consumers' attitudes toward biotechnology in the EU (including Sweden) do not differ from other regions in the world. The common view that Europeans are more negative toward GMOs results from more negatively formulated questions and a greater focus on risks and ethics in the EU surveys. Whether real or perceived, consumer attitudes on biotechnology are important for the players within the Swedish food supply chain and largely determine whether the technologies will be allowed to play a major role in the development of more sustainable agricultural systems.

As acknowledged by some of the interviewed representatives, the current biotechnology discourse may lose ground if, for example, public opinion changes. As noted above, the current discourse is very much framed in terms of the risks involved in cultivating GM varieties. Other common themes in the Swedish public debate on biotechnology and GMOs are: the naturalness and moral permissibility of 'playing God'; health issues related to the consumption of GM foods; and the alleged greediness of large biotech companies. A more balanced debate on biotechnology and GMOs could possibly come about with the help of academia, as suggested by one informant:

Researchers and research institutions should be clear about the facts. Somebody like [interviewee refers to a well-known Swedish researcher] has publicly announced several times ... and the environmental movement listens to him, that biotechnology is needed if we are to meet the challenges ahead. He has written about it in his books, but this message has to be announced more clearly: 'look, these are the facts,' so that populist arguments will not gain the upper hand in the debate (Representative of Lantmännen).

12.8   CONCLUSIONS

The results from both the study of the policy documents and the in-depth interviews reveal some interesting insights on how actors in the Swedish food supply chain perceive the concept of agricultural sustainability and the role of biotechnology.

Although the concept of agricultural sustainability lacks a formal definition, key actors in the Swedish food production chain have a shared understanding of the concept's core constituents. Through the interviews and our document study, we were able to identify three such constituents: commitment to environmental protection, commitment to ensuring intra- and inter-generational equity and a realization that sustainability involves several dimensions that need to be integrated in planning and decision-making concerning food production, processing and retailing.

In the interviewed organizations' view, there is nothing 'inherently' or manifestly unsustainable about biotechnology (including GMOs) as such. Particular applications of the technology must be assessed individually, taking into account the environmental, social and economic risks involved. At the same time, however, they reject food products containing GM-varieties in their present assortments. The perceived role of biotechnology in creating sustainable agricultural production systems can therefore rightly be described as ambiguous.

The essentially contested nature of the concept of agricultural sustainability (i.e., general agreement on the first-level meaning of agricultural sustainability, but less explicit consensus on how the concept should be put into practice, in particular what role biotechnology could play in creating sustainable agricultural production systems) renders the concept vulnerable to political 'hijacking.' In Sweden, the prevalent agricultural sustainability discourse, including the perceived role of biotechnology in creating sustainable agricultural production systems, has been shaped by consumer attitudes and pressure from strong environmental organizations to a significant extent. The sensitivity of anti-GM campaigns differs along the food production chain from production to retailing.

It is important to keep in mind that the aim of the present study was a modest one and that the collected empirical data are sparse. In the study, only a relatively small part of the food and retail market was covered. Among the actors that were not included in the study were large dairy and meat cooperatives/associations, such as Arla Foods and Swedish Meats. Moreover, the study focused exclusively on the Swedish market, which might differ from other European countries.

APPENDIX: INTERVIEW TEMPLATE

A1. Background

What is your role/function within the organization?

A2. Design and implementation of sustainability policies

Who initiates and designs the organization's policy documents concerning sustainability/sustainable development?
What factors/actors affect the organization's sustainability policies?
Does the organization prioritize any particular sustainability aspects/areas?
How are the sustainability policy documents used in the organization's daily work? Please give some examples.

A3. The concept of agricultural sustainability/sustainable agricultural production systems

How does the organization define the concept of agricultural sustainability/sustainable agricultural production systems?
How (in the organization's view) do the sustainability dimensions (ecological, economic, or social) affect one another?
Whose interests/needs should (in the organization's view) be given priority in creating agricultural sustainability/sustainable agricultural production systems?
Are there any goal, interest, or value conflicts among the actors in the field? If yes, please give some examples.
The global demand for agricultural products is expected to increase in the future due to population growth and changed consumption patterns. Is it possible to intensify agriculture while at the same time rendering agriculture more sustainable?

A4. The role of biotechnology

Does the organization work with issues related to biotechnology?
What is the organization's view on biotechnology (including but not limited to GMOs)?

Could biotechnology be used to increase agricultural sustainability? If so, how?
What are the potential risks and goal conflicts involved?
Can those goal conflicts be overcome? If so, how?
In the organization's view, is the attitude towards biotechnology and/or GMOs
changing? If yes, in what direction?

REFERENCES

1. Food and Agriculture Organization of the United Nations (FAO). The State of Food and Agriculture 2003–2004; FAO: Rome, Italy, 2004.
2. Hansson, S.O.; Joelsson, K. Crop biotechnology for the environment? J. Agric. Environ. Eth. 2013, 26, 759–770.
3. Wang, W.; Vinocur, B.; Altman, A. Plant responses to drought, salinity and extreme temperatures: Towards genetic engineering for stress tolerance. Planta 2003, 218, 1–14. [PubMed]
4. Thomson, J.A.; Shepherd, D.N.; Mignouna, H.D. Developments in agricultural biotechnology in Sub-Saharan Africa. AgBioForum 2010, 13, 314–319.
5. Fuchs, M. Plant resistance to viruses: Engineered resistance. In Desk Encyclopedia of Plant and Fungal Virology; Mahy, B.W.J., van Regenmortel, M.H.V., Eds.; Elsevier: Amsterdam, The Netherland, 2010; pp. 44–52.
6. Qaim, M. The economics of modified genetically modified crops. Annu. Rev. Resour. Econ. 2009, 1, 665–693.
7. Shiva, V.; Emani, A.; Jafri, A.H. Globalisation and threat to seed security: Case of transgenic cotton trials in India. Econ. Polit. Wkly. 1999, 34, 601–613.
8. Robinson, J. Ethics and transgenic crops: A review. Electron. J. Biotechnol. 1999, 2, 5–6.
9. Peters, C.J. Genetic engineering in agriculture: Who stands to benefit? J. Agricult. Environ. Eth. 2000, 13, 313–327.
10. Harwood, R.R. A history of sustainable agriculture. In Sustainable Agricultural Systems; Edwards, C.A., Lal, R., Madden, P., Miller, R.H., House, G., Eds.; Soil and Water Conservation Society: Ankeny, IA, USA, 1990; pp. 141–156.
11. Ikerd, J.E. Agriculture's search for sustainability and profitability. J. Soil Water Conserv. 1990, 45, 18–23.
12. Brklacich, M.; Bryant, C.R.; Smit, B. Review and appraisal of concept of sustainable food production systems. Environ. Manag. 1991, 15, 1–14.
13. Crews, T.; Mohler, C.; Power, A. Energetics and ecosystem integrity: The defining principles of sustainable agriculture. Am. J. Altern. Agric. 1991, 6, 146–149.
14. Crosson, P. Sustainable agriculture. Resources 1992, 106, 14–17.
15. Fresco, L.O.; Kroonenberg, S.B. Time and spatial scales in ecological sustainability. Land Use Policy 1992, 9, 155–168.
16. Smit, B.; Smithers, J. Sustainable agriculture: Interpretations, analyses and prospects. Can. J. Reg. Sci. 1993, 16, 499–524.
17. Francis, C.A.; Youngberg, G. Sustainable agriculture—An overview. In Sustainable Agriculture in Temperate Zones; Francis, C.A., Flora, C.B., King, L.D., Eds.; John Wiley & Sons: New York, NY, USA, 1990; pp. 1–23.

18. Hansen, J.W. Is agricultural sustainability a useful concept? Agric. Syst. 1996, 50, 117–143.
19. Gallie, W.B. Essentially contested concepts. Proc. Aristot. Soc. 1955, 56, 167–198.
20. Thompson, P.B. The Agrarian Vision: Sustainability and Environmental Ethics; The University Press of Kentucky: Lexington, KY, USA, 2010.
21. Baker, S. Sustainable Development; Routledge: London, UK; New York, NY, USA, 2006.
22. Lafferty, W.M.; Meadowcroft, J. Introduction. In Implementing Sustainable Development: Strategies and Initiatives in High Consumption Societies; Lafferty, W.M., Meadowcroft, J.M., Eds.; Oxford University Press: Oxford, UK, 2000; pp. 1–22.
23. Jacobs, M. Sustainable development as a contested concept. In Fairness and Futurity: Essays on Environmental Sustainability and Social Justice; Dobson, A., Ed.; Oxford University Press: Oxford, UK, 1999; pp. 21–45.
24. Connelly, S. Mapping sustainable development as a contested concept. Local Environ. 2007, 12, 259–278.
25. Hart, H.L.A. The Concept of Law; Clarendon Press: Oxford, UK, 1961.
26. Rawls, J. A Theory of Justice; Clarendon Press: Oxford, UK, 1972.
27. Swedish Board of Agriculture. Sveriges handel med jordbruksvaror och livsmedel 2012; Swedish Board of Agriculture: Jönköping, Sweden, 2013.
28. LRF. LRF—We Make the Country Grow; LRF: Stockholm, Sweden, 2013.
29. Delfi. Dagligvarukartan 2013. Available online: http://www.delfi.se/wp-content/uploads/Dagligvarukartan2013.pdf (accessed on 16 November 2013).
30. SCB (Statistiska centralbyrån). Hållbarhet i Svenskt Jordbruk 2012; Statistics Sweden: Stockholm, Sweden, 2012.
31. LRF. LRF's Policy för Hållbar Användning av Kemiska Växtskyddsmedel; Decision by the LRF Board 2010-03-16. LRF: Stockholm, Sweden, 2010.
32. LRF. LRF's Genteknikpolicy Samt Frågor och Svar; Decision by the LRF Board 2010-04-22. LRF: Stockholm, Sweden, 2010.
33. Lantmännen. Lantmännen's Code of Conduct; Approved by the Board of Directors on 27 February 2008. Lantmännen: Stockholm, Sweden, 2008; Available online: http://lantmannen.se/en/start/our-responsibility/reporting/management-approach/code-of-conduct/ (accessed on 2 December 2013).
34. ICA. ICA-Koncernens årsredovisning och hållbarhetsredovisning 2012; ICA: Stockholm, Sweden, 2012; Available online: http://reports.ica.se/ar2012sv/Materiale/Files/ICA+H%C3%A5llbarhetsredovisning+2012_opt.pdf(accessed on 2 December 2013).
35. United Nations. United Nations Global Compact: Corporate Sustainability in the World Economy; UN Global Compact Office, United Nations: New York, NY, USA, 2013.
36. Axfood. Sustainability Programme 2013/Final version; Axfood: Stockholm, Sweden, 2013.
37. Coop. Kooperativa förbundets policy för hållbar utveckling; Coop: Stockholm, Sweden, 2009.
38. Coop. KF Verksamhetsberättelse 2012; Coop: Stockholm, Sweden, 2013.
39. LRF. Spridning av avloppsslam i jordbruket–så här ser LRF på frågan; LRF: Stockholm, Sweden, 2013.
40. The International Council for Local Environmental Initiatives (ICLEI). The Local Agenda 21 Planning Guide; ICLEI and IDRC: Toronto, ON, Canada, 1996.
41. Bernauer, T.; Meins, E. Technological revolution meets policy and the market: Explaining cross-national differences in agricultural biotechnology regulation. Eur. J. Polit. Res. 2003, 42, 643–683.

42. Kurzer, P.; Cooper, A. What's for dinner? European farming and food traditions confront American biotechnology. Compar. Polit. Stud. 2007, 40, 1035–1058.

43. Godfray, H.C.J.; Beddington, J.R.; Crute, I.R.; Haddad, L.; Lawrence, D.; Muir, J.F.; Pretty, J.; Robinson, S.; Thomas, S.M.; Toulmin, C. Food security: The challenge of feeding 9 billion people. Science 2010, 327, 812–818. [PubMed]

44. Fedoroff, N.V. The past, present and future of crop genetic modification. New Biotechnol. 2010, 27, 461–465. [PubMed]

45. Good, A.G.; Johnson, J.S.; de Pauw, M.; Carroll, R.T.; Savidov, N.; Vidmar, J.; Lu, Z.; Taylor, G.; Stroeher, V. Engineering nitrogen use efficiency with alanine aminotransferase. Can. J. Botany 2007, 85, 252–262.

46. Robinson, J. Squaring the circle? Some thoughts on the idea of sustainable development. Ecol. Econ. 2004, 48, 369–384.

47. Aerni, P. Is agricultural biotechnology part of sustainable agriculture? Different views in Switzerland and New Zealand. AgBioForum 2010, 13, 158–172.

48. Gunnarsson-Östling, U.; Edvardsson Björnberg, K.; Finnveden, G. Using the concept of environmental sustainability: Interpretations in academia, policy and planning. In Sustainable Stockholm: Exploring Urban Sustainability through The Lens of Europe's Greenest City; Metzger, J., Rader, O., Eds.; Routledge: New York, NY, USA; London, UK, 2013; pp. 51–70.

49. Hess, S.; Lagerkvist, C.L.; Redekop, W.; Pakseresht, A. Consumers' Evaluation of Biotechnology in Food Products: New Evidence from a Meta-Survey. In Proceedings of the Agricultural & Applied Economics Association's 2013 AAEA & CAES Joint Annual Meeting, Washington, DC, USA, 4–6 August 2013.

Beyond Climate-Smart Agriculture: Toward Safe Operating Spaces for Global Food Systems

Henry Neufeldt, Molly Jahn, Bruce M. Campbell,
John R. Beddington, Fabrice Declerck,
Alessandro De Pinto, Jay Gulledge,
Jonathan Hellin, Mario Herrero, Andy Jarvis,
David Lezaks, Holger Meinke, Todd Rosenstock,
Mary Scholes, Robert Scholes, Sonja Vermeulen,
Eva Wollenberg, and Robert Zougmoré

13.1 INTRODUCTION: HISTORY OF THE CONCEPT OF CLIMATE-SMART AGRICULTURE

If trends in human diet and waste in food systems remain unchecked, food production would have to increase by about 70% to feed an estimated 9 billion people by 2050, with unprecedented consequences for the environment and society. The food price spikes of recent years have reinforced awareness of obvious links between political and economic stability and food security. As a consequence, agricultural development is now the focus of renewed attention in both the research and policy communities. In describing tensions between maximizing global agricultural productivity, increasing resilience of agricultural systems in the face of climate change and mitigating greenhouse gas (GHG) emissions from agriculture, the term climate-smart agricultural development was first used in 2009 [1,2]. A year later, at the First Global Conference on Agriculture, Food Security and Climate Change at the Hague, the concept of climate-smart agriculture (CSA) was presented and defined as agriculture that "sustainably increases productivity, enhances resilience, reduces/removes greenhouse gas emissions, and enhances achievement of national food security and development goals" [3]. This definition represented an attempt to set a global agenda for investments in agricultural research and innovation, joining the agriculture, development and climate change communities under a common brand.

Drawing on this original framing, CSA has been applied to diverse aspects of agriculture, ranging from field-scale agricultural practices to food supply chains and food systems generally. Beyond agricultural practices and outcomes, a wide array of institutions, policies, finance, safety nets, capacity-building and assessment have all been identified as enabling CSA. Following the Second Global Conference on Agriculture, Food Security and Climate Change in Hanoi in 2012, the recently published Climate-Smart Agri- culture Sourcebook further advanced the concept with the intention of benefiting primarily smallholder farmers and vulnerable people in developing countries [4].

Building on this brief outline of the concept, in this article we first lay out major implications and shortcomings of what CSA means in practice. We then describe the challenges we are facing in assessing our trajectories toward long-term safe operating spaces of social-ecological systems for humanity within planetary and local boundaries and suggest an agenda for immediate action required to step up to meet the challenges.

13.2 CLIMATE-SMART AGRICULTURE ENCOMPASSES VIRTUALLY ANY AGRICULTURAL PRACTICE

Although in principle only agricultural practices that en- compass all components of CSA should be branded as "climate-smart," the term has been used very liberally because it is unclear how the different dimensions interact. Therefore, virtually any agricultural practice that improves productivity or the efficient use of scarce resources can be considered climate-smart because of the potential benefits with regard to food security, even if no direct measures are taken to counter detrimental climate effects. In addition, virtually any agricultural practice that reduces exposure, sensitivity or vulnerability to climate variability or change (for example, water harvesting, terracing, mulching, drought-tolerant crops, index insurances, communal actions) are also climate-smart because they enhance farmers' ability to cope with weather extremes. Likewise, agricultural practices that sequester carbon from the atmosphere (for example, agroforestry, minimum tillage), reduce agricultural emissions (for example, manure management, biogas plants, reduced conversion of forests and rangeland) or improve resource use efficiency (for example, higher productivity crop and livestock breeds, improved crop management and animal husbandry) can all be considered climate-smart because they contribute to slowing the rate of climate change. CSA has been a powerful concept to direct a focus on the climate change–agriculture nexus and has united the agriculture, climate change and development communities under one brand. Almost any agricultural practice or outcome currently qualifies as climate-smart, however, suggesting that CSA is a triple-win for all without regrets, losers and trade-offs. Thus, CSA can easily be appropriated for a wide range of even conflicting agendas.

13.3   WHAT CLIMATE-SMART AGRICULTURE FAILS TO ENCOMPASS

By recognizing links between our choices in agricultural systems and outcomes related to food systems in human dimensions, the incorporation of food security as an imperative for CSA differentiates it from concepts such as sustainable intensification [5] and ecoefficiency [6]. Balancing priorities at the intersections of food security, adaptation and mitigation, however, always occurs in the context of region-specific conditions and cultures. Why should resource-poor farmers invest in agricultural practices that may reduce emissions if there are few if any immediate benefits related to food or water security? ("It's hard to be green when you are in the red.") CSA, as currently conceived and implemented, fails entirely to recognize different actors, incentives and interactions between different (but related) provisioning demands for food, water, energy, materials and ecosystem services.

Furthermore, the concept of CSA fails to consider possible impacts of agriculture on other ecosystem services, biodiversity conservation and broader social, political and cultural dynamics. Reducing GHG emissions or improving resilience may not always result in the best natural resource management outcomes if consequences include biodiversity loss, degradation of cultural heritage, increased social inequity or long-term ecosystem instability [7].

Finally, CSA has been defined to focus exclusively in developing countries because national food security and development goals have been implicitly and incorrectly understood as issues of importance only in the developing world [3]. This focus has engendered opposition from those who fear that some developed countries may insist on mitigation of agricultural GHG emissions as a condition of continued development aid. Food security, nutritional security and nutritional health are obviously not limited to the developing world; there is also a widespread prevalence of food insecurity in high-income countries, where there are different, but overlapping, policy, governance and technical challenges [8]. With regard to the recent focus on smallholder farmers [4], the policy dialogue about CSA now systematically overlooks any impacts and opportunities connected to innovations and implications of large-scale agricultural practices for and in food systems in both developing and developed country contexts, further reducing the utility of the CSA framework.

In summary, the current framing of CSA gives no specific direction, no new science agenda, no ability to negotiate and prioritize contentious and conflicting agendas and no compelling reason to increase or shift investment, despite the monumental importance of these challenges in the coming decades. In fact, current agricultural practices are neither smart nor dumb. Our current agricultural and food systems are simply the manifestations of political, bio- physical, socioeconomic and other influences that lead to sustainable or unsustainable outcomes, depending on the perspectives, scales, valuations of trade-offs and time frames considered. In the aggregate, however, our current systems fall well outside any defensible concept of long- term safe operating space considered in human and/or environmental terms [9,10]. Without radical interventions and innovations to curb fundamentally extractive processes toward the renewal of the resources upon which agricultural productivity depends, we stand only to slip further away [11-14].

This recognition provides a strong mandate for agricultural systems that better meet human and environmental needs. Although major improvements in food security and livelihoods through agricultural development have been achieved, often this has occurred at the expense of nutritional health and environmental sustainability, thereby eroding the very foundations of our long-term capacity to care for ourselves. Under current default development pathways, food systems often arise in such a way that large populations remain food-insecure while other populations begin to suffer from the pathologies of over- or malnutrition. Although equity issues dictate clear differences in responsibilities between developed and developing countries, agricultural systems that will lead to the desired outcomes of improved food security and dietary health remain common goals for our global community.

Recent reports[a] have set forth specific principles and recommendations to improve the sustainability of agriculture and food systems that explicitly address various threats, including that of climate change [15]. None of these reports, however, moved beyond incremental improvements to specify in any detail a future state in which we commit ourselves to a food-secure world within planetary or local boundaries over the short or long term. Recently, the Commission on Sustainable Agriculture and Climate Change synthesized a vast array of literature on agriculture, food systems, food and nutritional security, dietary health, adaptation to climate change and mitigation of agricultural GHG emissions into a series of recommended policy actions [10,16]. In its report, the commission extended the concept of "safe operating space" beyond the original framing, which focused on biophysical attributes of the planet, to include social- ecological systems related to human welfare, agriculture and food security [9,10,17].

In our view, a safe operating space for agricultural and food systems represents a set of conditions that demonstrably better meets human needs in the short and long term within foreseeable local and planetary limits and holds ourselves accountable for outcomes across temporal and spatial scales. In our view, agriculture and food systems are climate-smart when it can be shown that they bring us closer to safe operating spaces.

Although well intentioned and potentially costly, the current mode of incremental improvement may still fall well short of achieving safer spaces. For instance, some argue that we are already able to produce enough food to feed a worldwide human population of 9 billion, especially under scenarios of improved dietary health, reduced waste and loss and diversified, intensified production systems [18]. Although this view is valid and important, we still do not know whether, even after such major shifts, our food systems would be in long-term balance with our natural resources base. Along the way, we may cross tipping points that demarcate permanent transitions to new states that will become apparent only when it is too late to turn back.

Improvements in the management of agricultural systems that bring us significantly closer to safe operating spaces (however we learn to define these conditions) will require transformational changes in governance, management and use of our natural resources that are underpinned by enabling political, social and economic conditions. This is a major challenge in itself, considering that investments in agricultural development have often yielded unintended detrimental social and environmental consequences on various spatial and temporal scales [19,20].

As a coordinated international attempt to address such issues, Rio+20 member states recently reaffirmed in the outcome document "The Future We Want" their commitments regarding "the right of everyone to have access to safe, sufficient and nutritious food" [21]. India and Mexico, for example, have moved to enshrine the right to food in law and are seeking means by which to implement such policy effectively [22,23]. Although the recognition of a human right to food security places human welfare and humanitarian values at the center of development, the short- and long-term social, economic, political and environmental effects of such commitments remain unclear. For instance, how will these efforts affect a "land-degradation neutral world" that these countries committed themselves to in the same document?

To answer such a question, we need to have processes in place that can provide relevant insights into issues such as the following. Can these more holistic approaches be integrated into research and development, informing a robust representation of conditions on the ground in near real time and more informative tools looking forward? Can we integrate approaches and insights derived from diverse sources to predict, mitigate and innovate regarding food security and nutritional health issues in the face of climate change? What are the specific boundaries of safe spaces? How do we deal with ambiguity and uncertainty across scales and priorities? How will we describe these boundaries and how they move dynamically as the trajectories of food systems evolve? Can we identify synergistic, transformational changes that may vault us to more stable and secure food systems across scales? What governance mechanisms are needed to ensure that the benefits and costs of well-grounded choices and their positive and negative consequences are shared as equitably as possible? How will we know whether our collective investments in the future are bringing us closer to safer spaces? Can we reach a condition which we can objectively defend as a safer space at any scale before reaching critical tipping points and thresholds in Earth and human systems? How will we know whether the changes are sufficiently bold so that, by the middle of this century, we can avoid the recognition we now face with virtual certainty that our best intentions simply have not been good enough?

13.4   WHAT SHOULD WE DO NOW IN RESEARCH?

Finding adequate responses to questions such as those in the preceding paragraph, as well as many more connected to reaching long-term safer spaces as our collective target, will require transformational changes in our commitment toward the future we want across all of society, including our science agenda. It means changing how we fund and evaluate agricultural research, how we evaluate agricultural practices and how we describe relevant parameters of human conditions linked to our choices in agriculture, natural resources management and food systems.

To develop an operational characterization of the safer operating spaces, working definitions of food security, resilience and mitigation are essential. It is also essential that these concepts are defined across the stakeholder dimension in addition to the spatial and temporal dimensions, as preferences and priorities vary significantly across institutional structures. The operational viability CSA is dependent on the resolution of these barriers (definitions and synthesis across scales) and a correct representation of trade-offs that truly informs decision-makers. The science agenda for the 21st century must improve our ability to recognize and achieve long-term safer spaces across scales for agriculture and food systems. Key areas of innovation in support of such a science agenda will not be restricted to, but will include:

- *Discovery, testing and implementation of mechanisms across scales that allow for adaptive management and adaptive governance of social-ecological systems essential for long-term human provisioning:* Adaptive management and governance will afford the capacity, protocols and processes to learn from mistakes and successes, including both anticipated and undesired outcomes.
- *Development of integrated metrics of safe space that are practical and meaningful for decision-making by relevant communities in near real time:* Indicators, proxies and other attributes of agricultural social-ecological systems that provide relevant feedback to stakeholders are required to monitor, evaluate and appraise changes in systems over space and in time, allowing for better decision-making and providing milestones for adaptive management.
- *Systematic gathering and integration of quality data and information to generate knowledge in time frames and at scales relevant for decision-making through analytical tools, models and scenarios:* Describing the consequences of our provisioning demands and choices in human and ecological terms requires the integration of high-quality data into knowledge for improved decision-making that will increasingly be collected, filtered, analyzed and interpreted by using automated self-learning algorithms to transform the vast amounts of data into useful information. Drawing on technological and computation innovations already in place as well as implementing

strategic funding investments would help to bridge the gap between the developing and developed scientific communities.

• *Establishment of legitimate and empowered science policy dialogues that frame post–disciplinary science agendas on local, national and international scales:* Dialogues and roundtables between relevant stakeholders that scientifically test decisions at the interface of diverging interests of business, environment and civil constituencies in often contentious topical areas can improve outcomes and help identify scientifically credible interpretations of long-term safe operating spaces in the context of a changing climate and growing environmental and societal changes.

Shifting our agricultural and food systems toward development trajectories that with greater certainty can be defined as safe is what we understand as being truly climate-smart. Although the areas of innovation described above will not take us to safer spaces for humanity without massive investments in sustainable natural resource management, a transformation of global food systems and ambitious low emissions development pathways, they will improve our ability to predict whether our investments in future agricultural and food systems can be considered climate-smart with greater certainty.

ENDNOTE

[a]Also see Commission on Sustainable Agriculture and Climate Change, *Achieving Food Security in the Face of Climate Change*; World Bank, *World Development Report*; Organization for Economic Cooperation and Development (OECD), Green Growth Strategy for Food and Agriculture and Green Growth Knowledge Platform (http://www.oecd.org/greengrowth/ greengrowth knowledgeplatform.htm); United Nations Environment Program (UNEP) and International Water Management Institute (IWMI), *Ecosystems for Water and Food Security*; Food and Agriculture Organization of the United Nations (FAO), *Climate-Smart Agriculture Sourcebook and How to Feed the World in 2050*; Foresight, *Report on Global Food and Farming Futures*; International Assessment of Agricultural Knowledge, Science and Technology for Development (IAASTD), *Synthesis Report*; United Nations High Level Task Force on the Global Food Security Crisis (http://www.un.org/en/issues/food/taskforce/); Millennium Ecosystem Assessment Reports (http://www.unep.org/ maweb/en/index.aspx); World Economic Forum (WEF), *Realizing a New Vision for Agriculture: A Roadmap for Stakeholders*.

REFERENCES

1. Food and Agriculture Organization of the United Nations: Food Security and Agricultural Mitigation in Developing Countries: Options for Capturing Synergies. Rome; 2009.

2. Food and Agriculture Organization of the United Nations: Harvesting Agriculture's Multiple Benefits: Mitigation, Adaptation, Development and Food Security. Rome; 2009.

3. Food and Agriculture Organization of the United Nations: "Climate-Smart" Agriculture: Policies, Practices and Financing for Food Security, Adaptation and Mitigation. Rome; 2010.

4. Food and Agriculture Organization of the United Nations: Climate-Smart Agriculture Sourcebook. Rome; 2013.

5. Tilman D, Balzer C, Hill J, Befort BL: Global food demand and the sustainable intensification of agriculture. Proc Natl Acad Sci USA 2011, 108:20260–20264.

6. Hershey CH, Neate P: Eco-Efficiency: From Vision to Reality. Cali, Colombia: International Center for Tropical Agriculture (CIAT); 2013.

7. McCarthy N, Lipper L, Mann W, Branca G, Capaldo J: Evaluating synergies and trade-offs among food security, development and climate change. In Climate Change Mitigation and Agriculture. Edited by Wollenberg E, Nihart A, Tapio-Biström ML, Grieg-Gran M. London: Earthscan; 2012:39–49.

8. Coleman-Jensen A, Nord M, Andrews M, Carlson S: Household Food Security in the United States in 2010. Washington, DC: Economic Research Service, US Department of Agriculture; 2011.

9. Rockström J, Steffen W, Noone K, Persson A, Chapin FS 3rd, Lambin EF, Lenton TM, Scheffer M, Folke C, Schellnhuber HJ, Nykvist B, de Wit CA, Hughes T, van der Leeuw S, Rodhe H, Sörlin S, Snyder PK, Costanza R, Svedin U, Falkenmark M, Karlberg L, Corell RW, Fabry VJ, Hansen J, Walker B, Liverman D, Richardson K, Crutzen P, Foley JA: A safe operating space for humanity. Nature 2009, 461:472–475.

10. Beddington J, Asaduzzaman M, Clark M, Fernández A, Guillou M, Jahn M, Erda L, Mamo T, Van Bo N, Nobre C, Scholes R, Sharma R, Wakhungu J: Achieving Food Security in the Face of Climate Change: Final Report from the Commission on Sustainable Agriculture and Climate Change. Copenhagen: CGIAR Research Program on Climate Change, Agriculture and Food Security (CCAFS); 2012. Available at: http://ccafs.cgiar.org/sites/default/files/assets/docs/ climate_food_commission-final-mar2012.pdf (accessed 15 August 2013).

11. United Nations Environment Program: Global Environment Outlook 5: Environment for the Future We Want. Nairobi: Author; 2012.

12. Millennium Ecosystem Assessment: Ecosystems and Human Well-Being: Synthesis. Washington, DC: Island Press; 2005.

13. Stern N: The Economics of Climate Change. Cambridge, UK: Cambridge University Press; 2007.

14. Bernstein L, Bosch P, Canziani O, Chen Z, Christ R, Davidson O, Hare W, Huq S, Karoly D, Kattsov V, Kundzewicz Z, Liu J, Lohmann U, Manning M, Matsuno T, Menne B, Metz B, Mirza M, Nicholls N, Nurse L, Pachauri R, Palutikof J, Parry M, Qin D, Ravindranath N, Reisinger A, Ren J, Riahi K, Rosenzweig C, Rusticucci M, et al: Climate Change 2007: Synthesis Report. Contribution of Working Groups I, II and III to the Fourth Assessment Report of the Intergovernmental Panel on Climate Change. In Edited by Core Writing Team, Pachauri RK, Reisinger A. Geneva: Intergovernmental Panel on Climate Change; 2007. Available at: http://www.ipcc.ch/pdf/ assessment-report/ar4/syr/ar4_syr.pdf (accessed 15 August 2013).

15. McKenzie F, Ashton R: Safeguarding and Enhancing the Ecological Foundation of Agricultural and Food Systems to Support Human Well-being: Bridging the Implementation

Gap. Nairobi: United Nations Environment Programme. Available at: http://www.geosci. usyd.edu.au/documents/fiona1. pdf (accessed 15 August 2013).

16. Beddington JR, Asaduzzaman M, Clark ME, Fernández Bremauntz A, Guillou MD, Howlett DJB, Jahn MM, Lin E, Mamo T, Negra C, Nobre CA, Scholes RJ, Van Bo N, Wakhungu J: Agriculture: What next for agriculture after Durban? Science 2012, 335:289–290.

17. Ostrom E: A general framework for analyzing sustainability of social-ecological systems. Science 2009, 325:419–422.

18. Foley JA, Ramankutty N, Brauman KA, Cassidy ES, Gerber JS, Johnston M, Mueller ND, O'Connell C, Ray DK, West PC, Balzer C, Bennett EM, Carpenter SR, Hill J, Monfreda C, Polasky S, Rockström J, Sheehan J, Siebert S, Tilman D, Zaks DPM: Solutions for a cultivated planet. Nature 2011, 478:337–342.

19. Leith P, Meinke H: Overcoming adolescence: Tasmania's agricultural history and future. Griffith Rev 2013:39. Available at: http://griffithreview. com/edition-39-tasmania-the-tipping-point/overcoming-adolescence (accessed 15 August 2013).

20. Tábara JD, Chabay I: Coupling human information and knowledge systems with social-ecological systems change: reframing research, education, and policy for sustainability. Environ Sci Policy 2012, 28:71–81.21.

21. United Nations General Assembly: The Future We Want, Rio+20 Outcome Document. Resolution adopted by the General Assembly at the 123rd plenary meeting 27 July 2012. A/RES/66/288. Available at: http://unstats.un. org/unsd/broaderprogress/pdf/GA%20 Resolution%20-%20The%20future% 20we%20want.pdf (accessed 15 August 2013).

22. Kasmin A: Delhi pushes through right-to-food programme as hunger persists. Financial Times 3 July 2013.

23. Office of the High Commissioner for Human Rights, United Nations Human Rights: Mexico: UN Expert welcomes constitutional recognition of the right to food. Available at: http:// www.ohchr.org/en/NewsEvents/Pages/ DisplayNews.aspx?NewsID=11491&LangID=E (accessed 15 August 2013).

Author Notes

Chapter 2

Acknowledgments
The author would like to thank Seonaidh McDonald for her comments on an earlier draft of this paper.

Authors' Contributions
The author gathered, transcribed, translated and interpreted the data herself. All work presented in this paper is her own derived from personal research. The Albanian data were gathered in the Albanian language and translated back to English for the purpose of the data analysis.

Conflicts of Interest
There are no conflicts of interest.

Chapter 2

Acknowledgments
Many thanks to Debra Moffett whose research greatly enhanced the original report from the Toronto Food Policy Council [1] on which this paper is based. Josée Johnston of the University of Toronto was also instrumental in helping us understand many issues related to the citizen-consumer phenomenon.

Conflict of Interest
The authors declare no conflict of interest.

Chapter 5
Research leading to these results received funding from the European Community's Seventh Framework Program (2007–2013) under grant agreement n 217280. Entitled 'Facilitating Alternative Agro-Food Networks' (FAAN), the project had five national teams, each linking an academic partner with a civil society organization partner. The UK team consisted of the Open University and GeneWatch UK, also with participation by the University of Lancaster. We thank our colleagues Helen Wallace and Becky Price from GeneWatch UK, and

Bronislaw Szerszynski from Lancaster University, for stimulating discussions about local food networks. More results, especially a Europe-wide summary report, can be found at the FAAN project website www.faanweb.eu

We thank staff at Cumbria Fells and Dales Leader for helpful comments towards refining an earlier draft of this article, as well as the many practitioners who cooperated with the Cumbria study. Also thanks to two anonymous referees of this journal.

Chapter 6

Acknowledgements
The authors would like to acknowledge the assistance of Louisa Winkler, Colin Curwen-McAdams, Brigid Meints and Steve Lyon as well as two anonymous reviewers for comments that have greatly strengthened this paper. Financial support for this research was provided by the Organic Farming Research Foundation and the Clif Bar Family Foundation's Seed Matter Initiative.

Chapter 7

Acknowledgements
The judgments and conclusions herein are those of the authors and do not necessarily reflect those of the U.S. Department of Agriculture. The authors are responsible for all errors.

Chapter 8

Acknowledgments
This research was partly supported by the Tyson Chair Endowment at the University of Arkansas and by a Korea University grant.

Author Contributions
This paper is the result of teamwork. Giovanna Sacchi developed the original idea, contributed to the research design and was responsible for data collection. She wrote paragraphs 2, 3, and 4. Vincenzina Caputo contributed to the data set up and data analysis. She wrote paragraphs 5 and 6. Rodolfo M. Nayga Jr. provided guidance and advice. All the authors jointly wrote paragraphs 1 and 7. All authors have read and approved the final manuscript.

Conflicts of Interest
The authors declare no conflict of interest.

Chapter 9

Acknowledgements
This study was supported by the EU grant no. KBBE-245058-SOLIBAM. Special thanks go to the case farmer for providing the data and allowing us to perform research at his interesting farm. Further, we acknowledge the contribution from other SOLIBAM participants, specifically Elena Tavella at the University of Copenhagen, who calculated the revenues.

Author Contributions
The study was designed by Mads V. Markussen in collaboration with all co-authors. Data was collected by Mads V. Markussen and Michal Kulak with assistance from Laurence G. Smith. All data analysis related to the emergy evaluation was done by Mads V. Markussen, and all data analysis related to LCA was done by Michal Kulak. All authors contributed to combining the analyses and interpreting the results.

Supplementary Materials
Supplementary materials can be accessed at: http://www.mdpi.com/2071-1050/6/4/1913/s1.

Conflicts of Interest
The authors declare no conflict of interest.

Chapter 10

Acknowledgements
The second author was supported by a Discovery grant from NSERC and research grants from Alberta Crop Industry Development Fund Inc. Much research reported herein was conducted by our research group, many of whom have moved on to brighter futures. These students and research associates include (and this is not a total listing) A. Navabi, R. Degenhardt, A. Kaut, H. Mason, T. Reid, L. Annett, and A. Nelson.

Chapter 11

Acknowledgments
This work was funded by NERC Knowledge Exchange Programme Grant NE/K001191/1. WJS is funded by Arcadia. TGB's input was funded by the UK Global Food Security Programme. We thank Richard Brazier, Humphrey Crick, Bridget Emmett, Corrina Gibbs, Rob Macklin, Sarah Mukherjee, Simon Potts

and Andy Richardson for contributions to the early stages of prioritisation. We also thank Stephanie Prior, Joscelyne Ashpole, James Hutchison, Hugh Wright and Richard German for help before and during the meeting, and for collating results during and after the workshop. Special thanks to Michel Kaiser for chairing sessions.

Conflict of Interest

The following authors represented the interests of their organizations in this process. We do not interpret this as a conflict of interest because the process was designed to take account of a wide range of interests, including those of commercial and campaigning organizations.

Angela Booth (AB Agri), Jan Bouwman (Syngenta), Chris Brown (Asda), Ian Crute (AHDB), Frances Dixon (Welsh Government), Caroline Drummond (LEAF), Andrea Graham (National Farmers Union), James Hallett (British Growers Association), Beth Hart (Sainsbury's), John Holland (Game and Wildlife Conservation Trust), Vanessa King (Unilever), Tom MacMillan (Soil Association), Daniel McGonigle (Defra), Carmel McQuaid (Marks and Spencer), Tim Nevard (Conservation Grade), Steve Norman (DowAgro), Catherine Pazderka (British Retail Consortium), Inder Poonaji (Nestle), Duncan Sinclair (Waitrose), Juliet A. Vickery (RSPB) and William Wolmer (Blackmoor Estate).

Chapter 12

Acknowledgments
The authors would like to thank the interviewed representatives of the five organizations participating in this study. The authors would also like to thank two anonymous reviewers for their valuable comments and suggestions. Special thanks to Lotta Rydhmer and Sven Ove Hansson for their careful reading of previous versions of the manuscript.

The study was supported by Mistra, the Swedish Foundation for Strategic Environmental Research (grant "Mistra Biotech"). Their support is gratefully acknowledged.

Author Contributions

All authors participated in the design of the study, in analysis of the data, and in writing of the paper. Karin Edvardsson Björnberg and Håkan Marstorp conducted the interviews. An external person was hired to transcribe the interviews. All authors read and approved the final manuscript.

Conflicts of Interest

The authors declare no conflict of interest.

Chapter 13

Competing Interests
The authors declare that they have no competing interests.

Authors' Contributions
HN and MJ are co–first authors. HN and MJ developed the outline and drafted the manuscript. BMC, FDC, ADP, JG, JH, MH, HM, MS, SV and EW contributed to the evolving outline. BC, JRB, FDC, ADP, JH, MH, AJ, DLZ, HM, TR, MS, RS, SV, EW and RZ provided input to the writing of the manuscript. All authors read and approved the final manuscript.

Acknowledgements
We would like to acknowledge support provided by the CGIAR Research Program on Climate Change, Agriculture and Food Security (CCAFS).

Index

Milton Keynes UK
Ingram Content Group UK Ltd.
UKHW022058141024
449569UK00031B/1682